National Security
and United States Policy
toward Latin America

National Security and United States Policy toward Latin America

LARS SCHOULTZ

PRINCETON UNIVERSITY PRESS

PRINCETON, NEW JERSEY

To Karina, Nils, and Jane

CONTENTS

LIST OF ILLUSTRATIONS

LIST OF TABLES

PREFACE

THIS is a book about the importance of Latin America to the security of the United States. In 1975, when I first began to interview policy makers in Washington, I knew nothing about this subject and, disillusioned by the experience of Vietnam, I consciously avoided learning anything about it. Instead, I studied human rights. By the time that study was completed, however, I had become convinced that human rights is a residual category in United States policy toward Latin America; it (along with economic development) is what policy emphasizes when there is no security problem on the horizon. If one wants to understand the core of United States policy toward Latin America, one studies security.

And so in 1980 I began some initial interviewing designed to identify the boundaries of such a study. It soon became apparent that U.S. policy makers are deeply divided in their beliefs about what constitutes a security threat from Latin America. In my earlier interviewing I had also found deep divisions over the issue of human rights, but these divisions tended to be tactical—disputes over which rights to emphasize and, especially, how best to promote respect for fundamental rights. Everyone agrees that policy makers should protect U.S. security, of course, but beyond that there is almost no consensus on even the most basic aspects of the subject. Specifically, there is no agreement on what might cause threatening instability in Latin America, nor is there agreement on what the actual consequences of the instability might be. This study therefore focuses upon policy makers' differing beliefs about these two core concerns: the causes and consequences of Latin American instability for U.S. security interests.

Because the focus is upon individuals' beliefs, the data for this study are, by their nature, "soft." They are of two types. First, there is the public record. Much of this record is printed, and it is cited in the footnotes and bibliography that follow. Other parts of the public record are oral, the feelings and beliefs expressed in the speeches and presentations of policy makers. I have often found the question-and-answer segments of congressional hearings to be helpful, particularly when they force relatively senior policy makers (who are typically unavailable for individual interviews) to defend themselves and their policies. Attending these hearings is my hobby.

Second, data have been obtained from personal interviews with policy makers. Although most of the formal interviewing occurred during the 1982 to 1983 academic year, data have been drawn from interviews conducted as early as November 1975 and as late as June 1986. The total N now exceeds 290, but that figure hides so much diversity in interviewing form and substance as to be almost meaningless. The most striking feature of these interview data are their variable quality. It really is not possible to compare in any numerical sense the value of data obtained while driving a member of Congress to the airport with a formal interview in the office of a deputy assistant secretary of state, a three-day conference with a dozen CIA officials, or a year-long academic exchange with a Special Forces officer responsible for opening the Army's Regional Training Center in Puerto Castilla, Honduras. All provide useful information. All help an interviewer understand the nature of U.S. security concerns in Latin America; some help understand policy makers' beliefs about these concerns.

This study employs a broad definition of the term "policy maker." The conventional definition—career personnel and the political appointees serving the administration currently in power—is too narrow for my purpose. Day-to-day turnover is rapid in foreign-policy staffing (the tenure of an assistant secretary of state for inter-American affairs is typically less than two years), and major shifts occur whenever one administration replaces another. By using the conventional definition, one takes a snapshot of the policy-making process that is often outdated before the film is developed. In a study that focuses upon the disagreements among policy makers that have been emerging over the course of at least four decades, it is much more useful to include all individuals who recently have had (that is, some Carter and Nixon administration officials) or who currently have a direct impact upon policy. This definition also permits the inclusion of a few "informal" participants, individuals who do not hold public office but who nonetheless influence policy. Thus for the purposes of this study both Walter Mondale and Jeane Kirkpatrick, for example, retained their status as policy makers after they ceased drawing government paychecks because both continued to influence U.S. policy directly. Similarly, a few representatives of interest groups who appear to have a direct influence upon policy are also considered policy makers. Some are former employees who, for one reason or another, appear to have maintained unusually close ties to the federal government's foreign policy bureaucracy. Their beliefs and perceptions directly influence policy. It is important to include them.

More difficult than identifying those who deserve the title of "policy maker" is the problem of determining policy makers' beliefs. The reader should be warned that this book surveys analytic terrain where the footing is far from certain. Much of the evidence is in the form of statements of belief and perception. For example, as evidence that some policy makers believe Communism causes instability, President Reagan will be quoted as saying that he believes "the Soviet Union underlies all the unrest that is going on"; similarly, as an example of the belief that poverty causes instability, senators Leahy and Pell will be quoted as saying that they believe "the large social and economic disparities which have dominated Salvadoran society bred this revolution."

There are at least two major problems with using this type of evidence. First, no one knows for sure if the person making the statement is expressing a core belief or simply making a political point before a receptive audience. Howard Wiarda, for example, argues that "no one really believes, despite frequent assertions to the contrary, that the Soviet Union is the prime cause of the upheavals" in Central America.[1] I disagree. My data indicate that in recent years a very large number of officials—people like Nestor Sánchez, Constantine Menges, Gordon Sumner, Roger Fontaine, Jesse Helms, Lewis Tambs, Otto Reich, and dozens and dozens of others—have agreed wholeheartedly with Mr. Reagan's statement. But I could be wrong. These officials could have been less than candid in their responses to my questions; they could have changed their minds; they could have said what they thought I wanted to hear; they could have said what they thought would get me out of their offices most quickly.

The second problem (and one that is exacerabated by using a broad definition of policy maker) is that no one can ever be sure that he or she has obtained a representative sampling of all diverse policy makers' beliefs. I have interviewed virtually any policy maker concerned with issues related to Latin America who would give me the time for an interview. Despite my best efforts to be systematic, I have found the basic fact of foreign policy interview-

---

[1] Howard J. Wiarda, ed., *Rift and Revolution: The Central American Imbroglio* (Washington, D.C.: American Enterprise Institute for Public Research, 1984), p. 18.

ing is that you take what you can get and then do your best with it. I always asked respondents for introductions to (or at least the names of) other officials who might help me clarify this or that aspect of inter-American relations. I always asked for the name of each official's predecessor, for individuals who have recently left a position are often the most eager to discuss their views and experiences. Every time I was in the State Department's main building I went by the Office of Central American Affairs in the Bureau of Inter-American Affairs, and I looked for new names on the little plates beside the door. If there was a new name—and there almost always was at least one newcomer—I would try to arrange an interview. In its own way all of this was, to my mind, reasonably systematic in the sense that I do not think I missed any significant viewpoint. But my more scientifically inclined colleagues would probably raise an eyebrow here and there, and the reader therefore deserves a warning.

Not all that follows is built on shaky ground, however. There are some indisputable facts as well. When a policy maker asserts, for example, that the United States needs strategic raw materials from Latin America, a researcher can investigate the specific relationship between U.S. demand and Latin American supply. Similarly, when a policy maker argues that access to South Atlantic sea lines are crucial to U.S. security, it is possible to determine exactly what goods pass through these sea lines and what the impact of their absence might be. I have greatly enjoyed this part of the research for this book. It has taken me into odd nooks and crannies in Washington—only the Coast Guard, for example, has any reliable information on the military significance of Navassa Island (the answer is that it has none)—and into the University of North Carolina's Davis Library, where I learned more than I ever wanted about such topics as the military uses of exotic minerals and the diplomatic niceties of transiting the Bosphorus in a submarine. I now know that in speaking to navy personnel one does not spell out the initials used to abbreviate Sea Line of Communication, SLOC, but rather says "slock"; that one should not assume that all CIA officials are ignorant reactionaries because many of them are neither; and that one should never question Luigi Einaudi's belief that the U.S. invasion of Grenada was the best thing that had happened to the island since nutmeg. The difficult days of writing were compensated by these discoveries. There is much I still do not know about many of these subjects—the secrecy that surrounds antisubmarine warfare was especially frustrating—and I keep forgetting the difference between a frigate and a destroyer, but with the help of my respondents, many noted in the text and many others who requested anonymity, and with the groundwork provided by earlier generations of researchers, all noted in the footnotes and the bibliography, I think I learned the most important facts about U.S. security interests in Latin America.

The focus, however, is not upon geostrategic facts but upon policy makers' beliefs. The physical dimensions of a new airport in Grenada are irrelevant to U.S. policy, for example. When policy makers argue that a long runway in the eastern Caribbean poses a threat to U.S. security, they are asserting a belief about how the runway will be used. When all is said and done, uninterpreted facts have an extremely limited impact upon policy. It makes almost no difference at all whether the existence of such a runway is, in fact, crucial to U.S. security. If senior officials believe it is crucial, then U.S. policy will reflect that belief, regardless of its accuracy.

Nearly everyone would agree, however, that facts *should* make a difference in the creation of United States policy toward Latin America. A policy based upon ignorance (or the willful misrepresentation of the truth) is by its nature highly susceptible to mistakes, and foreign policy mistakes kill people. If, as many policy makers argue, we must be prepared to fight and die in order to deny the Soviet Union access to military bases in Latin America, then it seems to me that we should know exactly how Moscow could use those bases to threaten U.S. security. In the following chapters the search for this knowledge leads us at times away from a singular emphasis upon policy makers' beliefs and into such diverse subjects as Argentine submarines and the Trans-Alaska pipeline. The overall purpose of this study, then, is not only to analyze policy makers' beliefs but to assess, where possible, their accuracy.

Although it is not always possible to make this assessment— Who really knows how the airport in Grenada would have been used?—readers will note a judgmental quality in some of the pages that follow. Whenever conscientious (if not infallible) scholarship has led me to a conclusion about a controversial policy issue, I have stated it. In every case, however, I have tried mightily to be fair: to gather, analyze, and present my data in an even-handed manner, to treat all but the most idiotic opposing positions with respect, and to acknowledge that legitimate differences of opinion will always be just that—legitimate—on any subject as complex as national security and United States policy toward Latin America.

ACKNOWLEDGMENTS

THE MONEY to support the research for this book was obtained from a variety of sources. A postdoctoral research grant from the Joint Committee on Latin American Studies of the Social Science Research Council and the American Council of Learned Societies, supplemented by funds from the University of North Carolina, provided me with the opportunity to spend the 1982/1983 academic year conducting the initial interviews in Washington. Three separate interviewing trips to Central America in 1981, 1984, and 1986 were funded by the Faculty Committee for Human Rights in Central America, the University of North Carolina, and the Ford Foundation. An interviewing trip to Grenada in 1984 was funded by the Institute of Latin American Studies of the University of North Carolina. The Rockefeller Foundation provided an invitation to spend part of the summer of 1983 at the Foundation's Study Center in Bellagio, Italy, where I was partially successful in overcoming the distractions of princely surroundings in order to hammer out the original drafts of several key chapters. From 1983 to 1986, the University Research Council of the University of North Carolina paid for about three dozen brief trips to Washington to conduct interviews. Finally, a research leave of absence funded by the University of North Carolina's Joseph M. Pogue Trust gave me the opportunity to complete the manuscript during the 1985/1986 academic year.

Like most professors, I do my share of muttering about insufficient pay and inadequate working conditions. Once in awhile, however, I am obliged to write a paragraph like the one immediately above, and then I feel a bit guilty about my complaining. A lot of people spent a lot of their money to help me write this book. In particular, the people of North Carolina have been uncommonly generous. I deeply appreciate their support.

Although I have been known to deprecate the role of academic administrators in furthering the scholarly enterprise, basic honesty requires me to acknowledge that in this case, at least, the administrators of the University of North Carolina have helped me considerably. From time to time during the past six years, one provost—Sam Williamson—one dean—Gillian Cell—and two department chairs—Dick Richardson and the late Jim Prothro—have not only helped me obtain much of the generous assistance mentioned above but have as well freed me from certain other time-consuming administrative responsibilities. I appreciate their understanding and cooperation.

I am also indebted to my wife, Jane Volland, and to several friends and colleagues who took the time to read all or part of this manuscript and make suggestions about how it could be improved. They are Bill Ascher, Enrique Baloyra, Martin Diskin, Patricia Weiss Fagen, Richard Fagen, Yale Ferguson, Jo Marie Griesgraber, Henry Landsberger, Bill LeoGrande, Chris Mitchell, Rose Spalding, Carlos Rico, José Simán, Richard Ullman, and, once again, Federico Gil, whose influence can be found on every page of this book.

Finally, I am grateful for the assistance and the friendship of two uncommonly supportive editors. As a social science editor at Princeton University Press, Sandy Thatcher once asked me if I was ever going to write a book. Thus encouraged, I wrote one. Now, again with his encouragement, I have written another. And now as editor-in-chief of the Press, Sandy continues to serve as the principal force behind Princeton's publications on Latin America. As perhaps the most competent and unquestionably the most pleasant editor in the publishing business, Cathy Thatcher has corrected hundreds of my errors and infelicities, greatly improving the pages that follow. Both the reader and I are in her debt; fortunately, only I know how much.

LIST OF ABBREVIATIONS

ACP	Asian-Caribbean-Pacific
ADA	Americans for Democratic Action
AID	Agency for International Development
ANS	Alaska North Slope
ARA	Bureau of Inter-American Affairs, Department of State (formerly Bureau of American Republics Affairs)
ASW	Antisubmarine Warfare
CEB	*Comunidades Eclesiales de Base*—Christian Base (or Grass-roots) Communities
CREM	Centro Regional de Entrenamiento Militar—Regional Military Training Center
CRS	Congressional Research Service, Library of Congress
CVBG	Aircraft Carrier Battle Group
DIA	Defense Intelligence Agency
DOD	Department of Defense
DSAA	Defense Security Assistance Agency
DWT	Deadweight Tons
ECLAC	Economic Commission for Latin America and the Caribbean (formerly Economic Commission for Latin America—ECLA)
ESF	Economic Support Fund
FMS	Foreign Military Sales
FSO	Foreign Service Officer
FTG	Fleet Training Group
GAO	General Accounting Office
IADB	Inter-American Defense Board
IADC	Inter-American Defense College
IAPF	Inter-American Peace Force (Dominican Republic)
IMET	International Military Education and Training
ISA	Office of International Security Affairs, Department of Defense
MAAG	Military Assistance Advisory Group
MAP	Military Assistance Program

MILGP	Military Group
MTT	Mobile Training Team
NSC	National Security Council
OAS	Organization of American States
OAPEC	Organization of Arab Petroleum Exporting Countries
OECS	Organization of Eastern Caribbean States
PL	Public Law
PL480	Public Law 83-480 (Food for Peace)
SIMA	Shore Intermediate Maintenance Activity
SLOC	Sea Line of Communication
SONUS	Sound Surveillance System
SOUTHCOM	United States Southern Command
SSBN	Strategic Nuclear Submarine
VLCC	Very Large Crude Carrier

National Security
and United States Policy
toward Latin America

INTRODUCTION

IN THE waning days of America's involvement in Vietnam, a weary Secretary of State was asked to discuss the forces that had brought about defeat. Henry Kissinger responded that the collapse had come at home, not in the jungles of Indochina. Too many citizens had come to oppose U.S. policy, he said, and "any foreign policy of the United States that is not based on public support, and above all on congressional support, will not have a firm foundation."[1] Lacking this support, two administrations had been ruined and a third was overseeing the forced withdrawal of American forces from Indochina.

Within a few years, similar opposition had developed to United States policy toward Central America. As disagreement grew, the Carter and Reagan administrations faced the herculean task of defending themselves at home while simultaneously implementing their policies with the tepid support of nervous regional allies. By the late 1970s, large numbers of U.S. citizens were working with extraordinary vigor to hamstring virtually every administration proposal on Central America. Within the policy-making process itself, each issue occasioned a major controversy, each victory was soured by concessions or threatened by a subsequent reversal. By the end of the Carter administration, many officials could not even talk reasonably to one another. Questions were raised about patriotism and loyalty, ominous questions that poisoned principled debate. In 1980, retired General Gordon Sumner, who would soon enter the Reagan administration in the State Department's Bureau of Inter-American Affairs, spoke for many policy makers when he lamented the existence of interest groups that "owe no loyalty to the United States and are ideologically committed to ideals and policies rejected by the American people. Two examples are Amnesty International and the Washington Office on Latin America. There are many more."[2] By 1983, dissent had be-

---

[1] Press conference, January 14, 1975.

[2] U.S. Congress, House, Committee on Appropriations, Subcommittee on Foreign Operations and Related Agencies, *Foreign Assistance and Related Programs Appropriations for 1981*, 96th Cong., 2d Sess., 1980, pt. 1, p. 284. Before retiring from active duty in 1978, Lt. General Sumner held the position of chairman of the Inter-American Defense Board. His retirement was prompted in part by his oppo-

come so widespread that President Reagan was forced to appoint a special commission to recommend an appropriate policy to pursue in Central America. Yet policy makers would not even declare an overnight truce so that they could read the Kissinger Commission's report; Senator Jesse Helms issued his rejection three days before the document was made public. By the mid-1980s, the sense of confrontation was palpable in Washington.

The Loss of Consensus

The evidence of dissension is everywhere. Among members of the attentive public, it lies in the explosion of interest group activity. One cannot help but be impressed by the dramatic proliferation of organizations dedicated to influencing U.S. policy, often by mobilizing support at the grass-roots level in communities across the country, a tactic unknown in the long history of inter-American relations. While included among the organizations opposed to the Reagan administration's policies were virtually all members of the human rights lobby—particularly those singled out as disloyal by General Sumner and a vigorous newcomer, Americas Watch—perhaps the most remarkable feature of the opposition was the participation of previously inactive religious organizations. In 1981 it was simply startling to watch the American Baptist Churches—no radicals they—denounce the administration's policy:

> Supporters of the current government in El Salvador have attempted to portray the opposition to that government as being only the extreme left. Such a generalization is completely at variance with the facts. The extreme leftist elements are indeed part of the opposition, as are the moderate left, the center, the academic community, labor unions, peasant groups, and vast numbers of average Salvadoran people. The centrists are gone from the government, most of them having fled into exile or hiding. The United States and the right-wing junta are aiding the elimination of the center as a political reality in El Salvador. Just at the time we should be midwifing the birth of

sition to the 1978 Panama Canal treaties. Upon retirement he accused the Joint Chiefs of Staff of being "ignorant men" who were "taking what their staff is telling them and what their political masters are telling them." *Army Times*, June 12, 1978, pp. 1, 22. During the Reagan administration, Sumner served as Special Advisor to the Assistant Secretary of State for Inter-American Affairs.

a new center, we are instead arming the terrorist right more thoroughly.[3]

As we will see in the chapters that follow, most of the church groups that opposed the Reagan policy did so because of their members' first-hand experiences in Latin America, experiences that were often so strongly offensive that they created a passionately committed opposition.

Meanwhile, as some U.S. citizens traveled to Nicaragua as part of Witness for Peace, an organization committed to halting U.S. support for the anti-Sandinista guerrilla units, other U.S. citizens had formed groups to provide aid and, on at least one occasion, to fight alongside the Contras. In September 1984, two members of an organization called Civilian Military Assistance (CMA) were killed when their helicopter was shot down over Nicaragua during an attack on the government's military base at Santa Clara. Why were they there? An angry CMA leader told one reporter: "The U.S. and [Latin] America deserve a victory over communism. I'm not talking about Grenada, which was a small operation. I'm talking about a major victory in which the Communists can be the ones for a change who have to go home and lick their wounds."[4] Here, too, there was impassioned commitment.

At the time the two CMA volunteers were killed, retired General John Singlaub and his Council for World Freedom were organizing a "Freedom Fighters Ball and Banquet" in Dallas, and in Washington, D.C., the *Washington Times*, a paper controlled by Reverend Moon's Unification Church, had launched a Nicaraguan Freedom Fund to raise money for the Contras. Other pro-Contra U.S. interest groups included the Air Commando Association, which transported material from private aid groups to the Contras; Americares Foundation, which sent "Mercylifts" to right-wing groups in Central America; the Caribbean Commission, an organization that sought to "maintain, promote, and strengthen the free enterprise system in the western hemisphere to prevent totalitarian infiltration"; the Christian Broadcasting Network of Reverend Pat Robertson, which sent aid to Contra camps through

---

[3] U.S. Congress, House, Committee on Foreign Affairs, Subcommittee on Inter-American Affairs, *U.S. Policy toward El Salvador*, 97th Cong., 1st Sess., 1981, p. 296. The speaker was the Reverend Robert W. Tiller, Washington representative of the American Baptist churches, who testified before Congress on the basis of official denomination policy adopted by the general board of the church.

[4] James L. Pate, "CMA in Central America: The Private Sector Suffers Two KIA," *Soldier of Fortune* 10 (January 1985): 123.

"Operation Blessing"; the Council on National Policy, an organization that included such conservative luminaries as Jesse Helms and Joseph Coors, which sponsored speaking tours for Contra leaders; the Friends of the Americas, which sent "Shoe Boxes for Liberty" to Nicaraguans in Honduras, storing them in a Contra safe house in Tegucigalpa; the Knights of Malta, an organization involving such public figures as J. Peter Grace and William Casey that styles itself after the Catholic soldier-monks of the crusades, which lent moral and political support to the Contra effort; *Soldier of Fortune* magazine, which helped the Contras obtain weapons; and World Medical Relief, provider in an earlier era of medical supplies to CIA mercenaries in Laos, which sent similar supplies to the CIA's forces in Honduras.[5] Never before in the two-century history of our republic had so many voluntary organizations addressed any issue of inter-American relations.

The lack of consensus is most evident in Congress, for there legislators must periodically cast votes on a variety of aspects of U.S. policy. For two decades after World War II these votes indicated a remarkable Cold War consensus: U.S. participation in the United Nations was approved in 1945 by a 98 percent majority, the Truman Doctrine was accepted in 1947 by a 73 percent majority, the Marshall Plan in 1948 by 80 percent, NATO in 1949 by 86 percent, the U.S.-Korean Mutual Defense Treaty in 1954 by 93 percent, the Cuban-U.S. trade embargo in 1961 by 100 percent, and, of course, the Gulf of Tonkin resolution in 1964 by 98 percent in the Senate and 100 percent in the House.[6]

But then the consensus began to crack in Congress, and since the 1970s votes on issues of inter-American relations have regularly thrown both houses into uncommonly bitter debates. There will probably never be a better illustration of this dissensus than the April 1985 votes on aid to the Nicaraguan Contras. After weeks of debate on and off the floor, the vote indicated that Congress had split into two equal parts: in the Senate, 46 opposed, 53 in favor; in the House, 215 opposed, 213 in favor.[7] These and sim-

---

[5] Inter-Hemispheric Education Resource Center (Albuquerque, N.M.), *Bulletin* 2 (Fall 1985): 1, 4; *Washington Post*, May 3, 5, and 7, 1985; Pate, "CMA in Central America."

[6] Ole R. Holsti and James N. Rosenau, *American Leadership in World Affairs: Vietnam and the Breakdown of Consensus* (Winchester, Mass.: George Allen and Unwin, 1984), p. 221; *Congressional Record* 107 (August 18, 1961): 16,292.

[7] The vote cited in the House was one of several during the final deliberations on April 24. This particular vote, the closest, was also the most important. It was on a

ilar votes in 1986 tell only half the story, however. Committee and floor debates have become increasingly acrimonious, exposing raw nerves and deepening cleavages. A member's stand on Latin America has now become an identifying device, a litmus test for friends and adversaries alike. History shows no example of a more profound disagreement in Congress over any aspect of United States policy toward Latin America.[8]

The dissensus has profoundly affected the executive branch, where fundamental disagreements have traditionally been minimal on issues of inter-American relations. In that type of environment all points of view were tolerated, if not welcomed. The Ford-Kissinger State Department, for example, gave much responsibility for United States policy toward Latin America to an entire litter of mainstream Democrats: in late 1975, William D. Rogers, Albert Fishlow, Tom Farer, and Joseph Grunwald all held political appointments in the Republican administration. The Carter administration followed this pattern as well.[9] Mr. Carter awarded liberals a few high profile positions of influence—Andrew Young as Ambassador to the United Nations, Patricia Derian as Assistant Secretary of State for Human Rights and Humanitarian Affairs—balanced by the appointment of moderates such as Cyrus Vance and Warren Christopher to all senior positions and to most positions with responsibility for Latin American policy. Three moderate career foreign service officers (FSOs)—Terence Todman, Viron "Pete" Vaky, and William Bowdler—served as Mr. Carter's assistant secretaries of state for inter-American affairs, and the young, hawkish Robert Pastor was the Latin Americanist on the Carter National Security Council staff. It is true that extremely conservative officials had to trim their sails a bit, but they maintained their positions during the Carter administration.

This type of executive branch staffing can only be effective if

---

White House-sponsored amendment to provide $14 million in direct nonmilitary aid to the Contras. In this case as in others, the limited voting options (essentially "yea" or "nay") often mask some dimensions of the disagreement over U.S. policy, but they serve well to underscore the lack of consensus.

[8] For a detailed discussion of the lack of consensus in Congress during the first Reagan administration, see I. M. Destler, "The Elusive Consensus: Congress and Central America," in *Central America: Anatomy of Conflict*, ed. Robert S. Leiken (New York: Pergamon Press, 1984), pp. 319-335.

[9] For a contrasting view that suggests the Carter administration was fairly homogeneous, see I. M. Destler, Leslie H. Gelb, and Anthony Lake, *Our Own Worst Enemy: The Unmaking of American Foreign Policy* (New York: Simon and Schuster, 1984), pp. 97, 118-119.

there is a broad agreement on the basic contours of policy. Given a consensus on basics, divergent views on secondary subjects such as tactics and priorities are probably quite functional; competition encourages the consideration of fresh ideas. Absent both a consensus on fundamental questions and strong leadership, divergent views make policy impossible to create or implement.

It was Mr. Carter's misfortune to enter office after the consensus on basics had disappeared but before its disruptive effect on policy had been identified. As a result, the administration's policy came to be characterized by drift and indecision. It is arguable that had he been a stronger leader, President Carter would have taken a hand in forging a narrow but functional consensus. Critics may be correct when they accuse Mr. Carter of refusing to lead, of vacillating on tough issues. On U.S. policy toward El Salvador, for example, one senior White House aide complained to Raymond Bonner of the *New York Times* that Carter "was a little wishy-washy, always reluctant to come down hard on the Pentagon and their insistence on more military aid."[10] The aide's implication is that Mr. Carter wanted to reject the Pentagon's requests but could not muster the fortitude to do so. There is no reason to make this assumption. A more plausible explanation is that the President perceived a complex situation in El Salvador because he received advice from a variety of perspectives. Then, in keeping with what we know about his temperament, Mr. Carter attempted to fine-tune U.S. policy to accommodate this complexity and, lacking the appearance of decisiveness, he failed to communicate a coherent policy.

Whatever the case, policy making on issues involving Latin America degenerated into a form of bureaucratic guerrilla warfare during the Carter administration. Thus while one special envoy (Ambassador William Jorden) was in Managua telling Anastasio Somoza that he need not pay attention to the Deputy Assistant Secretary of State for Human Rights, Mark Schneider, because Schneider was, after all, "a horse's ass," Mr. Schneider was cantering around Washington halting one form of aid after another because of Somoza's human rights abuses.[11] Many foreign policy officials accept these antics as part of the push and pull of

[10] Raymond Bonner, *Weakness and Deceit: U.S. Policy and El Salvador* (New York: Times Books, 1984), p. 208.

[11] Ambassador Jorden's taped conversation with Somoza in November 1978 is transcribed and printed in Anastasio Somoza as told to Jack Cox, *Nicaragua Betrayed* (Belmont, Mass.: Western Islands, 1981), p. 323.

Washington politics, but the principal members of the incoming Reagan administration were determined not to make policy in the same fashion. They argued that this disorganized behavior cost the United States dearly in terms of its credibility. How was Somoza to know whom to believe? More important (but less to the point and certainly less central to Reagan policy makers), how many Nicaraguans would be alive today had Somoza known in 1977 that the United States would eventually cease supporting his crumbling regime? On Somoza and a hundred and one other issues, the Carter administration appeared to damage U.S. interests because it had no coherent policy toward Latin America. And it had no policy because it could not manage its own internal disagreements.

Recognizing the breakdown in consensus, President Reagan's team sought to minimize conflict within the executive branch by permitting only one point of view to be represented among the individuals making U.S. policy toward Latin America. According to one observer, the administration conducted "one of the most thorough purges in State Department history, unlike anything since the ouster of 'China hands' during the witch-hunts for 'Communists' by Senator Joseph McCarthy in the early 1950s."[12] The terms "purge" and "witch-hunt" are clearly too strong to capture accurately the Reagan approach to foreign policy staffing, but an attempt was clearly made to simplify policy making by eliminating divergent policy perspectives.

The initial signals came quickly. A few days before Ronald Reagan's inauguration, the Assistant Secretary of State for Inter-American Affairs, William Bowdler, a career foreign service officer, was told to be out of the State Department building by the time Mr. Reagan was sworn in. By tradition, high-ranking FSOs enter retirement at a department ceremony, during which they are presented with the flag that had stood in their last office. Mr. Bowdler, who had risen through the ranks during a thirty-year career to become ambassador to South Africa, Guatemala, El Salvador, and finally Assistant Secretary of State, received his flag by mail in a manila envelope.[13] Bowdler's dismissal was followed almost immediately by the recall and subsequent forced retirement of Robert White as ambassador to El Salvador and, somewhat later, by the premature retirement of Lawrence Pezzullo, another career offi-

[12] Bonner, *Weakness and Deceit*, p. 244; Destler, Gelb, and Lake, *Our Own Worst Enemy*, pp. 97, 99.

[13] *Wall Street Journal*, August 29, 1984, p. 19; George Gedda, "A Dangerous Region," *Foreign Service Journal* 60 (February 1983): 18.

cial who had been U.S. ambassador to Nicaragua. These diplomats were not simply removed from their posts; they were in effect fired from the Foreign Service. Lower-ranking career officials with ties to the policies of the Carter administration searched frantically for the closets left vacant by emerging conservatives—peripheral roles such as diplomats-in-residence at universities or glorified clerks in the office of the inspector general. James Cheek, a foreign service officer who used the State Department's dissent channel in the early 1970s to voice his fear about the consequences of U.S. friendliness to Somoza, was exiled to Nepal. According to George Gedda, "virtually all of the career diplomats who guided Central American policy during the Carter years either saw their careers set back or found themselves, like Bowdler, out of the Foreign Service altogether."[14]

In sum, wherever one looks in Washington—at the explosion of interest group activity, at the acrimony of congressional debates, at the Reagan "purge" of executive branch personnel—there is evidence of strong dissension over United States policy toward Latin America. We find ourselves in an era where disagreements are so profound that administration officials feel obliged to conduct something resembling a purge in order to make effective policy. The fact that this tactic did not work for the Reagan administration but served merely to transfer the locus of disputes to other policy-making arenas demonstrates not only the vigor of American pluralism but the depth of dissensus. Losers refuse to yield; instead, they find another battleground.

[14] Gedda, "A Dangerous Region," p. 18. Gedda has been the Associated Press's State Department reporter since 1968. An early product of the Peace Corps (1962-1964 in Venezuela), Gedda produces perhaps the most insightful regular coverage of U.S. policy toward Latin America.

One effect of the Reagan approach to executive branch staffing was to split the remaining officials into warring factions, in this case the "hardline" conservatives versus the "soft" conservatives. This internecine squabbling came to a head during the first half of 1983, when National Security Adviser William Clark convinced President Reagan to send Ambassador Jeane Kirkpatrick on a ten-day fact-finding trip to Central America. She returned with a proposal that the United States cease negotiations in favor of increased military pressure in both El Salvador and Nicaragua. When this proposal was opposed by the "softliners" in the Department of State, Mr. Reagan removed Assistant Secretary Thomas Enders and Ambassador Deane Hinton. The personnel change signalled not only a victory by the hardliners but a defeat for Secretary of State Shultz and the Department of State. For a broader discussion of this issue, see Barry Rubin, *Secrets of State: The State Department and the Struggle over U.S. Foreign Policy* (New York: Oxford University Press, 1985), especially p. 225.

INTRODUCTION

BELIEFS AND PERCEPTIONS

Contemporary disputes are the result of changes in policy makers' beliefs and perceptions. This book is an analysis of these changes. The focus is upon three factors that dominate the policy process: simplicity, stability, and security. Specifically, the thesis is that United States policy toward Latin America is largely determined by the manner in which one fact, instability, is simplified when it is perceived by U.S. policy makers; this simplification, in turn, is governed by a series of beliefs about the causes of instability and its consequences for U.S. security. Not long ago almost all officials held similar beliefs about instability, and the result was a consensus; today Washington policy makers have sharply differing beliefs about instability, and the result is dissensus.

One principal reason for these disagreements is that instability is itself a difficult phenomenon to evaluate. Although some analysts possess more knowledge than others, in any specific case it is always difficult to determine the precise mix of factors that causes instability in Latin America, and no one is so prescient as to be certain of the consequences of instability for U.S. security. But, unlike the rest of us, Washington policy makers have to make some judgment about instability, even if their judgment is that instability is irrelevant to U.S. interests. In fact, some officials will often argue that there is no reason to worry, that instability in Latin America does not threaten the United States, while others will be extremely concerned about the potential impact of instability. The important point here, however, is that policy makers must reach some conclusion about instability. That is their job. But—and here is the problem—since it is difficult to determine the cause of instability, and since no one can predict with certainty what the effects of instability might be, policy makers' conclusions must be based upon their judgment. They have to look at the available information and, using their beliefs about a host of factors—from the thought process of Central American peasants to the intended use of an unfinished airfield in Grenada—reach a conclusion. Policy makers used to reach the same conclusion because they used to hold the same beliefs. Now policy makers hold differing beliefs, and more often than not their conclusions are extremely dissimilar.

In other words, the "fact" of instability never speaks for itself; only when this fact is interpreted does it obtain meaning. This, of course, is the fundamental message of cognitive psychology: what is perceived depends upon the perceptual apparatus, particularly the belief system, of the perceiver. The most insightful analysts of an earlier era have always underscored this aspect of foreign policy. Walter Lippmann, for one, noted that "the facts we see depend on where they are placed, and the habits of our eyes"; reality to all of us, he suggested, "is a combination of what is there and of what we expected to find."[15] Similarly, Louis Halle wrote a generation ago that "what the foreign policy of any nation addresses itself to is the image of the external world in the minds of the people who determine the policy of that nation. That image may approximate the reality more or less closely, but at best it can never be quite the same thing. And it is generally different in fundamental respects."[16]

This book explores these linkages, first, between policy makers' beliefs and their perceptions of reality and, second, between policy makers' perceptions and the policies they create. Following Holsti, Jervis, George, and others, the gist of the argument is that each policy maker's perceptions "are filtered through clusters of beliefs or 'cognitive maps' of different parts of his social and physical environment. The beliefs that compose these maps provide the individual with a more or less coherent way of organizing and making sense out of what would otherwise be a confusing array of signals picked up from the environment by his senses."[17]

[15] Walter Lippmann, *Public Opinion* (New York: The Free Press, 1965), pp. 54, 76.

[16] Louis J. Halle, *American Foreign Policy: Theory and Reality* (London: George Allen and Unwin, 1960), p. 316.

[17] The quotation is from Ole Holsti, "Foreign Policy Viewed Cognitively," in *The Structure of Decision: The Cognitive Maps of Political Elites*, ed. Robert Axelrod (Princeton, N.J.: Princeton University Press, 1976), p. 19. Holsti's essay serves as an excellent introduction to a subtle theoretical literature that helps to clarify a number of aspects of United States policy toward Latin America. For further studies see in particular George's discussion of belief systems and the operational codes they produce in Alexander L. George, "The 'Operational Code': A Neglected Approach to the Study of Political Leaders and Decision-Making," *International Studies Quarterly* 13 (June 1969): 190-222; Axelrod's exploration of cognitive maps in *Structure of Decision*, especially chapters 1 and 3; Jervis's discussion of the impact of perceptions on policy in Robert Jervis, *Perception and Misperception in International Politics* (Princeton, N.J.: Princeton University Press, 1976); Robert Jervis, "Hypotheses on Misperception," *World Politics* 20 (April 1968): 454-479; and, in general, Ole R. Holsti, "Individual Differences in 'Definition of the Situation,' " *Journal of Conflict Resolution* 14 (September 1970): 303-310; John D. Steinbruner, *The Cybernetic Theory of Decision* (Princeton, N.J.: Princeton University Press, 1974); Chris Lamb, "Belief Systems and Decision Making in the *Mayaguez* Crisis," *Political Science Quarterly* 99 (Winter 1984-1985): 681-702; William A. Scott,

No one would assert that instability is the only subject of relevance to United States policy toward Latin America; rather, the following chapters argue that instability is the *principal* subject and that when it becomes widespread it is the only subject that can totally dominate foreign policy decision making. Similarly, no one would argue that beliefs and perceptions are the only factors that determine policy. But as the chapters that follow will argue, a clear understanding of policy makers' beliefs about instability are absolutely crucial to any explanation of United States policy toward Latin America.[18]

The Evolution of Postwar Beliefs

In the years immediately following World War II, a consensually held belief system—Cold War internationalism—kept debates over United States policy toward Latin America within well-defined boundaries.[19] Virtually everyone focused upon keeping the

---

"Psychological and Social Correlates of International Images," in *International Behavior: A Social Psychological Analysis*, ed. Herbert Kelman (New York: Holt, Rinehart and Winston, 1965). An initial review/synthesis is Michael J. Shaprio and G. Matthew Bonham, "Cognitive Process and Foreign Policy Decision-Making," *International Studies Quarterly* 17 (June 1973): 147-174.

[18] It should be noted that for our purposes a "correct" perception is unimportant. As Jervis notes, "sometimes it will be useful to ask who, if anyone, was right; but often it will be more fruitful to ask why people differed and how they came to see the world as they did." *Perception and Misperception*, p. 29. Our goal is to understand U.S. policy, and that is not dependent upon knowing which (if any) policy makers possess an accurate understanding of the "real" causes and consequences of instability in Latin America. For a discussion of perceptual inaccuracies, see Whitehead's study in which he observes that "perceptions in Washington of the Central American situation are what shape America's isthmian policies, and those perceptions have a surprisingly loose relationship to local realities." Laurence Whitehead, "Explaining Washington's Central American Policies," *Journal of Latin American Studies* 15 (November 1983): 358.

[19] For a discussion of the role of beliefs and perceptions in the creation of this consensus, see Deborah Welch Larson, *Origins of Containment: A Psychological Explanation* (Princeton, N.J.: Princeton University Press, 1985).

For discussions of the breakdown of the Cold War internationalist consensus, see Charles W. Kegley, Jr., and Eugene R. Wittkopf, "Beyond Consensus: The Domestic Context of American Foreign Policy," *International Journal* 38 (Winter 1982-1983): 77-106; Michael Mandelbaum and William Schneider, "The New Internationalisms: Public Opinion and Foreign Policy," in *Eagle Entangled: U.S. Foreign Policy in a Complex World*, ed. Kenneth A. Oye, Donald Rothschild, and Robert J. Lieber (New York: Longman, 1979), pp. 34-88; Ole R. Holsti, "The Three-Headed Eagle: The United States and System Change," *International Studies Quarterly* 23 (September 1979): 339-359; Michael A. Maggiotto and Eugene R. Wittkopf, "Amer-

Soviet Union and Communism out of the hemisphere. Policy disagreements were restricted to debates over important but secondary issues—the appropriate reaction to the expropriation of U.S.
investments, for example—and to conflicts over tactics, timing,
and budget priorities. Differences of opinion on these secondary
issues were resolved in the atmosphere of civility that used to
characterize the homogeneous U.S. foreign policy establishment.
These gentlemen were pleased to listen with respect to the views
of others, to offer alternative perspectives, and then to compromise quickly by splitting the difference, for the difference was always very small.

Disagreements were minor because virtually all policy makers
shared two core beliefs: they believed that instability in Latin
America was caused by Communist adventurism and, as a result
of this, they believed that instability, left unchecked, was a threat
to U.S. security. The policy that flowed from these two beliefs was
a straightforward extrapolation of the global policy of containment: stop Communist adventurism in the Western Hemisphere
by demonstrating that Communist-inspired instability will be
met with armed resistance. The high-water mark of this consensus occurred in the mid-1950s and is best exemplified by U.S. efforts to overthrow the Arbenz government in Guatemala. Looking
back through the available documents, it is evident that policy
makers perceived instability in Guatemala as little more than a
transfer of European instability to the Western Hemisphere.[20]
Guatemala was the next Greece or Czechoslovakia.

---

ican Public Attitudes toward Foreign Policy," *International Studies Quarterly* 25
(December 1981): 601-632; Ole R. Holsti and James N. Rosenau, "Vietnam, Con-
sensus, and the Belief Systems of American Leaders," *World Politics* 32 (October
1979): 1-56; Holsti and Rosenau, *American Leadership in World Affairs*; Stanley
Hoffmann, *Primacy or World Order: American Foreign Policy since the Cold War*
(New York: McGraw-Hill, 1978), pp. 14-17.

[20] In the 1970s, three authors used the Freedom of Information Act to force the
declassification of a large number of documents related to U.S.-Guatemalan relations during the period 1944 to 1954. These documents from the National Ar-
chives, the Department of State, the Department of the Navy, and the Federal Bureau of Investigation were combined with materials from the Eisenhower and
Truman presidential libraries and from the Dulles brothers' papers at Princeton
University to produce two uncommonly illuminating books: Richard H. Immerman, *The CIA in Guatemala: The Foreign Policy of Intervention* (Austin: University of Texas Press, 1982), and Stephen C. Schlesinger and Stephen Kinzer, *Bitter
Fruit: The Untold Story of the American Coup in Guatemala* (Garden City, N.Y.:
Doubleday, 1982). Immerman's bibliography is a fine guide to the documents.

By the mid-1950s, the Cold War stranglehold was beginning to weaken on issues involving Latin America.[21] In the 1952 presidential campaign, Dwight Eisenhower had accused the Truman administration of ignoring Latin America and promised to end the neglect. In mid-1953, the President's highly respected brother, Milton, conducted a month-long fact-finding tour of South America and returned home convinced that the region's grinding poverty and stunted economic development were the principal issues the United States needed to address. In Eisenhower's report, references to poverty were on almost every page, while comments about Communism were few in number and always linked to poverty: Communists would take advantage of instability caused by poverty. Eisenhower's fundamental message was that "a tremendous social ferment exists today throughout Latin America."

> Leaders of the nations to the South, recognizing that too many of their people are desperately poor, that widespread illiteracy is a handicap to progress, that educational and health facilities are woefully inadequate, and that improvement calls for capital for machinery, tools, highways, schools, hospitals, and other facilities, look to the United States for help. . . . They want greater production and higher standards of living, and they want them *now*.[22]

This message was reinforced by two prominent Republican senators, Homer Capehart and Bourke Hickenlooper, who made separate fact-finding trips to Latin America in the early 1950s. Both were impressed by the poverty of the region; both noted what soon

---

[21] Latin America was not unique. Two studies published in 1963 indicated, for example, that a significant schism had developed over both arms control and nuclear war. See Robert A. Levine, *The Arms Debate* (Cambridge, Mass.: Harvard University Press, 1963), p. 49, and Arthur Herzog, *The War-Peace Establishment* (New York: Harper and Row, 1963), pp. 5, 256.

[22] Milton S. Eisenhower, *United States-Latin American Relations: Report to the President* (Washington, D.C.: Department of State, 1953), pp. 7-8. The twenty-three-page report, an uncommonly bland document, is reprinted in the Department of State *Bulletin*, November 23, 1953. For a much better (but later) understanding of Milton Eisenhower's views on Latin America, see his *The Wine Is Bitter: The United States and Latin America* (Garden City, N.Y.: Doubleday, 1963). The President's brother was always concerned about what he called "ultra-nationalism" in Latin America and the way Communists and ultra-nationalists use one another to gain advantage from the United States. In his 1963 work he wrote that Latin Americans know that "the way to get action out of the United States is to stir up trouble or threaten to turn to Moscow for help" (p. 215).

came to be known as "the revolution of rising expectations." Hickenlooper tentatively endorsed the Bolivian revolution of 1952, and Capehart used his influence on the Senate Banking Committee to engineer legislation to permit Export-Import Bank loans for Latin American development projects. By the mid-1950s, there was a hairline fracture in the Cold War establishment; policy makers disagreed on the causes of instability.

Nowhere is this fracture more apparent than in policy makers' evaluations of Vice President Richard Nixon's disastrous goodwill tour of Latin America in 1958. Nixon faced angry demonstrators in several cities during his eighteen-day, eight-nation trip, and at his final stop in Caracas the limousines carrying the Vice President and his wife were attacked by an angry crowd, some members of which smashed the cars' windows, spat on the Vice President, and attempted to drag him from his car. Nixon escaped to the American embassy, but the demonstration was so alarming that he cut short the last stop on his itinerary and flew to Puerto Rico.

Once the Vice President was back in the United States, the inevitable congressional investigation began. The printed record indicates how policy makers had come to disagree about instability. The initial administration witnesses came before the Senate Committee on Foreign Relations to voice their opinion that the attacks on Mr. Nixon were caused by a Communist conspiracy. CIA director Allen Dulles took the lead, telling senators that the attacks were planned after the fortieth anniversary meeting of the Communist party in Moscow, held in 1957, which adopted a strategy of "hit the weak points" that was "worldwide in scope."[23] For its part the State Department reported on

> considerable evidence that the demonstrations in the various countries visited by the Vice President followed a pattern in South America and were Communist-inspired and staged. Slogans on the banners carried by students and others were similar in the different countries: "Little Rock," "Guatemala," "Yankee Imperialism," "Wall Street Agents," "McCarthyism," "Colonialism," "Nixon, Go Home," were among those repeated. The tactics were much the same with young students, urged on by older persons leading the activities. Intelligence reports from Latin American capitals also support

[23] U.S. Congress, Senate, Committee on Foreign Relations, *Executive Sessions of the Senate Foreign Relations Committee (Historical Series)*, vol. 10, 85th Cong., 2d Sess., 1958, p. 34. Made public November 1980.

the conclusion of a leading Communist role in the demonstrations.[24]

At first glance the two senators most experienced with the Latin American policy of the United States seemed to agree: Homer Capehart characterized the attacks upon Nixon as "a 100 percent Russian penetration," and Bourke Hickenlooper said the incidents were part of a "world-wide pattern of Communist stimulus."[25]

But lurking inconspicuously in an occasional comment was clear evidence of differing perceptions among policy makers. Senator J. William Fulbright noted the existence of economic problems in the region,[26] and Senator Capehart contradicted his own analysis when he observed that "this whole problem down there is divided into two parts. It is the agitation on the part of the Russians and Communists to stir up trouble, and then there is on the part of the Latin Americans a genuine dissatisfaction with particularly our trade policies."[27] Under questioning, the State Department revised its perception that the anti-Nixon demonstrations in Venezuela were only "Communist inspired"; specifically, the administration agreed that some Venezuelans were upset over Washington's support of the country's dictator:

> We gave General Jimenez [*sic*] a medal of the Legion of Merit in 1954. The presentation was made by our Ambassador at the time on November 12, 1954, in a ceremony in the Embassy in Caracas. . . . Adverse reaction in the United States and among Latin American liberal circles was taken into consideration by the Embassy at that time when it made its recommendation. Having in mind the apparent solidity of the Perez Jimenez Government, the virtual lack of any organized opposition, the corresponding belief that the proposed action would not have any practical effect on the Venezuelan political situa-

[24] Ibid., pp. 221, 251. The speaker was Robert Murphy, Deputy Undersecretary of State for Political Affairs.

[25] Ibid., pp. 205-206. The only administration official not to blame the demonstrations on Communism was the U.S. ambassador to Argentina who, years after the event, nonetheless argued in his memoirs that the hostility shown Mr. Nixon in Buenos Aires was not an indicator of broadly based animosity but merely the work of "a handful of extremists and political punks." Willard L. Beaulac, *A Diplomat Looks at Aid to Latin America* (Carbondale: Southern Illinois University Press, 1970), p. 7.

[26] *Executive Sessions of the Senate Foreign Relations Committee*, 10: 262-265, *supra* n. 23.

[27] Ibid., p. 206.

tion, the importance to our national security of additional petroleum concessions to increase reserves, and the danger of
the development of an antagonistic attitude toward United
States investments, the advantages of the decoration appeared
to outweigh the disadvantages.[28]

The often unarticulated implication of these statements was a belief that instability in Latin America, including violence directed
against the United States, is not caused only by an international
Communist conspiracy. The seed of postwar dissensus had
sprouted.

But because it was only a small Latin American sprout, nearly
everyone missed seeing it. Once Mr. Nixon was safe, Washington's focus returned to Europe and the Soviet Union where the
Cold War consensus was still in full bloom. Writing in 1961, journalist Fred Cook lamented "that there is virtually no debate, not
even on the most awesome issues that have ever confronted mankind. We wrap our increasingly militaristic society in the folds of
the flag and the emotional words of a patriotism that requires no
thought." [29]

Meanwhile, back in Latin America the political ground continued to shake. There is no need to measure individually the impact
of each salvo in the barrage of explosive events, particularly the
Cuban revolution, that shattered forever the simple bonds of Cold
War internationalism linking U.S. policy makers to one another.
Historian Charles Griffin was among the first to identify the fundamental disagreement that had developed over United States policy toward Latin America:

> It is no secret to anyone that Latin America is in a state of con-
> fusion, effervescence, and revolutionary agitation. Many
> Americans seem to regard the problem as one created by
> Communist propaganda and subversive activity, and assume
> that conditions would be basically satisfactory were it not for
> the trouble-making of the Kremlin and the Chinese Reds. Ac-
> cording to this view, anti-American riots and outbursts are
> presumed to be almost exclusively the result of Communist

[28] Ibid., p. 249. Again, the State Department spokesperson was Robert Murphy.
[29] Fred J. Cook, "Juggernaut: The Warfare State," *The Nation*, October 28, 1961,
p. 282. This entire issue of *The Nation* was devoted to Cook's analysis of the mili-
tary-industrial complex; it was a seminal critique that activists used throughout
the 1960s to buttress their attacks upon the militaristic orientation of U.S. foreign
policy.

influence. There is another view of Latin America, also widely accepted in this country, according to which the area as a whole is suffering from massive poverty, ill-health, and illiteracy—so intensified by the rapid growth of population as to produce an explosive situation.[30]

Starting with the common perception of "confusion, effervescence, and revolutionary agitation" in Latin America, by the early 1960s, U.S. policy makers clearly held conflicting explanations of why Latin Americans engage in "anti-American riots and outbursts." Over the course of the 1950s, a substantial group of policy makers came to believe that, while Communist adventurism was not unimportant, poverty was the fundamental cause of instability. They believed that many Latin Americans had been swept up by the revolution of rising expectations; tired of a life of poverty, the poor had begun to voice their complaints. To policy makers who believed this, instability represented an understandable desire to escape the deprivation into which most Latin Americans had been born.

The best way to conceptualize this evolution in policy makers' beliefs is as a continuum that arrays policy makers along a line according to their beliefs about the cause of instability. Early in the 1950s, no continuum was necessary; virtually all policy makers agreed that Communist adventurism caused instability in Latin America. A decade later, poverty was accepted by many officials as the cause of instability, while still others (like senators Capehart and Hickenlooper) were in the middle, either because they varied in their interpretation of each instance of instability or, more frequently, because they believed in multiple causality.

The result of this dispersal of policy makers' beliefs about the cause of instability was the compromise that we have come to associate with John Kennedy's administration: since there were two views on the cause of instability, there were two cures. One was a new economic aid program, the Alliance for Progress, designed to end instability by reducing the level of poverty in the region. The other cure was new or expanded counterinsurgency programs: the Army's Special Forces—the Green Berets—and AID's Office of Public Safety, designed to find and destroy Communist insurgents.

It is somewhat difficult to picture these Kennedy policy makers

[30] Charles Griffin, "On the Present Discontents in Latin America," *Vassar Alumnae Magazine* 48 (1963): 12.

as they sat down to decide on an appropriate policy toward instability in Latin America. One group of officials wanted to shoot the
Latin Americans responsible for instability because they were believed to be Communists; the other group wanted to provide these
Latin Americans with a Food for Peace shipment. The solution, of
course, was to gloss over this very fundamental difference by simply asserting that the Communists and the poor comprised two
identifiably separate groups. Shooting a Communist was concep-
tually isolated from feeding the poor, and the two groups of policy
makers went about their respective tasks.

The glue holding these two sides together was a common perception of the *consequences* of instability: all policy makers
agreed that, if allowed to continue, instability would be exploited
(if not caused) by Communists. This instability, therefore, would
lead to a threat to U.S. security. In other words, regardless of the
differences among policy makers over the *causes* of instability,
everyone in Washington retained a strong consensus on the negative consequences of instability for U.S. security. The result in the
1960s and 1970s was an uneasy compromise; when disagreements
erupted over the causes of turmoil, policy makers avoided the dispute by falling back upon their continuing accord over the deleterious consequences of instability, regardless of its origins.

Meanwhile, the evolution in policy makers' beliefs continued as
the United States emerged from its prewar isolationist cocoon.
The explosive growth of colleges and universities across the country coincided with the Cuban revolution, the Alliance for Progress, and a general interest in the subject of Third World development. Thousands of citizens began to study Latin America and
subjects related to social change and instability. At the same time,
large numbers of U.S. citizens went to live and work in Latin
America, not as relatively wealthy businesspeople, but as Peace
Corps Volunteers and as church workers. They spent time with
the poor, and they developed their own beliefs about the causes
and especially the consequences of instability. And, not inciden-
tally, they also witnessed the debacle of Vietnam, the seminal foreign policy event of the postwar era.

For whatever reasons (and there surely were many), the stage
was set for a further split among U.S. policy makers when instability flared in Central America in the late 1970s. In recent years a
second cleavage has appeared in addition to the old Communism/
poverty dispute. This new cleavage separates those who believe
instability is a threat to U.S. security from those who believe it is

not. As we will see, members of this third group of policy makers believe, first, that instability will lead to reforms that will end, or at least lessen, deprivation and injustice in Latin America. And, second, they believe that U.S. security interests would be enhanced by the existence of Latin American social and political systems that provide greater equity in the distribution of political and economic resources. These policy makers perceive this reform-motivated instability in its most positive sense; no broken eggs, no omelettes. They do not believe instability is an actual or potential threat; in interviews they are obviously uninterested in discussing either security threats in general or Communist adventurism in particular. They want to talk about the need for social reform.

Everyone in Washington recognizes that reforms aimed at rapidly decreasing deprivation and injustice would, by definition, be revolutionary in most of Latin America. These reforms would require the breakdown of the existing structure of socioeconomic privilege and the fitful creation of another that, with luck, would be more equitable and just. Members of this third group of policy makers recognize that the period of breakdown and reconstruction would be chaotic. Instability, even for a brief period, is greatly feared by most other officials because they believe that it is either (a) an indicator of Communist adventurism or (b) an opportunity for Communists to seize control of a chaotic situation and threaten the United States. This third group of policy makers, conversely, perceives instability as the price of progress not only for Latin Americans but for the long-term interests of the United States. This group is new, and if it persists, it threatens to change forever the face of United States policy toward Latin America.

This group is also relatively small. As already noted, its members went into exile or into deep cover during the Reagan years. But even during the Carter era, unabashed members of this group—officials like Mark Schneider, Patricia Derian, and Andrew Young—were few; most of those who tended in this direction found it inappropriate to tend too distinctly. In Congress, members of this group are more evident—senators Harkin, Kennedy, Dodd, and Representatives Barnes, Gejdenson, Moakley, and Studds are examples, as are a relatively large number of staffers, some of whom occupy important positions in the policy-making process. In general, however, there are few members of this group on Capitol Hill, if not so few as across town in Foggy Bottom. In the 1980s, not many policy makers argued that instability in Latin

America is desirable because it offers the opportunity for needed change.

Nonetheless, the following chapters demonstrate that this third group does, in fact, exist, and that there are several reasons to identify and analyze its members' beliefs. First, there are more members than meet the eye in Washington. Although forced to adopt a low profile in order to accommodate the tenor of the Reagan foreign policy, they did not leave town. Just as the "poverty people" had to await the death of John Foster Dulles and then the Kennedy victory in 1960 in order to obtain full membership in the post-McCarthy foreign policy establishment, so the members of this third group are waiting their turn. My interviews suggest that those with the most to lose—midcareer foreign service officers— are waiting the most quietly. They might have to wait forever, but then again they might not.

Second, this group is significant because of its mushrooming presence on the fringes of the policy-making process. Here we enter a gray area, where it is often difficult to separate genuine policy makers, on the one hand, from mere participants in the policy-making process, on the other. To be sure, the representatives of Amnesty International, Americas Watch, the Washington Office on Latin America, and especially the host of church groups interested in Latin America do not make policy, and (with rare exceptions) they therefore do not constitute policy makers by any reasonable definition of the term. But it is impossible to make sense out of U.S. policy toward Latin America without acknowledging their impact—their growing impact—upon policy. Without an understanding of these participants, for example, the close congressional votes on aid to the Nicaraguan Contras in 1985 and 1986 are simply inexplicable. The beliefs of these Washingtonians now count heavily in the construction of U.S. policy toward Latin America, and that is why they are sometimes included, always with a warning that we are discussing "participants" rather than formal policy makers in the chapters that follow.

Third, this group is significant because the scholarly community of Latin Americanists has almost wholly deserted the other two groups, thus providing the members of the third group with the ammunition they need to dominate the intellectual debate over what is actually happening in Latin America. Those of us who teach courses on inter-American relations find it increasingly difficult to create a balanced reading list for our students, one that includes rigorous scholarship from the three basic political camps in Washington. Most undergraduates can spot this imbalance instantly, and some accuse their professors of permitting personal political preferences to influence their selection of readings; in the following chapters, readers will find this imbalance to be particularly acute if they contrast the scholarly underpinnings of Chapter 2 (on poverty) and Chapter 3 (on Communist adventurism). In truth, however, professors of inter-American relations, like policy makers, are constrained by the availability of sound scholarship. Where, for example, is the scholarly competition for William Durham's *Scarcity and Survival*, Merilee Grindle's *State and Countryside*, Nora Hamilton's *Limits of State Autonomy*, Cynthia McClintock's *Peasant Cooperatives and Political Change*, or Peter Evans' *Dependent Development*? Not in the antiempirical, impressionistic musings of a Jeane Kirkpatrick or the extemporaneous current events reports of a Robert Leiken; nor, with exceptions, is it to be found in the soporific pages of mainstream foreign policy journals. The third group of policy makers can draw for support upon the largely (but far from entirely) apolitical research of scholars who have won hands down the intellectual debate over the causes and, to a lesser extent, the consequences of instability in Latin America. This is not to argue, of course, that the possession of sound scholarship is a prerequisite for victory in policy debates, much less that the pen is mightier than the sword, but only to note that the side with researchers supplying sound empirical analysis has an advantage in most policy debates, even in Washington.

In short, it makes sense to include in the analysis this third group of policy makers and related participants in the policy-making process, despite the fact that the group's members are a distinct minority in policy-making circles. Not to include this group would be to ignore strong evidence of an emerging position that is already exerting a substantial influence upon policy and, in my view, has an intellectual capacity to become a dominant participant in the closing years of the twentieth century. On the other hand, this group's political power in the 1980s has been substantially less than that of the other two groups and as a result its presence will be less obvious in the chapters that follow, for this is a book about how policy makers' beliefs influence what U.S. national security policy actually is, not how they influence what it might (or might not) become.

In any event, this is where we are in the mid-1980s. Over the course of about three decades, U.S. policy makers have undergone an intellectual evolution, splitting a strong consensus into three fragments. In the years immediately following World War II, almost all policy makers believed that Communist adventurism caused instability in Latin America. There was consensus. Then some policy makers began to believe that poverty caused instability. This second group continued to believe, however, that Communists would take advantage of instability and thereby threaten the United States. A decade passed, and the evolution continued as some policy makers, all of whom believed poverty is the cause of instability, came to believe that the consequence of instability is not to provide an opportunity for Communist adventurism but rather to provide an opportunity for reforms that can, in time, facilitate the creation of a more mature, stable security relationship with the United States. A second cleavage now lies across the first, further fragmenting the political spectrum and making compromise extremely difficult and perhaps impossible. The 1980s are therefore unlike the 1960s, when all U.S. policy makers feared instability, regardless of its cause, because of its potential negative consequences for U.S. security. Now there are three groups of policy makers involved in the making of United States policy toward Latin America, each clearly separated from the others by the distinctive manner in which its members perceive the causes and consequences of "confusion, effervescence, and revolutionary agitation"—instability—in Latin America.

The chapters that follow analyze the contemporary implications of the evolution of policy makers' beliefs about the causes and consequences of instability in Latin America. As mentioned previously, they focus upon three crucial values that orient the policy-making process: simplicity, stability, and security.

The first three chapters discuss policy makers' beliefs about *stability*. Chapter 1 examines the general concept of instability and its influence upon U.S. policy toward Latin America during the past decade. Chapters 2 and 3 focus upon policy makers' beliefs about the two principal causes of instability in Latin America: poverty and Communist adventurism.

The next five chapters focus upon policy makers' beliefs about the consequences of instability for U.S. *security*. Here there are two themes. One focuses upon strategic access: many policy makers believe that Latin America provides the United States with access to resources that are necessary for U.S. security. Instability, these officials suggest, can lead to a disruption of access. These resources are grouped into three categories: access to essential raw materials (Chapter 4), access to military bases and military support (Chapter 5), and access to vital maritime routes (Chapter 6). Throughout most of the twentieth century the determination to retain access to these three types of resources has constituted a fundamental rationale for United States policy toward Latin America. Today, however, significant differences have developed among policy makers over the importance of strategic access to the region.

The second theme focuses upon strategic denial: many policy makers believe that Latin America contains resources that the United States may not want but that we cannot permit others— specifically the Soviet Union—to have. These resources can be grouped into two categories: military bases (Chapter 7) and a vaguely defined but absolutely crucial category that can be labeled the balance of world power (Chapter 8). On these two issues of strategic denial, policy makers have come to divide extremely sharply. As we will see, nothing is more divisive than the debate over Latin America's role in the global balance of power.

The only one of the three values that does not receive a separate set of chapters is *simplicity*; its role in policy making will be woven into discussions throughout this volume. Here, however, it is important to make explicit the meaning of simplicity in foreign policy making. Specifically, it is necessary to discuss, first, how simplicity serves both as a value—something policy makers strive to achieve—and a tactic—a weapon policy makers use to defeat their opponents. And second, it is crucial to recognize that simplicity does not mean simplistic, and that someone who simplifies is not necessarily simple-minded.

SIMPLICITY

Everybody simplifies reality, including Washington policy makers who create United States policy toward Latin America and social scientists who write books that array complex human beings on two-dimensional continua. The focus here, however, is upon the former group of individuals who must manage an unusually strong tension between the human tendency to simplify reality, on the one hand, and the complexities of both Latin American instability and U.S. security, on the other.

The simplification begins with senior officials, who tend to ignore most of the countries of the region. That simplifies their world considerably. These senior officials are extraordinarily busy people, and under normal circumstances they do not have time for the relatively unimportant places in Latin America. They have to concentrate upon relations with the Soviet Union, the nations of the Middle East, Japan, China, our major NATO allies, Mexico, and occasionally Brazil. Alongside these concerns, Uruguay and El Salvador simply do not matter very much.

Instead of spending time on relations with these less important places, senior policy makers assign lower-ranking officials to watch over U.S. interests in Latin America. Under normal circumstances these Latin Americanists do their job in much the same way that art museum guards watch over their paintings. They are essentially guardians; their job is to protect what they are given. Most of the time this task is challenging but fairly unexciting, like grading term papers or filing an income tax return. Every so often, however, instability flares. Someone (or some group) starts running around the galleries of the Latin American museum, readjusting the paintings, talking in loud voices, and generally disturbing the desired atmosphere of calm. When this occurs, the guards sound the alarm, senior officials come running, and the museum analogy ends.

As we will see, the senior officials want to know the answer to two and only two questions: 1) What is the cause of the instability? 2) What are the consequences of the instability for U.S. security? In addition to this simplification senior officials also ask for simple answers. Lower-level officials know that busy people are asking the questions; there is no time for a seminar on inter-American relations or political development. "Like Kissinger, Brzezinski . . . could explain problems and ideas simply, put them in words that politicians could use in speeches and television appearances, give them a clever phrase to catch a headline. He understood their dilemmas."[31] So do lower-level officials, nearly all of whom could be coaxed into accepting the position of National Security Adviser; the policy-making environment rewards those who can simplify. The result, as Laurence Whitehead has observed, is that "Washington policy debates are, in general, conducted on the basis of capsule statements about complex and unfamiliar foreign conditions."[32] This is one of the principal forms of

[31] Destler, Gelb, and Lake, *Our Own Worst Enemy*, p. 95.
[32] Whitehead, "Explaining Washington's Central American Policies," p. 358.

simplification that will be found throughout the following pages.[33]

Lower-level policy makers also engage in three additional forms of simplification. First, these officials tend to simplify in the way they define the position of their opponents. Read carefully how an old hand at bureaucratic battles, Assistant Secretary of State Thomas Enders, explained instability in Central America. First, there is complexity:

> Central America has deep political divisions, among nations as well as within them. It suffers severe economic troubles, with the world recession devastating economies already weakened by high oil prices and internal inefficiencies. And it is fragmented by social tensions, with population growth straining public services and popular aspirations outrunning the historically possible.
>
> But [and here comes the simplification] the tangle of violence that has taken so many lives traces directly to the clash of two polar approaches to these problems. One is the way of the violent right—to ignore socioeconomic problems and, when that proves impossible, to shoot the messengers of despair. The other is the way of the violent left—to magnify injustices and provoke confrontations so as to rationalize shooting their way to power.[34]

From the first paragraph, it seems clear that Mr. Enders does not believe—and, indeed, I have never met a policy maker familiar with Central America who believes—that the instability plaguing the region in the 1980s was simply the clash of the violent right and the violent left. Everyone believes it is a complex situation. Yet in the second paragraph he defined the combatants as no more than the polar right shooting the messengers of despair and the violent left shooting its way into power. Then, having simplified the position of his opponents by identifying them as extremists, Mr. Enders offered the administration's proposal: a third, complex alternative, the Duarte government. Mr. Enders thereby positioned Mr. Duarte as the "reasonable" alternative. He proposed a policy

[33] Capsule statements—summaries—are not necessarily simplifications. For some officials they represent the distillation of carefully reasoned, complex arguments. As we will see, however, this is not often the case in the making of United States policy toward Latin America.

[34] Speech to the Commonwealth Club of San Francisco, August 20, 1982.

that recognized complexity; he identified the other positions as simplistic and therefore unreasonable.

What Mr. Enders did is what policy makers will never admit to doing but what, in fact, they do every day: he simplified his reality by dismissing as simplistic the position of his opponents. First, he separated the Duarte government from the violent right—no easy task in a country where many argue that the military, over which Mr. Duarte had nominal control, *is* the violent right. Second, by saying that the insurgents in El Salvador represented no more than violent leftists shooting their way to power, he freed himself from having to negotiate with them. Not surprisingly, Mr. Duarte, who became something of a policy maker himself during his numerous trips to Washington, endorsed the Enders simplification. Of his principal opponent on the right, Duarte was asked: "Are you calling D'Aubuisson a Nazi Fascist?" His response: "I am absolutely convinced of that." Of the opposition on the left, specifically the Mothers of the Disappeared, Duarte remarked: "Have you seen who they are? There are two types. There are those with white scarves on their heads, who are not mothers but, rather, Marxist activists, and they're the ones who run the show."[35] The choices were simplified to three: Nazi Fascists (the violent right), Marxist activists (the violent left), and José Napoleón Duarte (the reasonable alternative).

The tactic of simplifying the position of one's opponents is itself one of the most frequently observed simplifications in Washington. It is the way policy makers disagree with one another. Career foreign service officers are particularly frequent users of this approach when dealing with outside criticism. Critical analyses are never "wrong" but "oversimplified." Outsiders are never "mistaken"; rather, they do not understand the complexities of the situation being discussed, generally because they have not seen the cable traffic. Because everyone agrees that simplicity distorts reality, to accuse opponents of simplification is to dismiss them without having to deal with the substance of their arguments.

Second, policy makers simplify by asking a narrow set of questions about inter-American relations. Simplification begins when policy makers select an issue and then make that issue the totality of Latin America's reality. President Carter quite clearly made human rights the only subject of broad discussion in the first two

---

[35] Marc Cooper and Greg Goldin, "Playboy Interview: José Napoleón Duarte," *Playboy* 31 (November 1984): 72, 73.

years of his presidency; similarly, instability in Central America became the only issue of significance to most policy makers during the final months of the Carter administration and throughout the Reagan administration. The historical record indicates, moreover, that whenever instability flares in Latin America, the decks are cleared of all other issues.

In the case of instability the simplification continues when policy makers focus upon two, and only two, aspects of the problem: its causes and consequences for U.S. security. This narrow focus reflects in part the demands of senior policy makers, but there is no rule that keeps lower-level officials from thinking about other aspects of the problem. Yet most policy makers do not ask, for example, *whether* instability has any significant consequences for U.S. security. They assume that it does. In a sense, this complicates officials' lives; they must pay attention to everything. On balance, however, this assumption of importance simplifies policy making because it permits officials the freedom of never having to answer the question of significance. This simplification will appear repeatedly in the chapters that follow.

The process of simplifying instability then proceeds one final step, as most policy makers adopt simplified views of the available alternatives. To Mr. Enders, who asserted (and perhaps believed) that instability in El Salvador was caused by the polar right shooting the messengers of despair while the violent left was shooting its way to power, there was no alternative but Mr. Duarte. His focus, like that of almost all his colleagues, is on the search for moderation. In selecting this focus, policy makers tend to make several simplifying assumptions, not all of which are deserving of uncritical acceptance. First is the assumption—probably correct in many cases—that the center is good, that extremes are bad, and therefore that moderation is a desirable feature of political life. Second—and more questionable—is the assumption that the center exists in each Latin American society and will thrive if it simply can be found and nurtured by the United States. Mr. Duarte, for example, was found in Venezuela, where he had been living for seven years. Third—and highly suspect—is the assumption that Latin American politics are fundamentally unlike politics in the United States in the sense that moderation results *not* from a balance of left and right but rather from the elimination of both the left and the right (conveniently labeled the "polar" right and the "violent" left) and the placement of political power in the hands of those who remain: the moderates in the center.

Perhaps the greatest irony of inter-American relations is that the complex dynamics underlying the moderation of the U.S. political system is rejected by most policy makers' approach to Latin America. Moderation, which in the United States is the result of compromises hammered out in the course of extraordinarily complex political battles, is the price Washington officials attempt to charge as an entrance fee for participation in political life in Latin America. In El Salvador in the 1980s, the right had to agree to reforms that they opposed with every fiber of their being; the left had to agree to accept the existing structure of socioeconomic privilege, modestly altered by a half-hearted agrarian reform. Those who refused to become moderates were excluded—witness the fate of Roberto D'Aubuisson—as the Salvadoran political spectrum was condensed, simplified.

Third, policy makers simplify by permitting urgency to determine what is important. At a symposium in late 1984 on democracy in Latin America, the Deputy Director of Policy Planning in the State Department's Bureau of Inter-American Affairs (ARA), Michael Skol, took the floor to comment: "I'd like to look at something I think has not been sufficiently emphasized, and that is the incredible weight of stereotypes, particularly the stereotype of Latin America's capacity for democracy. I will deal in benchmarks, in headlines here. In this case, of course, what I am doing is repeating the error of relying on stereotypes, because I do not have the time for anything else."[36] Other participants in the discussion seemed to have plenty of time; many of the professors talked interminably. And Mr. Skol did not seem to be going anywhere. Yet he had to use the shorthand of stereotypes. The pressure of time forced him to simplify.

Washington policy makers are extremely busy people. It is difficult for them to find the time to talk with an inquisitive researcher, and when an appointment is made it often has to be rescheduled or shortened. Sitting down in a junior or middle-level policy maker's office, researchers accustomed to hiding in their library carrels will be surprised by the hum of activity: incoming telephone calls, secretaries bearing memoranda, colleagues popping their heads in the door for a quick word. One cannot help but notice the mountain of paper that apparently awaits reading or processing of some sort. At the end of an interview, after the re-

[36] Skol's comments are printed in *AEI Foreign Policy and Defense Review* 5 (Winter 1985): 41-43.

searcher has collected the necessary data and is heading for the door, an official will frequently comment that the research project seems interesting and ask to see a copy of any publication that might be produced. Waving an arm across the sea of white that covers most desks in Washington, policy makers often close an interview by remarking: "As you can see, we don't have much time to do your kind of work."

Policy makers' working hours are occupied by a truly extraordinary array of immediate demands. Many of these demands are fairly routine but urgent in the sense that there is always an approaching deadline. In the morning most officials concentrate on preparing for the afternoon; in the evening they prepare for the next morning. If there is a meeting, the papers prepared for the meeting must at least be skimmed. If Senator Nunn's aide has requested the name of the individual in Asunción who is responsible for the Paraguayan government's purchase of widgets from an Atlanta manufacturer, a cable must be sent to the embassy and an answer provided before the aide complains to a more senior official or, worst of all, before the assistant secretary receives a call from The Senator Himself.

Many of the demands are unexpected. If there was a riot last night in Lima, Washington must learn if there will be another today; instability is important. If there was a guerrilla clash in Quetzaltenango, Washington must learn what group did it, how they did it, when or whether they can be expected to do it again, and what, if anything, the United States should do about it. If there is misery, injustice, deprivation, and oppression in Paraguay, well, that's no crisis, nothing unusual. Misery has always existed in Paraguay. Officials believe that injustice and deprivation can eventually lead to instability, but there is no instability now in Paraguay. Until there is, no one in Washington will be able to devote much time to the issue of injustice and deprivation in that country. The reason why, as Henry Kissinger observed long ago, is that "problems are dealt with as they arise. Agreement on what constitutes a problem generally depends on an emerging crisis which settles the previously inconclusive disputes about priorities."[37] As Mr. Kissinger's comment suggests, permitting urgency to dictate the focus of policy simplifies the lives of policy makers; it elimi-

[37] Henry A. Kissinger, "Domestic Structure and Foreign Policy," in *International Politics and Foreign Policy*, rev. ed., ed. James N. Rosenau (New York: The Free Press, 1969), p. 268.

nates one of the great dilemmas of the human race: what to do next. In foreign policy making one does not have to decide; one picks the most urgent task and, in the case of Latin America, that means the case with the most instability.

In discussions of U.S. foreign policy this issue is typically framed as a conflict between a short-term and a long-term perspective. With policy makers' eyes confined to the near horizon, "short-term realism usually prevails over long-term realism in the decision-making process," writes Robert Johansen, "not because of its greater validity, but because of its greater urgency in the minds of decision-makers."[38] When the short term wins, urgency has been awarded the power to determine the attention given any particular issue of international relations. And because so many issues have been declared urgent, little time remains for examining broader, less urgent (and extremely complex) issues such as misery, injustice, deprivation, and oppression—issues that might, in the end, have a determining impact on U.S. security interests in Latin America.

Social science, too, rewards simplicity. Other things equal, anyone who can explain U.S. policy with two continua (poverty/communism and threat/no threat) is more likely to be noticed than someone who needs a factor matrix printed on a fold-out page. As I developed my interpretation, I would discuss it with friends in Washington—some policy makers, some analysts—all of them perceptive observers. Most agreed that the two continua were a useful if rough way of describing contemporary policy disputes, but all argued that *they* could not be positioned on the continua. They moved around, depending on the specific circumstances of each instance of instability, or they needed more than two dimensions. They said I was being too simplistic.

I cannot concede the point. Unquestionably, to classify policy makers solely on their beliefs about two issues is crude. And, also unquestionably, the reality perceived by some policy makers is much more subtle. But the evidence presented in the following chapters suggests that it is not unrealistic to think of *most* officials as arrayed along two continua, from those who believe poverty is the only cause of instability to those who believe Communist adventurism is the only cause and from those who perceive a

[38] Robert C. Johansen, "Toward an Alternative Security System," World Policy Paper No. 24 (New York: World Policy Institute, 1983), p. 2; David Scott Palmer, "Military Governments and U.S. Policy: General Concerns and Central American Cases," *AEI Foreign Policy and Defense Review* 4 (1982): 28.

severe threat to U.S. security in any act of Latin American instability to those who perceive no threat whatsoever.

In my interviews with policy makers, I asked a broad variety of
questions on a single theme: "Why is Latin America important to
U.S. security?" I received a bewildering array of responses. But as
I sat and pondered these responses, a pattern emerged. The respondents clustered along two continua. These continua represent what policy makers believe about the importance of Latin
America to U.S. security *abstracted to represent central tenden-
cies*. It is the pattern that I found in the interviews, the documents,
the speeches and debates. Specifically, the data indicate that there
is only one crucial focus—instability—that there are only two
questions raised about instability—its cause and its consequences—and that *most* policy makers begin to answer these
questions by protesting complexity and end by depositing themselves fairly clearly on some point on the two continua. Not every
policy maker fits on a single point, but most do; and the act of
identifying one's point is only the last in a series of simplifications. Two dimensions will never capture all of a complex reality,
but here, I think, it captures most of what divides policy makers
today as they struggle to make United States policy toward Latin
America.

· 1 ·

INSTABILITY IN LATIN AMERICA

"Since the early nineteenth century, the primary interest of the United States in Latin America has been to have the area be a peaceful, secure southern flank."[1] Thus begins the brief discussion of Latin America in the most widely adopted national security textbook of our time. Few U.S. citizens recognize how difficult it is for policy makers to be given these obvious but vague goals—peace and security—and told to convert them into concrete public policy. It is a staggering task. No U.S. citizen knows, for example, how to make Bolivia peaceful and secure. Frustrated with attempts in this direction, in 1959 one embassy official in La Paz threw up his arms and voiced the despair that afflicts many foreign service officers who seek peace and security in Latin America. We had given $129 million in aid to Bolivia since 1953, he complained, and "we don't have a damn thing to show for it. We're wasting money. The only solution to Bolivia's problems is to abolish Bolivia."[2]

In a casual conversation in early 1983, another beleaguered U.S. official, this time in San Salvador, was told about his predecessor's

[1] Amos A. Jordan and William J. Taylor, Jr., *American National Security: Policy and Process* (Baltimore: The Johns Hopkins University Press, 1981), p. 436.

[2] Publication of this remark in *Time* magazine's Latin American edition (March 2, 1959) had the predictable result of decreasing both Bolivian peace and U.S. security: anti-American demonstrations in La Paz and other Bolivian cities led to the hasty destruction of sensitive files and the hurried evacuation of the U.S. embassy. The derrogatory comment also led to problems for Claire Booth Luce, wife of *Time* publisher Henry Luce and the nominee for U.S. ambassador to Brazil. In an era when wives were presumed to agree with everything their husbands (and, should they happen to own one, their husband's magazine) said about foreign affairs, Ms. Luce had difficulty overcoming charges that she shared her spouse's alleged insensitivity to Latin America. It might be added that Ms. Luce was capable of creating problems for herself, particularly with Senator Wayne Morse, chairman of the Senate Foreign Relations Subcommittee on Western Hemisphere Affairs, for whom she held scant affection. Shortly after surviving the confirmation battle, she remarked that her difficulties with the U.S. Senate originated "some years back when Senator Morse was kicked in the head by a horse." Confirmed as ambassador to Brazil on April 28, 1959, she resigned three days later. U.S. Congress, Senate, Committee on Foreign Relations, *Executive Sessions of the Senate Foreign Relations Committee (Historical Series)*, vol. II, 86th Cong., 1st Sess., 1959, made public 97th Cong., 2d Sess., 1982, especially p. 247.

frustration in Bolivia and asked what he thought about the idea of abolishing El Salvador. "There's no comparison," he responded; "we've got a handle on stability here."[3]

In the latter official's response was a statement about the process U.S. policy makers use to create a peaceful, secure southern flank. Unable to implement the vague general goals of "peace" and "security," Ambassador Hinton had subconsciously replaced them with something more concrete, something that could be implemented: stability. He and his colleagues had determined what *means* would best promote the macrogoals of peace and security and then converted that means into the goal of U.S. policy. This conversion process is not peculiar to inter-American relations. "Because most important goals (e.g. security, high influence) are too general to provide guidelines for actions," Jervis notes, "actors must establish subgoals (e.g. strong alliances, military perponderance) that are believed to contribute to the higher ones." Subgoals—more accurately, means—are thereby transformed into ends, and they "often come to be valued for their own sakes, especially when their attainment requires a great deal of time, effort, and attention." Particularly in large bureaucracies the mere act of doing something (ensuring stability) in time tends to become divorced from the goals (peace and security) that the act was intended to achieve. In this sense inertia determines policy; as John Barth once observed, "processes persisted in long enough tend to become ends in themselves."[4]

In the aftermath of World War II, the United States passed through a brief period during which it felt obliged to commit whatever resources were necessary to ensure stability throughout the non-Communist world. Billions of dollars and tens of thousands of soldiers' lives were invested in securing the peace in such faraway places as Southeast Asia. After the sobering experience of Vietnam, however, America's ambitions moderated. Today any official who asserted that U.S. peace and security are dependent upon the maintenance of stability in, say, Burma, would be dismissed as

[3] Interview with Ambassador Deane Hinton, San Salvador, January 5, 1983.

[4] Robert Jervis, *Perception and Misperception in International Politics* (Princeton, N.J.: Princeton University Press, 1976), pp. 411-412; John Barth, *The Floating Island* (New York: Avon Books, 1956), p. 59. Holsti, too, notes that "means have a way of becoming ends" in the implementation of foreign policy. Ole R. Holsti, "The Study of International Politics Makes Strange Bedfellows: Theories of the Radical Right and the Radical Left," *American Political Science Review* 68 (March 1974): 235.

an antiquated Cold Warrior. But any official who *failed* to make such an assertion in 1962 would have been accused of misunderstanding the fundamental nature of U.S. security interests. Time passes, and the early postwar conception of stability as a global requirement of U.S. peace and security has been undergoing a substantial reappraisal.

It is a quiet reappraisal, manifested less by open confrontation and debate than by simply dropping out of the discussion. In 1983, members of Congress yawned, doodled, and kibitzed with aides as Secretary of State Shultz, pleading for a larger foreign aid budget, used a tired metaphor to express an exhausted idea: "The fault line of global instability runs strongly across the continents of the Third World. This instability is inimical to our security."[5] It was a boring message from a tedious speaker to an audience numbed by forty years of investing heavily in Third World stability. By the 1980s, the purses of few taxpayers could be pried open by citing something called the fault line of global instability. The "bear any burden, pay any price" language that symbolized the Kennedy commitment to stability has, for better or for worse, become a curious relic of an overly ambitious era. Now policy makers have more modest goals. Peace, operationalized by a subgoal—the absence of instability in countries ruled by governments friendly toward the United States—is perceived as a requirement of U.S. security in a more limited number of places.[6] For most U.S. policy makers, Latin America remains one such place.

[5] Prepared remarks before the House Committee on Foreign Affairs, February 16, 1983. This type of rhetoric is completely bipartisan. For a similar statement by a Carter administration official, see U.S. Congress, House, Committee on Foreign Affairs, *Foreign Assistance Legislation for Fiscal Year 1982*, 97th Cong., 1st Sess., 1981, pt. 2, pp. 42-43.

[6] Perhaps it is unnecessary to point out that there is no intrinsic value in stability. To U.S. policy makers, most of the time stability is good, but sometimes it is bad, and on occasion (such as, for example, when one has been assigned to a tour of duty in Montevideo) stability is neither good nor bad but just boring. Throughout most of the 1980s, for example, the United States sought to promote peace and security in Latin America by encouraging stability in El Salvador and instability in nearby Nicaragua. In the mid-1970s, Representative Michael Harrington coined a new verb, "to destabilize," to describe U.S. policy toward Allende's Chile. And in a *much* earlier era, complained Prince Metternich, the United States perversely encouraged instability throughout Latin America: "In fostering revolutions wherever they show themselves, in regretting those which have failed, in extending a helping hand to those which seem to prosper," U.S. officials "lend new strength to the apostles of sedition, and re-animate the courage of every conspirator." Metternich's 1824 letter, which was written in reaction to the Monroe Doctrine, is translated into English and reprinted in Dexter Perkins, *The Monroe Doctrine, 1823-1826*

In the chapters that follow it will be argued that the high value most policy makers place upon stability in Latin America is not a peculiar feature of the Cold War era. The quest for stability has been the basis for U.S. policy toward Latin America for nearly two centuries.[7] The early twentieth century, for example, was not unlike today. Kane tells us that "the constant state of political disorder in the Caribbean was generally regarded by American military strategists with something approaching horror."[8] To Theodore Roosevelt, "the specter of German aggression in the Caribbean or elsewhere in Latin America became a veritable nightmare with him. He was absolutely convinced that the Kaiser would one day start trouble somewhere in this hemisphere."[9] Over time the precise *form* of the threat has changed (from the Holy Alliance to German imperialists to Fascists to Communists), but the *nature* of the threat has remained fairly constant. This nature was identified in 1904 by President Roosevelt as "a general loosening of the ties of civilized society"—instability.[10]

(Cambridge, Mass.: Harvard University Press, 1927), p. 167. Overall, it should be emphasized that there are relatively few examples of the United States encouraging instability in Latin America. Since early in the nineteenth century Washington policy makers have overwhelmingly agreed that, other things being equal, the road to U.S. peace and security is paved with stability in Latin America.

[7] The term "instability" is subject to a variety of definitions, and, as luck would have it, the one used by most social scientists is not the same as the one used by policy makers. Within academic circles political instability "somehow means the absence of violence, governmental longevity, the absence of structural change, legitimacy, and effective decision making." Leon Hurwitz, "Contemporary Approaches to Political Stability," *Comparative Politics* 5 (April 1973): 463. This is too broad a definition for policy makers. Their definition of instability (and, therefore, my definition here) focuses upon violence. It has been stated most precisely by the Feierabends as "the degree or the amount of aggression directed by individuals or groups within the political system against other groups or against the complex of officeholders and individuals and groups associated with them. Or, conversely, it is the amount of aggression directed by these officeholders against other individuals, groups, or officeholders within the polity." Ivo K. Feierabend and Rosalind L. Feierabend, "Aggressive Behaviors within Polities, 1948-1962: A Cross-National Study," *Journal of Conflict Resolution* 10 (September 1966): 250.

[8] William Everett Kane, *Civil Strife in Latin America: A Legal History of U.S. Involvement* (Baltimore: The Johns Hopkins University Press, 1972), p. 60.

[9] Quoted in Harold Sprout and Margaret Sprout, *The Rise of American Naval Power, 1776-1918* (Princeton, N.J.: Princeton University Press, 1942), p. 253. See also Henry Cabot Lodge, ed., *Selections from the Correspondence of Theodore Roosevelt and Henry Cabot Lodge, 1884-1918*, 2 vols. (New York: Charles Scribner's Sons, 1925), 1: 484, where Roosevelt notes "the evident intention of the German military classes to take a fall out of us when the opportunity arises."

[10] More recent than President Roosevelt's 1904 announcement of his corollary to the Monroe Doctrine is a 1953 analysis by the National Security Council that

37

Instability *per se* is not the issue, of course. Few U.S. policy makers would be concerned if Salvadorans or Guatemalans or Haitians spent their time shooting one another were it not for the fact that one possible consequence of this instability might be to provide hostile forces with the opportunity to seize territory in Latin America and then use it to threaten U.S. security. Thus geographic proximity is crucial in determining the high value U.S. policy makers place upon the maintenance of stability in Latin America. In a geopolitical sense Latin America is, as every president since James Monroe has repeatedly remarked, "our backyard." Some officials call Latin America our sphere of influence. Yet, just as the Soviets have discovered in Afghanistan and Poland, we too have found that the maintenance of a geographic sphere of influence is no easy task. What is ours is also ours to lose. That is why analysts in the Defense Intelligence Agency argue that any change in Latin America "will most likely work to our disadvantage."[11]

The emphasis U.S. policy makers place upon stability in Latin America is largely a reflection of the fact that the United States is king of the surrounding hemispheric mountain, and from the summit the only way to go is down. It may not be pleasant to be forced off the summit in Vietnam, but we can live with defeat in a far off corner of Southeast Asia. Latin America is an entirely different matter, however. "Central America is America. It's at our doorstep," President Reagan asserted in his speech to the nation in May 1984. Because of this geographic proximity, if we permit instability there it will "eventually move chaos and anarchy toward the American border."[12] This basic causal linkage—instability in Latin America causes a threat to United States security—is the cognitive bedrock of United States policy toward Latin America.

Not long ago Richard Barnet noted that nearly everyone who made U.S. foreign policy owned a piece of this rock. Writing in 1972, he found that to all policy makers "the real enemies are chaos and disorder."[13] If this were true today, there would be no

---

found "a wide-spread tendency toward immediate political change which produces instability" to be the first of "three basic problems which adversely affect our relations with Latin America." U.S. National Security Council, "U.S. Objectives and Courses of Action with Respect to Latin America," mimeographed, NSC Report 144, March 6, 1953, p. 3.

[11] U.S. Congress, House, Committee on Foreign Affairs, Subcommittee on Inter-American Affairs, *Impact of Cuban-Soviet Ties in the Western Hemisphere, Spring 1980*, 96th Cong., 2d Sess., 1980, p. 36.

[12] *Weekly Compilation of Presidential Documents* 20 (May 14, 1984): 676-677.

[13] Richard J. Barnet, *Roots of War* (New York: Atheneum, 1972), p. 74.

fundamental differences over an appropriate policy to pursue in Latin America. But, as the following chapters will demonstrate, it is not true. Because U.S. officials have come to perceive instability in quite different ways, they have naturally divided over what, if anything, needs to be done about it. The quiet reappraisal has reached our backyard.

The place to begin an analysis of this reappraisal is with U.S. policy makers' perceptions of the cause of instability in Latin America. Like medical doctors attempting to diagnose the ills of ailing patients, the first step to take is to listen as the patients describe their symptoms. That requires an initial focus upon U.S. policy toward Central America in the 1970s and 1980s. What officials saw in Central America tells us as much about the perceptual apparatus of U.S. policy makers as it does about the actual instability. Central America served as a focus of the policy debate, throwing into sharp contrast the conflicts that have come to plague U.S. policy.

INSTABILITY IN CENTRAL AMERICA: THE CARTER POLICY

When the policy makers appointed by President Carter sat down at their new desks in early 1977, Latin America was not on the agenda left behind by the departing Ford administration. Most of the instability that had existed in the Southern Cone was gone or, in the case of Argentina, going fast. Brazil, too, had quieted after a spate of political unrest earlier in the 1970s. In the Caribbean, Fidel Castro appeared to be occupied for the moment in rebuilding normal diplomatic relations with many of the region's governments. While this gave pause to some officials, particularly when they focused on Cuba's growing friendship with the Manley government in Jamaica, nearly everyone was relieved to note that there were no major groups of Castroite guerrillas active in Latin America. In Nicaragua there were the Sandinistas, of course, but in early 1977 they numbered perhaps one hundred and therefore showed absolutely no indication of being able to destabilize the Somoza dynasty. Two months before Jimmy Carter's election the Defense Intelligence Agency reported that "at this time, no insurgent group poses a serious threat to an existing government."[14]

[14] U.S. Congress, House, Committee on International Relations, Subcommittee on International Political and Military Affairs, *Soviet Activities in Cuba—Parts VI and VII: Communist Influence in the Western Hemisphere*, 94th Cong., 2d Sess., 1976, p. 107.

The absence of significant instability permitted a considerable amount of latitude in the making of United States policy toward Latin America, and the opportunity was seized by policy makers committed to increasing the prominence of human rights concerns in United States policy toward Latin America.

No one is certain exactly why Central America became one of the primary foci of the Carter human rights policy, but there are several plausible explanations. One is that the governments of El Salvador, Guatemala, and Nicaragua were, in fact, uncommonly repressive. Proponents of this explanation argue that these governments' violations of human rights were too gross to be overlooked. Another explanation is that the governments of El Salvador and Guatemala called attention to themselves in early 1977 by announcing that the United States could keep its military aid now that the receipt of assistance had been made dependent upon a State Department human rights report on each recipient.[15] A further explanation is that Mark Schneider, the Deputy Assistant Secretary of State for Human Rights, had a particularly soft spot in his heart for Central America, for he had spent two years in El Salvador as a Peace Corps volunteer. A final explanation is that many national security officials were reluctant to put up much of a fight with human rights activists over control of U.S. policy toward Central America, for they needed to husband their resources for more important battles over U.S. human rights policy toward Iran, the Philippines, and South Korea.

For whatever reasons—and it was probably some combination of all of these—human rights quickly came to dominate U.S. policy toward Central America. In one of his first appearances before Congress as Secretary of State, Cyrus Vance announced that the proposed military assistance budget for FY1978 had been reduced to exclude funds for El Salvador and Guatemala.[16] Eventually U.S.

---

[15] In 1976, when Congress rewrote Section 502B of the Foreign Assistance Act of 1961, it mandated the reports on all recipients of military aid. This revision of Section 502B was part of the International Security Assistance and Arms Export Control Act of 1976 (PL94-329). In 1979, Congress rewrote the reporting provisions for the *economic* aid program to require State Department human rights reports on all members of the United Nations. See Section 504 of the International Development Cooperation Act of 1979 (PL96-53), which amended Section 116 of the Foreign Assistance Act of 1961.

[16] Brazil, the other Latin American country that had rejected U.S. military aid, was also included on the Secretary's list. A fourth Latin American country, Uruguay, also balked at the report, but Congress had already terminated military aid to that country, so the Carter administration eliminated its small economic aid program.

economic assistance to El Salvador, Guatemala, and Nicaragua was halted, decreased, or redirected in order to bypass agents of the offending governments. A parade of U.S. officials flew in and out of Central America to communicate directly the concern of the United States over continued reports of gross violations of fundamental human rights. Within less than a year instability had increased dramatically throughout most of the region.

Although the growing instability in Nicaragua quickly drew the attention of the Carter administration's national security officials, initially there was little they could do. Throughout late 1977 and early 1978, the administration was deeply committed to an exhausting uphill battle for Senate confirmation of the Panama Canal treaties. Under the stern direction of the President—the only firm guidance he ever provided on U.S.-Latin American relations—no one in the administration was permitted to raise the issue of unrest in Central America, for to do so would risk upsetting already jittery senators and, in particular, revealing the active support Panamanian strongman Omar Torrijos was providing the Nicaraguan insurgents.[17] All administration officials did their best to assuage any concern on the part of Congress. No one mentioned the mounting level of violent opposition to the Somoza government.

With the treaties safely through the Senate, the growing fears over Nicaraguan instability became more apparent. By mid-1978, the Sandinistas had emerged as a popular force. Crowds of applauding citizens lined the road to the airport in August as Edén Pastora led his forces out of Managua after successfully capturing the legislative assembly and extracting humiliating concessions from Anastasio Somoza. In September 1978, the Carter administration acknowledged that the Somozas had to go, and it stopped the military assistance pipeline.[18]

From that moment on the growing instability in nearby El Sal-

[17] Studies of President Carter's active supervisory role, which contrasted starkly with his lack of leadership on other issues of inter-American relations, have been published by two participants in the treaty process: George D. Moffett, III, *The Limits of Victory: The Ratification of the Panama Canal Treaties* (Ithaca: Cornell University Press, 1985), and William J. Jorden, *Panama Odyssey* (Austin: University of Texas Press, 1984).

[18] No new Foreign Military Sales (FMS) commitments had been permitted since September 1977, but the pipeline of committed but undelivered FMS aid remained open for an additional year. International Military Education and Training (IMET) aid was not discontinued until February 1979. U.S. Congress, House, Committee on Foreign Affairs, Subcommittee on Inter-American Affairs, *United States Policy towards Nicaragua*, 96th Cong., 1st Sess., 1979, p. 39.

vador became an issue of U.S. national security. With one Central American nation already engulfed in a full-scale civil war, the Carter administration was not prepared to let another revolutionary situation develop. But after nearly two years of publicly denouncing El Salvador's human rights abuses, few officials were prepared to reverse U.S. policy and support the Romero government's indiscriminantly repressive attempts to defeat the rapidly growing guerrilla movement. While leftist in orientation, the insurgents had firmly rejected cooperation with El Salvador's Moscow-affiliated Communist party; moreover, the rebels had been active for years before either Cuba or the Soviet Union would lend so much as verbal support. Here, argued many policy makers, was a clear example of indigenous instability. Before it could be exploited by the Communists and become a threat to U.S. security Washington needed to encourage reforms that would alleviate the basic cause of the instability: poverty.

While most officials in Washington agreed that reforms were essential to end the growing instability, in San Salvador the Romero government remained intransigent, rebuffing all suggestions for reform. Increased pressure from the United States brought only increased resistance. Then in July 1979, the Somoza regime collapsed. The handwriting was on the wall in El Salvador: reform now or revolution later, perhaps not that much later. It was therefore with considerable relief that the Carter administration observed the overthrow of the Romero government in November by a group of army officers, some of whom appeared committed to reform.[19] A few days later a military-civilian government was created that included several prominent civilian reformists. At last there were leaders in El Salvador with whom the United States could work to halt instability.

The cornerstone of the U.S.-Salvadoran effort was a comprehensive agrarian reform. It consisted of three phases. Phase I was to nationalize all large landholdings (over 500 hectares) and convert them into cooperatives owned and operated by landless peasants. Phase II was to divide and redistribute moderately large landhold-

[19] In his memoirs the U.S. ambassador strongly denies any participation by embassy personnel. Nevertheless, the verb "observed" clearly understates the level of U.S. involvement in the coup. If nothing else, the Carter administration unmistakably communicated its dislike of Romero's government; the precise steps that the administration took, if any, remain secret. For the ambassador's story, see Frank J. Devine, *El Salvador: Embassy Under Attack* (New York: Vantage Press, 1981), especially p. 140.

ings (100 to 500 hectares). These parcels constituted the dominant pattern of ownership in the major coffee-producing regions and therefore were the very heart of the oligarchy's economic structure. The third phase, known as the "land-to-the-tiller" program, was to provide tenant farmers with titles to the land they were already working. If fully implemented, the agrarian reform would have altered substantially the existing structure of land ownership in El Salvador, a country whose economy is dominated by agriculture. To many U.S. policy makers comprehensive land reform was perceived as an absolutely essential step toward the alleviation of structural poverty and therefore instability in El Salvador.

Two additional reforms were designed to complement the agrarian reform. Commercial banks were to be nationalized, as was the export of El Salvador's three major agricultural products: coffee, cotton, and sugar. Both reforms reflected the view that continued private control of these industries would provide economic elites with sufficient power to block the agrarian reform. For the agrarian reform to be effective, money was needed to provide credit and other support services to new landowners, particularly in the initial years of the reform. As in the United States, El Salvador's commercial banks had traditionally been the source of agricultural operating capital, and there was little faith in Washington that El Salvador's private bankers would loan peasants money to plant crops on land that until recently had belonged to the shareholders of the banks.

The nationalization of agricultural exports was designed primarily to "rationalize" the use of export earnings. Under the reformed system the hard currencies earned by exporting coffee, cotton, and sugar would no longer pass directly into the producers' hands; rather they would flow first into the nation's Central Bank. Dollars, pounds, and marks would be retained by the bank, and the producers would be paid in *colones*. In this way the government would accumulate foreign currencies, which could be purchased with *colones* by importers if the products they wished to import were determined by the government to meet the development needs of the country. The goal was to end the tendency for export earnings to be used for luxury consumption and foreign real estate investments. A secondary motivation was to disable one major vehicle for capital flight, the underinvoicing of exports.[20]

[20] A discussion of various aspects of the reforms is included in U.S. Congress, House, Committee on Foreign Affairs, Subcommittee on Inter-American Affairs,

While some U.S. officials were working with the Salvadoran government to implement the reform program, other policy makers were trying to cope with the postrevolutionary instability in Nicaragua. They were not very successful. Try as they might, Carter administration officials could not develop a unified interpretation of the Sandinista-led revolution. This was attributable in large measure to the confusion that accompanied the fall of the Somoza regime; the situation was extremely fluid and therefore difficult to interpret. More significant, however, was the fact that the Carter administration was staffed by officials who differed widely—sometimes extraordinarily widely—in their perceptions of instability in Latin America. In general, the administration hedged its bets on Nicaragua, particularly on the key question of Communist involvement. Two months after the fall of Somoza, Assistant Secretary of State Vaky announced the administration's view: "Classically, Cuba sees Nicaragua's agony as a chance to advance its own interests. Cuban support for the Sandinista cause has been indirect, but has recently increased. The possibility that the particular guerrilla factions Cuba has supported and helped arm could come to exert significant political leverage is cause for concern." But, he continued, "we cannot lose sight of the basic issue. This is fundamentally a Nicaraguan crisis. And Cuba is not the only or even the most important of the supporters of the anti-Somoza rebellion."[21]

One major goal of the Carter administration, then, was to reduce to a minimum the power of "the particular guerrilla factions Cuba has supported and helped arm"—the Sandinistas. At first the Sandinistas were perceived as an insignificant force that just happened to be in the right place at the right time. According to the Defense Intelligence Agency,

---

Presidential Certification on El Salvador, vol. 1, 97th Cong., 2d Sess., 1982. For the State Department's understanding of the meaning of the banking and export industry nationalizations, see p. 487. For the most complete analysis of the agrarian reform, see Martin Diskin, "Agrarian Reform in El Salvador," in *Latin American Agriculture: Structure and Reform*, ed. William Thiesenhusen (Winchester, Mass.: Allen and Unwin, 1987), pp. 107-157. See also Roy L. Prosterman, Jeffrey M. Riedinger, and Mary N. Temple, "Land Reform and the El Salvador Crisis," *International Security* 6 (Summer 1981): 53-74; David Browning, "Agrarian Reform in El Salvador," *Journal of Latin American Studies* 15 (November 1983): 399-426; Laurence R. Simon, James C. Stephens, Jr., and Martin Diskin, *El Salvador Land Reform, 1980-1981: Impact Audit with 1982 Supplement* (Boston: Oxfam America, 1982).

[21] *United States Policy towards Nicaragua*, pp. 33-34, *supra* n. 18.

the Sandinistas were a small group of Castroite insurgents that had been badly mauled by the Government forces over the years. Since they had never shown much promise, Cuban aid had always been meager. Until 1974 the Sandinistas had little hope for victory. But their successful operation that year increased their prestige and apparently convinced Castro that the situation was ripe for revolution. With the assistance of (security deletion)—which had differing reasons for seeking the overthrow of the Somoza government—Cuba provided training, funds, arms, and advice to bolster the Sandinistas until they became a viable core around which other anti-Somoza forces coalesced.

In response to a specific question from a member of Congress, a Defense Intelligence Agency (DIA) official remarked that "of course, the Cubans were the driving force, if we could call it that, behind the Sandinistas."[22]

The administration's strategy was to downplay these allegations of Cuban involvement and attempt to construct a cautious but positive relationship with the Sandinista-led government. "The course of the Nicaraguan revolution will . . . depend in no small measure on how the U.S. perceives it and relates to it," argued Assistant Secretary Vaky.

We might write it off as already radicalized and beyond redemption, but that would surely drive the revolution into deeper radicalization. Or we can work with those leaders who seek profound socio-political change but one which safeguards individual rights and democratic procedures. This latter course is the one most likely to achieve an outcome compatible with our own interests and which is most likely to be of mutual benefit. It is the course which the United States seeks to follow.[23]

[22] *Impact of Cuban-Soviet Ties*, pp. 19, 28, 40, *supra* n. 11. The "security deletion" is probably all or some combination of Costa Rica, Mexico, Panama, and Venezuela. The "successful operation" is a reference to the December 1974 attack upon the home of Somoza's minister of agriculture, which occurred during a farewell party for the U.S. ambassador, Turner Shelton. Although the attack occurred after the ambassador had left the party, the Sandinistas managed to capture a number of prominent Nicaraguan government officials. They were freed only after Somoza met several demands, including the release of captured Sandinistas.

[23] *United States Policy towards Nicaragua*, p. 80, *supra* n. 18. See also the similar statement by Secretary of State Vance, *Washington Post*, September 28, 1979.

A major multinational relief effort to aid the war-ravaged country began immediately after the fall of Somoza. U.S. participation was substantial. In the two and a half months between July 17 and September 30, 1979, the Carter administration delivered $26.3 million in quick-disbursing aid, primarily in food ($13.3 million) and medical supplies. Other gestures of friendship quickly followed. On September 24, three members of the Nicaraguan government (Daniel Ortega, Sergio Ramírez, and Alfonso Robelo) breakfasted at the White House with President Carter and then lunched with members of the Senate Foreign Relations Committee. In November, Representative Dante Fascell, a key moderate on issues involving Latin America, led a group of members of Congress to Nicaragua. His favorable report provided substantial support for the Carter request for a special authorization of $140 million in economic and military aid to Latin America, $75 million of which was economic aid earmarked for Nicaragua.

No one seemed to be particularly enthusiastic about the aid proposal; rather, most supporters voiced only a fear of what might happen if the aid were not provided. The position taken by the Council of the Americas, a business lobby that had worked mightily to obtain Senate support for the Panama Canal treaties in 1978, was typical. As the president of the Council advised House members, "I am not necessarily bullish on the fact that it isn't going to go down the drain. I believe, though, however, in making errors of commission rather than omission, and I think it could guarantee it going down the drain if it wasn't passed."[24] With that kind of faint praise the aid bill faced such an uncertain future that the House leadership had to agree to a secret session on February 28, 1980—only the second such session in history—during which reluctant members were scared just enough to vote "yes" by a State Department/CIA briefing on the extent of Soviet and Cuban influence in Nicaragua. Majority leader Jim Wright emerged from the session convinced of the need to "make a fight for Nicaragua" rather than leave the country "to Cuba's voracious appetite." Indeed, the debate in both houses of Congress centered on the issue of whether Nicaragua was salvageable or had already been lost to Communism. Aid opponent Jesse Helms told his Senate colleagues that "what we are going to decide today is whether the

[24] U.S. Congress, House, Committee on Appropriations, Subcommittee on Foreign Operations and Related Agencies, *Foreign Assistance and Related Programs Appropriations for 1981*, 96th Cong., 2d Sess., 1980, pt. 1, p. 293.

U.S. taxpayers' money is going to be used to support a Communist regime. We would see U.S. aid going for Marxist central planning, confiscation of agricultural land and industrial plants and ensconcing a whole cadre of loyal Communists in high-paying government jobs." After an unusually difficult seven-month struggle, Congress authorized the $75 million.[25]

In addition to airing congressional concern over the direction of the Nicaraguan revolution, the debate also provided nervous members of Congress with an opportunity to cover themselves against the election-year charge that they were giving taxpayers' money to a Communist government. As a result, the authorization had attached to it a series of restrictive provisions unprecedented in the history of the U.S. aid program. The most important provision focused upon Sandinista support of guerrillas in El Salvador: the president was required to certify that the Nicaraguan government was not cooperating with, harboring, or otherwise aiding "terrorist organizations."[26]

President Carter made this certification in September 1980, despite evidence that there was a flow of arms from Nicaragua to the guerrillas in El Salvador. The administration based its decision upon the lack of proof of official government involvement in the

[25] PL96-257. Two additional months were needed to convince Congress to appropriate the $75 million as part of the FY1980 supplemental appropriations bill, PL96-304.

[26] Additional congressional stipulations included (1) a requirement that the Secretary of State produce a semi-annual report on the status of human rights, political pluralism, freedom of the press and assembly, freedom of religion, and freedom of labor to organize and bargain collectively in Nicaragua, (2) a requirement that the President encourage free elections in Nicaragua, (3) a requirement that aid be terminated in the event of gross violations of internationally recognized human rights, (4) a requirement that aid be terminated in the event the government violated "the right to organize and operate labor unions free from political oppression," (5) a requirement that aid be terminated in the event the government engaged in "systematic violations of free speech and press," (6) a requirement that aid be terminated if Soviet, Cuban, or other foreign combat forces were stationed in Nicaragua, (7) a prohibition on aid to any Nicaraguan educational institution that housed, employed, or was made available to Cuban personnel, (8) a requirement that at least 60 percent of the aid be used to assist the private sector, (9) a requirement that loan money spent abroad be spent in the United States, and (10) a requirement that 1 percent of the aid money be used to publicize the fact that the United States was providing aid. A much larger number of proposed restrictions was defeated, typically by a narrow margin. Senator Jesse Helms was able to accumulate the fewest number of votes—33—for his proposal that would prohibit disbursal of the $75 million until the Nicaraguans had adopted the U.S. Bill of Rights.

arms shipments.[27] At the same time, however, the administration did not protest Congress's decision to refuse military aid to Nicaragua. In its FY1981 aid proposals the Carter administration had asked for authorization to reinitiate the military assistance programs with $5 million in credit sales and $494,000 in military training. The requests were deleted by both the House and Senate. The lopsided vote in the House—267 to 105 against any military assistance—was more an indicator of election-year fears than a sign of congressional opposition to the Sandinista government. As it turned out, these fears were well-grounded.

INSTABILITY IN CENTRAL AMERICA: THE REAGAN POLICY

The Caribbean basin was in turmoil when Ronald Reagan became President. In Nicaragua, Sandinista leaders had solidified their control over the government and were attempting with varying degrees of success to manage the instability that inevitably accompanies a major social revolution. In El Salvador, the U.S.-backed government of José Napoleón Duarte had managed to turn back the guerrilla's "final offensive," but a long period of armed struggle seemed inevitable. Continued U.S. support for the Duarte government seemed questionable, for the Salvadoran military continued its long practice of terrorizing the population, and civilian elites persisted in efforts to block implementation of the agrarian reform program. In Guatemala, guerrilla units were growing increasingly bold, and in the heavily populated northwest highlands it appeared that the Indian population was shaking off centuries of political passivity, joining the rebellion against the government of General Romeo Lucas García. There were no active insurgencies in Honduras or Costa Rica, but in both countries severe economic woes boded ill for the future.

On the other side of the Caribbean was Grenada, an island ministate that few of the newly appointed policy makers could locate without the help of an atlas. Ruled since 1979 by something called a People's Revolutionary Government, Grenada had become the site of increasingly hostile anti-American rhetoric and, worse yet, a Cuban-assisted project to build a world-class airport, the purposes of which had become a matter of considerable anxiety in

---

[27] For a discussion of this issue, see U.S. Congress, House, Permanent Select Committee on Intelligence, Subcommittee on Oversight and Evaluation, *U.S. Intelligence Performance on Central America: Achievements and Selected Instances of Concern*, 97th Cong., 2d Sess., 1982, especially pp. 6-7.

Washington. Stretched across the Caribbean between Central America and Grenada were a dozen other nations with economies devastated by the global recession. And, as luck would have it, situated in the middle of all this instability was Fidel Castro's Cuba, for over two decades the most annoying burr under the U.S. foreign policy saddle. Now, after several years of focusing its attention upon Africa, Cuba was active in Central America. Reagan administration officials may have differed on Cuban's exact role in the region, but all agreed that Fidel Castro, smiling through his cigar smoke, was once again up to no good.

Looking at this instability, officials of the new administration perceived a casebook example of Communist adventurism. "Many of our citizens don't fully understand the seriousness of the situation," President Reagan said, "so let me put it bluntly: There is a war in Central America that is being fueled by the Soviets and the Cubans. They are arming, training, supplying, and encouraging a war to subjugate another nation to communism, and that nation is El Salvador. The Soviets and the Cubans are operating from a base called Nicaragua. And this is the first real Communist aggression on the American mainland."[28]

Why had Central America become the target of Communist expansionism? To begin with, many policy makers believed that the critical countries in Central America—Guatemala, El Salvador, Honduras, and Nicaragua—were afflicted by a "banana republic" political culture, in particular a tendency to resolve disputes with violence, that facilitated instability. Into this environment came the worldwide economic recession of the late 1970s, causing severe distress in virtually every sector of the Central American economies. The effects of plummeting prices for agricultural exports were exacerbated by skyrocketing energy costs and soaring interest rates, which increased the burden of servicing foreign debts that had been contracted at floating interest rates. As per capita gross domestic product fell and unemployment rates rose to include perhaps a third or more of the labor force, the stage was set for Communist adventurism.

"Weakness attracts the predator," intoned Undersecretary of State James Buckley. "A failure to achieve viable economies, credible defenses, and stable political institutions makes these less developed nations inviting targets for subversion."[29] The Director of

[28] Speech to the nation, July 18, 1983.

[29] U.S. Congress, House, Committee on Foreign Affairs, *Foreign Assistance Legislation for Fiscal Year 1983*, 97th Cong., 2d Sess., 1982, pt. 2, pp. 9-10.

the Defense Security Assistance Agency (DSAA), General Ernest Graves, warned how these economic problems were providing "opportunities for exploitation." "In Latin America," Graves elaborated, "the Soviets, principally through their Cuban surrogates, have seized the opportunity to expand their influence and to foster anti-American sentiment."[30] Many other officials and analysts had a similar perception of the cause of instability: Communism "is going to exploit any weakness"; economic discontent will "invite Cuban adventurism"; Cuba has "manipulated these frustrations"; and "Cuba, the Soviet Union, and the Sandinistas are actively exploiting these troubles."[31]

These officials noted that the Soviets had given little attention to Latin America during most of the 1970s. The CIA reported that until late in the decade "it was generally recognized by the Soviets that the Western Hemisphere was within the U.S. sphere of influence and the likelihood of gaining any significant positive results in Latin America, the Caribbean, and Central America were remote, particularly where the risk of confrontation with the United States was a very strong possibility."[32] The Cuban strategy prior to about 1978, reported Assistant Secretary of State Enders, was "to develop state-to-state relations."[33] But Communist encouragement of instability had lulled, not ended. "This brief respite from leftist inroads in the Caribbean basin will be of very short duration," reported the DIA. "Havana remains determined to exploit the social and economic ills of the region."[34]

Most of these new officials blamed the Carter administration for encouraging new Soviet-Cuban efforts at destabilization. Some believed the Soviets took advantage of what they thought to be the

[30] *Foreign Assistance Legislation for Fiscal Year 1982*, pt. 7, pp. 76-77, *supra* n. 5.

[31] The quotations are, in order, from retired Colonel Samuel Dickens, a lobbyist for the American Legion, in *Foreign Assistance and Related Programs Appropriations for 1981*, pt. 1, p. 360, *supra* n. 24; the State Department's John Bushnell in *Foreign Assistance Legislation for Fiscal Year 1982*, pt. 7, p. 72, *supra* n. 5; Roger Fontaine, Cleto DiGiovanni, Jr., and Alexander Kruger, "Castro's Specter," *Washington Quarterly* 3 (Autumn 1980): 16; and Assistant Secretary of State Langhorne A. Motley, prepared statement before the House Appropriations Subcommittee on Foreign Operations, March 27, 1984. Fontaine served briefly on the Reagan administration's National Security Council (NSC) staff; DiGiovanni once worked for the CIA in Latin America.

[32] *Impact of Cuban-Soviet Ties*, p. 47, *supra* n. 11.

[33] Thomas O. Enders, "The Central American Challenge," *AEI Foreign Policy and Defense Review* 4 (1982): 9.

[34] *Impact of Cuban-Soviet Ties*, p. 32, *supra* n. 11.

left-leaning bias of the Carter human rights policy. Others said the Soviets simply had a lucky break—the 1979 coup in Grenada—and the Carter administration, by refusing to react with firmness, encouraged further Soviet adventurism. Mr. Reagan thought it was the Panama Canal treaties. In 1977, he wrote that ratification would create "the picture of a United States that bows to demands" and thus encourage tests of Washington's resolve. "Precisely where and when the next demand will come we cannot say, but come it will."[35] To analysts in the Heritage Foundation, a Washington think-tank with close ties to the Reagan administration, the Soviets were encouraged because the Carter administration had been so easily duped into helping overthrow Somoza: "The FSLN was defeated in the field, but the dispatch of U.S. arms to Panama for trans-shipment to the Sandinistas, U.S. interdiction of Israeli, South African and Argentine weapons destined for Somoza and the U.S. embargo of munitions for the Nicaraguan National Guard, delivered the Nicaraguan nation to the Communist camp."[36]

Whatever the cause, by the beginning of the 1980s many policy makers had come to believe that the Soviets, to quote a CIA official, "now estimate that perhaps there is something to be gained by making some additional investments."[37] The DIA asserted that the Cubans, acting as Soviet surrogates, "are the driving organizational force behind the entire insurgency movement in the area" of Central America.[38] Assistant Secretary of State Enders, who cited 1978 as the beginning of "the surge in Cuban and Soviet support for organized violence in Central America," charged that "Cuba is now trying to unite the radical left, commit it to the use of violence, train it in warfare and terrorism, and attempt to use it

[35] Ronald Reagan, "The Canal as Opportunity: A New Relationship with Latin America," Orbis: 21 (Fall 1977): 556-557.

[36] "Soviet Penetration of the Caribbean," mimeographed, Heritage Foundation National Security Record No. 22, June 1980, p. 2.

[37] Impact of Cuban-Soviet Ties, p. 47, supra n. 11; U.S. Congress, House, Committee on Foreign Affairs, Subcommittees on International Security and Scientific Affairs and on Inter-American Affairs, U.S. Arms Transfer Policy in Latin America, 97th Cong., 1st Sess., 1981, p. 10.

[38] Impact of Cuban-Soviet Ties, p. 37, supra n. 11; U.S. Joint Chiefs of Staff, "United States Military Posture for FY1982," printed in U.S. Congress, House, Committee on Armed Services, Hearings on Military Posture and H.R. 2614 and H.R. 2970, 97th Cong., 1st Sess., 1981, pt. 1, p. 109; Thomas O. Enders, "Building the Peace in Central America," mimeographed, speech to the Commonwealth Club of California, San Francisco, August 20, 1982.

to destroy existing governments and replace them with Marxist-Leninist regimes on the Cuban model. . . . the evidence is now overwhelming."[39] The section on Latin America in the 1982 annual report of the Department of Defense began with the accusation that characterized perceptions within the Reagan administration: "Today, it is the Soviet empire that poses the challenge to this hemisphere by intruding with military and political means wherever the opportunity arises. . . . convincing evidence of Cuban subversion has surfaced in virtually every Caribbean Basin country."[40]

Nicaragua, the first target, had already been hit by Soviet fire, and as a result many officials had voiced substantial opposition to the Nicaraguan government long before they entered office with the Reagan administration. This was more than a simple anti-Sandinista plank in the Republican party platform.[41] During the 1979 floor debate over economic aid, for example, Representative Robert Bauman called Nicaragua "a new beachhead for communism in the Americas."[42] In early 1980, the administration's future special advisor on Latin America, retired General Gordon Sumner, argued before Congress that "the Sandinistas are blood brothers of the Marxist militants—red terrorists, who are holding this nation hostage today in Teheran."[43]

The Reagan administration initially adopted the position that while the Sandinistas were Communists, they could not yet muster the strength to destroy their non-Communist opposition. To support these non-Sandinistas, in early 1981 the administration asked Congress to approve $35 million in economic aid to Nicaragua for FY1982, all earmarked for the Nicaraguan private sector—groups that, according to Acting Assistant Secretary Bushnell, "have held on for more than a year as a strong force against those who would establish a totalitarian state."[44] Secretary of State Haig argued that the money was needed to support Nicara-

[39] Enders, "The Central American Challenge," p. 9.

[40] U.S. Department of Defense, *Annual Report to the Congress, Caspar W. Weinberger, Secretary of Defense, Fiscal Year 1983* (Washington, D.C.: GPO, 1982), pp. II-21-2, II-26.

[41] The 1980 Republican party platform stated: "We deplore the Marxist Sandinista takeover of Nicaragua. . . . We oppose the Carter administration aid program to the government of Nicaragua."

[42] *Congressional Quarterly 1979*, p. 263.

[43] *Foreign Assistance and Related Programs Appropriations for 1981*, pt. 1, p. 282, *supra* n. 24.

[44] *Foreign Assistance Legislation for Fiscal Year 1982*, pt. 7, p. 71, *supra* n. 5.

guans seeking "moderate outcomes. One of the great problems in deciding a future course with respect to Nicaragua is not to make their prospects for success aggravated or worse, even though it is clear that the main direction of Nicaraguan policy is in the hands of the extreme left."[45]

Then in early April 1981, the administration officially gave up its effort to save Nicaragua from Communism with a carrot and turned to the stick. President Reagan refused to extend Mr. Carter's September 1980 certification that Nicaragua was not assisting terrorist organizations; the State Department announced instead that the Nicaraguan government "was involved in acts of violence in El Salvador."[46] This automatically forced a halt to the disbursement of the final $15 million of the $75 million authorized during the Carter administration.[47] From this time forward U.S. policy

[45] U.S. Congress, Senate, Committee on Foreign Relations, *Foreign Assistance Authorization for Fiscal Year 1982*, 97th Cong., 1st Sess., 1981, pp. 37, 420.

[46] Ibid., p. 420.

[47] There is some controversy over the exact date that the United States halted aid to Nicaragua. One member of Mr. Reagan's National Security Council staff, Constantine Menges, said that the aid cutoff occurred almost immediately, on January 22, 1981. Constantine C. Menges, "Central America and the United States," *SAIS Review* 1 (Summer 1981), p. 27. This is almost assuredly incorrect. The announcement of the aid suspension on April 1, 1981, did not coincide with the actual decision, which probably occurred sometime in February. For the various public statements on this issue, see *Foreign Assistance Authorization for Fiscal Year 1982*, pp. 448, 453, *supra* n. 45; U.S. Congress, Senate, Committee on Appropriations, *Foreign Assistance and Related Programs Appropriations, Fiscal Year 1982*, 97th Cong., 1st Sess., 1981, pt. 1, pp. 272, 274; U.S. Congress, Senate, Committee on Appropriations, *Foreign Assistance and Related Programs Appropriations, Fiscal Year 1983*, 97th Cong., 2d Sess., 1982, pt. 1, p. 22.

The timing of the aid cutoff assumes significance in any debate over the question of who did what to whom first, a question that will probably never be resolved to everyone's satisfaction. Nicaraguans can cite a series of hostile gestures, beginning in late 1979 with statements about the Carter aid request and including the hostile language in the Republican party platform, to demonstrate that the first diplomatic shots were fired by the United States. For the Nicaraguan perspective, see Centro de Documentación de INIES, *Cronología de las relaciones Estados Unidos-Nicaragua, 1979-1984* (Managua: Editorial de Ciencias Sociales, 1985). Reagan administration appointees and many career foreign service officers were adamant in their contrary belief that, to quote Deputy Assistant Secretary of State James Michel, "the United States abandoned 'constructive engagement' only *after* it became apparent that this policy was not compelling the Sandinistas to end their support for the Salvadoran insurgency." James Michel, "Defending Democracy," in *Central America and the Western Alliance*, ed. Joseph Cirincione (New York: Holmes and Meier, 1985), pp. 54-55. See also the comment by ARA's Michael Skol in *AEI Foreign Policy and Defense Review* 5 (Winter 1985): 41-43. Mr. Reagan made asser-

toward Nicaragua was designed to undermine the Sandinista government. Options were debated throughout the summer and fall of 1981, and then in November President Reagan approved National Security Decision Directive 17, which initiated the CIA's covert war against Nicaragua.[48]

The rhetoric also escalated. Beginning in April 1981, the administration depicted the Sandinistas as dependent upon Cuba and the Soviet Union. In 1982, Assistant Secretary of State Enders told Congress that "the Nicaraguan state remains the preserve of a small Cuban-advised elite of Marxist-Leninists." Immediately after the fall of Somoza, he said, democracy had been possible. But then "a hard core of Marxist-Leninist ideologues began to consolidate a monopoly of force with Cuban assistance."[49] By 1983, Secretary of State Shultz was referring to Nicaragua's leaders as "a handful of ideologues, fortified by their Cuban and Soviet-bloc military advisors";[50] a year later President Reagan was calling the Sandinista rule "a Communist reign of terror."[51] The Soviet Union and Cuba had won the first round in the fight for control of Central America, these officials admitted, but only because the Reagan administration had not controlled U.S. policy when the contest began in the late 1970s. The rest of the struggle in Central America would be different, they vowed.

As in the case of Nicaragua, the new administration quickly changed its predecessor's policy toward El Salvador. By 1979 many

---

tions similar to those of Skol and Michel in his address to the nation on May 9, 1984. The reported assertion by another Deputy Assistant Secretary of State, Craig Johnstone, that the United States did not become hostile until April 1982 "because there was no flexibility on the Nicaraguan side" is belied by the evidence. Johnstone is quoted in Roy Gutman, "America's Diplomatic Charade," *Foreign Policy*, no. 56 (Fall 1984), p. 11.

[48] For the most insightful analysis of this decision and its potential consequences, see Richard H. Ullman, "At War with Nicaragua," *Foreign Affairs* 62 (Fall 1983): 39-58. For additional information, see *New York Times*, April 7, 1983, p. 1; *Foreign Assistance Authorization for Fiscal Year 1982*, pp. 420, 448, 453, *supra* n. 45.

[49] Enders, "Building the Peace in Central America," pp. 1, 4.

[50] George Shultz, "Struggle for Democracy in Central America," speech before the Dallas World Affairs Council and the Chamber of Commerce, Dallas, April 15, 1983.

[51] Speech to the nation, May 9, 1984. For an excellent example of the Reagan administration's overall view of revolutionary Nicaragua, see the article by Deputy Assistant Secretary of Defense (and career CIA official) Nestor D. Sánchez, "Revolutionary Change and the Nicaraguan People," *Strategic Review* 12 (Summer 1984): 17-22.

Carter policy makers agreed that Cuba and the Soviet Union were encouraging instability in El Salvador, but they downplayed the importance of this encouragement. Typical of this view was the position of the Latin Americanist on President Carter's National Security Council staff: "It is pointless to dispute the contention that the Cubans and Soviets have a stake in the outcome or a hand in the current turmoil," he wrote soon after leaving office. "However, to leap from a description of this support network to a conclusion that Moscow and Havana are running the show in El Salvador is a mistake. . . . The war in El Salvador is not between the Soviet Union and the United States; it is among Salvadorans."[52] This was precisely the view of the DIA, which, as we have seen in the case of Nicaragua, was hardly willing to adjust its views to the Carter policy. Cuba's role in the insurgency in El Salvador was "different from the more prominent position it exercised in Nicaragua," a DIA representative told Congress in 1980. "The war in El Salvador is largely 'locally grown.' "[53]

Opposition to this position had developed long before Mr. Reagan's electoral victory. In 1978, retired General Gordon Sumner characterized Mr. Pastor and other Carter Latin Americanists as members of "the radical left" who were permitting "the creation of a permissive strategic environment in which Communist Cuba has flourished." He warned: "The Cubans are literally and figuratively cutting our throats. . . . It is a strategic disaster of major proportions."[54] The El Salvador affiliate of the arch-conservative American Association of Chambers of Commerce in Latin America actively lobbied in Washington against the administration's policies. Its representatives warned repeatedly that the problem in El Salvador was Cuban adventurism, which would only be exacerbated by the proposed "socialist" reforms of the Carter administration.

Although the Reagan administration gave early verbal support for the reforms,[55] many analysts who moved into the administra-

---

[52] Robert A. Pastor, "Three Perspectives on El Salvador," *SAIS Review* 1 (Summer 1981): 39.

[53] *Impact of Cuban-Soviet Ties*, p. 29, *supra* n. 11.

[54] U.S. Congress, House, Committee on International Relations, Subcommittee on Inter-American Affairs, *Arms Trade in the Western Hemisphere*, 95th Cong., 2d Sess., 1978, p. 307. At the time he made this statement, General Sumner was a representative of the ultra-conservative Council on Inter-American Security.

[55] See, for example, the administration's statement in *Foreign Assistance Legislation for Fiscal Year 1982*, pt. 7, pp. 82-83, and pt. 2, p. 160, *supra* n. 5.

tion were already on record as outspoken critics. For example, Roger Fontaine, Mr. Reagan's initial replacement for Robert Pastor, had written soon after the reforms were announced in 1980 that the "pro-U.S. military government in El Salvador has been replaced by a center-left government supported by the U.S. embassy" and was bringing "the country to near economic ruin by desperate and sweeping reforms."[56] At the same time the Heritage Foundation published a study calling for cancellation of Phase II. The author, former CIA official Cleto DiGiovanni, wrote that in El Salvador "it is impossible to measure how happy the campesinos are, but in conversations this observer had privately and individually with about a dozen of them in late August, not one felt as much job and personal security under the cooperative system as under the previously private management."[57] The Western Hemisphere Task Force of the American Legion was equally critical. "It is inconsistent and wrong for the United States to foster 'Communist' or 'socialist' solutions to a nation's problems," wrote Legion representative Samuel Dickens. "The United States should not promote economic reforms for other countries, such as agrarian reform, nationalization of banks, and governmental control over exports, which are inimical to our own free enterprise and capitalist marketing system." Anyone who would suggest that these reforms "be implemented here would be run out of the country on a rail. At best they are socialist and at worst they are Communist."[58]

The Reagan administration faced a dilemma. On the one hand, its core supporters were clamoring for an end to U.S. encouragement of reform in El Salvador. On the other hand, the administration's support of reform was absolutely essential if it wished to retain congressional approval for military aid to fight the guerrillas. The solution was to tell each side what it wanted to hear. This task fell first to John Bushnell, a career official who served briefly as acting Assistant Secretary of State for Inter-American Affairs. To supporters of reform he spoke enthusiastically about progress in land redistribution. He told some members of Congress that the land reform "has meant that these people, who were so desperate,

[56] Fontaine, DiGiovanni, and Kruger, "Castro's Specter," p. 26.

[57] Cleto DiGiovanni, Jr., "U.S. Policy and the Marxist Threat to Central America," Heritage Foundation *Backgrounder*, no. 128, October 15, 1980, pp. 8, 12.

[58] U.S. Congress, House, Committee on Foreign Affairs, Subcommittee on Inter-American Affairs, *U.S. Policy toward El Salvador*, 97th Cong., 1st Sess., 1981, pp. 224, 211.

who were farming a few acres, who had large families which could not possibly make ends meet continuing in that way, whose kids were the best recruitment source for the guerrillas, now have hope." He remarked that Phase I "has been completed. The power has shifted" and that "this also has happened" with the land-to-the-tiller program. At the time Bushnell failed to mention Phase II.[59] When asked about this omission by another congressional committee, his response assuaged fears that the reform would be aborted: Salvadorans "don't want to get in a situation where production goes down very sharply, so they feel they cannot press ahead at this moment with the second phase until the first and the third phase are further along and they don't have to devote so much of their human resources—technical assistance, cadestral and what have you—to it."[60]

To opponents of the reforms Bushnell spoke a different language: there was no need to be alarmed; the reforms were not the Reagan administration's idea of good public policy. "Senator Helms, let me say on both banking and on agrarian reform, or on the nationalization of some parts of foreign trade, they are not policies which the United States pursues around the world. We are not in favor of nationalizing banks. In fact, we think that is a bad policy. . . . But it is the program that the government [of El Salvador] has. It is a sovereign government. It is under a threat, an attack, a massive attack by the worldwide Communist network, and we are supporting it against that."[61]

[59] *Foreign Assistance Authorization for Fiscal Year 1982*, p. 442, *supra* n. 45; *Foreign Assistance Legislation for Fiscal Year 1982*, pt. 7, p. 83, *supra* n. 5.

[60] Bushnell's statement is in *Foreign Assistance Legislation for Fiscal Year 1982*, pt. 7, p. 92, *supra* n. 5. On April 27, 1982, the Salvadoran Constituent Assembly repealed the law permitting the government to expropriate property for the purpose of reform, thereby ending permanently the reform effort. By July 1982, Bushnell's replacement, Thomas Enders, was telling Congress that Phase II was only "a proposal which was drafted by the Duarte government" that "was not backed by the United States at any point. It has now been explicitly set aside by the constituent assembly, but this is not the setting aside of a reform that was underway. This was a reform that was never born." *Foreign Assistance and Related Programs Appropriations, Fiscal Year 1983*, pt. 1, pp. 1,678, 1,664, *supra* n. 47. At about the same time, Roy Prosterman, a conservative land reform specialist who acted as an advisor to the Salvadoran agrarian reform institute, asserted that other action by the Constituent Assembly "raises virtually insuperable impediments to the continued viability and implementation of the vital Phase III of the land reform program." U.S. Congress, Senate, Committee on Foreign Relations, *U.S. Policy in the Western Hemisphere*, 97th Cong., 2d Sess., 1982, p. 216.

[61] *Foreign Assistance Authorization for Fiscal Year 1982*, p. 448, *supra* n. 45. By

It is difficult to determine the precise reason for opposition to the reforms, although part of it was clearly ideological. In 1980, for example, when Senator Helms was successful in writing into law a provision prohibiting U.S. financial assistance to fund either the land or banking reforms, his opposition was based upon the allegation that the reforms were "socialistic."[62] Four years later Helms issued an ideology-based preemptive broadside against agrarian reform as part of his overall attack upon the report of the Kissinger Commission:

> The confiscation of a productive economic system, and its dismemberment or collectivization, is an integral part of every Marxist regime. Even though it makes no sense economically, and usually results in a lower standard of living for the peasant who is its victim, agrarian reform is always a high ideological priority for those trying to destroy a traditional, evolving society. . . . Land reform has contributed more to the disintegration of the social and economic fabric of El Salvador than the terrorist attacks of the Communist guerrillas against the industrial infrastructure. The so-called land reform program dismantled the managerial and capital skills of the most advanced agricultural sector and that were earning the most foreign exchange [sic]. It has aggravated the tensions in society by uprooting families and communities, while at the same time thwarting the hopes of those to whom great promises were given.[63]

Like Senator Helms, Democratic Representative Clarence Long, chairman until 1986 of the powerful House Appropriations Subcommittee on Foreign Operations, held that the agrarian reform was ideologically flawed: "The part which just converts these huge estates into huge cooperatives is exactly what the Russians are doing. The Communists love that big cooperative deal."[64]

---

mid-1981, Senator Jesse Helms and his staff had concluded that Mr. Bushnell's two-sided approach to reform was an indicator of liberal inclinations, which it most certainly was not. In any event, Helms used his power as chairman of the Senate Foreign Relations Subcommittee on Western Hemisphere Affairs to block Bushnell's chances for an ambassadorship, and the hapless FSO was sent off to be deputy chief of mission in Buenos Aires.

[62] Section 721 of the International Security and Development Cooperation Act of 1980. This provision was repealed by the sunset clause in the following year's aid bill (Section 734 of PL97-113), but a new provision (Section 730) of the same bill then extended indefinitely the ban on U.S. aid to the Salvadoran reform effort.

[63] Press release, Office of Senator Jesse Helms, February 7, 1984.

[64] U.S. Congress, House, Committee on Appropriations, Subcommittee on For-

Senator Helms and Representative Long spoke for many policy makers in voicing ideological opposition to the nationalization of the banking and export service industries. The transfer of property from individual to state ownership was considered by many to be bad, regardless of its economic consequences. Similarly, although the agrarian reform transferred property from one set of individuals to another set of private owners and therefore was not "socialistic" in the normal sense of the word, the reform involved the temporary displacement of the free market while the government reallocated resources. Moreover, the agrarian reform placed restrictions on the amount of land that any individual could own, a fundamental violation of laissez-faire economic principles.

Some Washington policy makers also questioned the purely economic wisdom of the reforms. The Carter administration was always ambivalent about Phase II because of the balance-of-payments problems that could occur if, as many officials expected, coffee production were to decline. The fear was that the new peasant owners would plant corn to eat rather than coffee to export, crippling the nation's ability to import. During the Reagan administration, two State Department Latin Americanists, Craig Johnstone and Jon Glassman, drafted a cable that stated U.S. policy: "We want Phases I and III of the agrarian reform carried through as thoroughly and efficiently as possible. We favor implementation of Phase II only on a voluntary basis."[65] Their opposition, too, was based upon the fear of a decline in coffee production.[66] Indeed, virtually everyone who attacked the reforms made some remark similar to that of the American Legion: "We should stress our belief in the free enterprise system as the best economic system for improving man's material needs."[67]

Underlying both the ideological and the economic opposition, however, was yet a third objection to the reforms: many policy makers believe that poverty in Latin America is not structural,

---

eign Operations and Related Agencies, *Foreign Assistance and Related Programs Appropriations for 1983*, 97th Cong., 2nd Sess., 1982, pt. 1, p. 39.

[65] The confidential cable, approved by Assistant Secretary Enders and Charles Fairbanks, the Deputy Assistant Secretary of State for Human Rights and Humanitarian Affairs, was signed by Secretary of State Haig and sent to the embassy in San Salvador on May 22, 1983. For a brief discussion of the Carter administration's opposition to Phase II, see Raymond Bonner, *Weakness and Deceit: U.S. Policy and El Salvador* (New York: Times Books, 1984), p. 187.

[66] There is some evidence to suggest that these fears were not well-founded. See, for example, World Bank, "Land Reform," World Bank Paper—Rural Development Series, July 1974, pp. 21-22.

[67] *U.S. Policy toward El Salvador*, p. 224, *supra* n. 58.

and therefore structural remedies such as land reform are inappropriate. In the opinion of Senator Helms:

> El Salvador has done a good job in creating wealth and distributing it broadly, particularly in view of its immense population and slender resources. It ranks with the most mineral-rich, well-watered, and industrialized countries of Latin America in this regard, and even begins to approach the United States in certain significant respects. El Salvador can claim to have made progress, in the best sense of the term—progress for the people achieved through the actions of an enlightened and astute entrepreneurial class.[68]

Officials who believed that El Salvador was doing well "in creating wealth and distributing it broadly" would obviously see no reason for reform. If it ain't broke, don't fix it.

Unlike the subject of reform, there was never any question about where the Reagan administration stood with respect to Soviet and Cuban involvement in El Salvador. In February 1981, the State Department published a White Paper entitled "Communist Interference in El Salvador," as part of an effort to gain public support for increased military aid to the Salvadoran armed forces: "We feel that many Americans who oppose our policies do so by and large because they lack complete information about the real goals of the leftist insurgency and its connections with Cuba and other Communist countries," remarked Secretary Haig.[69] Briefly summarized the White Paper pictured a pro-Western government struggling to resist a small but fanatically dedicated group of Marxist insurgents, directed and supplied by Cuba and the Soviet Union—"another case of indirect armed aggression against a small Third World country by Communist powers acting through Cuba."[70]

---

[68] U.S. Congress, Senate, Committee on Foreign Relations, *Nomination of Robert E. White*, 96th Cong., 2d Sess., 1980, p. 3. Members of Senator Helms' staff were unable to identify the source of the Senator's information on the Salvadoran economy.

[69] *Foreign Assistance and Related Programs Appropriations, Fiscal Year 1982*, pt. 1, p. 80, *supra* n. 47.

[70] At the time the White Paper was released, the State Department also made available a set of captured documents that were said to constitute the evidence supporting the White Paper's claims. These documents are reproduced in Appendix A of Philip Agee, *White Paper Whitewash: Interviews with Philip Agee on the CIA and El Salvador*, ed. Warner Poelchau (New York: Deep Cover Books, 1981).

As it turned out, making these documents public proved to be a tactical mistake,

The White Paper on El Salvador was the first salvo in an ongoing public relations battle waged by the administration to demonstrate that instability in Central America was caused by the Soviet Union and its Latin American proxy, Cuba. A second major offensive began in February 1982 with a barrage of newspaper interviews and congressional testimony and was capped in mid-March by a coordinated CIA/DIA press briefing involving aerial photographs and, as the *coup de grâce*, the appearance at a Washington press conference of a Nicaraguan guerrilla captured in El Salvador. The guerrilla in question, Orlando José Tardencillas Espinosa, proved to be an unfortunate choice. Rather than testify as expected that he had received training in Cuba and been sent to El Salvador by the Sandinistas, Mr. Tardencillas indicated that he had never been abroad, had never been directed to do anything by the Sandinistas, and had decided by himself to fight against an unjust government in El Salvador. As Mr. Tardencillas continued with his lengthy testimony, one reporter asked him why he had agreed to fly to Washington from El Salvador. "They gave me an

---

for several enterprising journalists accepted the administration's suggestion that they inspect the documents, all of which were in Spanish along with the State Department's English translations. After some time a series of lengthy articles in the nation's best newspapers demolished the State Department's credibility on this issue. The articles pointed to (1) mistranslations that distorted the original Spanish, always with the effect of demonstrating greater Cuban and Soviet involvement than the originals would suggest; (2) a large number of unsupported inferences; and (3) a high probability of forgery of some documents and the uncertain origin of others. One former foreign service officer labeled the White Paper "a source of acute embarrassment to the administration, primarily revealing shoddy research and a fierce determination to advocate the new policy, whether or not the evidence sustained it." Wayne S. Smith, "Dateline Havana: Myopic Diplomacy," *Foreign Policy*, no. 48 (Fall 1982), pp. 161-162. Jorge Domínguez characterized the White Paper as "hardly a textbook case of anything"; to Raymond Bonner, it was "a textbook case of distortion, embellishment, and exaggeration." Jorge I. Domínguez, *U.S. Interests and Policies in the Caribbean and Central America* (Washington, D.C.: American Enterprise Institute, 1982), p. 33; Bonner, *Weakness and Deceit*, p. 256. For the original newspaper stories, see in particular John Dinges' initial analysis in the *Los Angeles Times*, March 17, 1981, and that of Robert Kaiser in the *Washington Post*, June 9, 1981; Jonathan Kwitny in the *Wall Street Journal*, June 8, 1981; and James Petras and Ralph McGehee in *The Nation*, March 28, 1981, pp. 353, 367-372 and April 11, 1981, pp. 423-425. For a game attempt to defend the White Paper, see Mark Falcoff, "The El Salvador White Paper and Its Critics," *AEI Foreign Policy and Defense Review* 4 (1982): 18-24. For an analysis that places the White Paper in perspective, see Walter LaFeber, "The Reagan Administration and Revolutions in Central America," *Political Science Quarterly* 99 (Spring 1984): 1-25.

option," he responded. "I could come here or face certain death." Walking out of the briefing room, State Department representative Dean Fischer sighed to one reporter, "You win some and you lose some."[71]

In the following years, the Reagan administration made repeated efforts to prove that the instability in El Salvador was caused by the Soviet Union. The State Department produced elaborate documents, every word of which received the concentrated attention of administration critics, looking for (and often finding) errors and inconsistencies.[72] After the March 1982 CIA briefing of the House Intelligence Committee, for example, a committee staff report acknowledged "the existence of convincing intelligence that insurgents rely on the use of sites in Nicaragua for certain HQ and logistical operations, and of persuasive evidence of Cuban and Nicaraguan support." But the report also noted that (1) "the statement [by the CIA briefing team] did not address the extent of control of the insurgency by non-Salvadorans, a point which remained subject to differing interpretations"; (2) "the presentation was flawed by several instances of overstatement and overinterpretation"; and (3) the CIA conveyed "the suggestion of greater certainty than warranted by the evidence."[73] Recalling the errors in the White Paper, a skeptical member of Congress asked not for a briefing but for access to the actual information used in the briefing. The response was that "the intelligence sources and methodology used to collect this evidence are highly confidential in nature. It would be impossible to release the information without jeopardizing those sources."[74]

Many observers interpreted this reluctance to offer proof of Soviet-Cuban activity as proof that it did not exist. In 1982, former

[71] New York Times, March 13, 1982, pp. 1, 5.

[72] Typical of these documents are: U.S. Department of State and Department of Defense, Background Paper: Nicaragua's Military Build-Up and Support for Central American Subversion (Washington, D.C.: Department of State and Department of Defense, July 18, 1984), and U.S. Department of State and Department of Defense, News Briefing on Intelligence Information on External Support of the Guerrillas in El Salvador (Washington, D.C.: Office of Public Diplomacy for Latin America and the Caribbean, Department of State, August 8, 1984). For critical assessments of the quality of these documents, see the sources cited in note 70, especially Bonner, Weakness and Deceit, pp. 256-264. See also U.S. Intelligence Performance on Central America, supra n. 27.

[73] U.S. Intelligence Performance on Central America, pp. 1, 8, supra n. 27.

[74] Foreign Assistance and Related Programs Appropriations for 1983, pt. 1, p. 126, supra n. 64.

ambassador Robert White asserted that "the United States is not producing the evidence" of Nicaraguan arms shipments to El Salvador "because they do not have it," and another former career diplomat, Wayne Smith, noting Secretary of State Haig's assertion that he had solid proof of Cuban arms shipments to Central America, wondered why "the administration never produced any such solid evidence."[75] These challenges were buttressed in mid-1984 by the revelations of a disgruntled former CIA analyst, David G. MacMichael. In a series of newspaper articles MacMichael charged that no substantial evidence existed to link Nicaragua to any arms flow into El Salvador.[76] At the same time MacMichael was making his assertions, news leaked out about the failure of U.S. naval exercises in the Gulf of Fonseca during May 1984, which were designed to intercept arms shipments from Nicaragua. The effort was so unproductive—"We haven't captured anything since early 1981," complained a U.S. official—that the United States dismantled its radar station on Tiger Island.[77] By the time the State Department provided additional evidence of shipments across the Gulf, one *Washington Post* reporter (Joanne Omang) had become so suspicious of administration claims that she shrugged the documents off as no more than "fuzzy videotape, translated documents and crude waterspotted maps."[78]

These Washington skirmishes often appeared to result in victory for the administration's critics, but appearances were deceptive. The skirmishes also changed in a subtle but important way the terms of the policy debate. The original question was: Is the instability in Central America caused by poverty or by Communist adventurism? Over time the question became: How much Communist adventurism exists? Poverty became irrelevant. What

---

[75] *Presidential Certification on El Salvador, 1: 248, supra* n. 20; Smith, "Dateline Havana: Myopic Diplomacy," pp. 163-164.

[76] *New York Times*, June 11, 1984; *Washington Post*, June 13, 1984. Asked about MacMichael's report, Salvadoran President Duarte responded that "the man is clearly a Marxist. . . . there are infiltrators everywhere." Marc Cooper and Greg Goldin, "Playboy Interview: José Napoleón Duarte," *Playboy* 31 (November 1984): 74.

[77] The radar station remained inactive only briefly. One official explained that it was reactivated because equipment scheduled to replace the island's facility (a ship?) was needed elsewhere. Another explanation offered by critics was that to dismantle the base was tantamount to admitting that no (or very few) arms were being shipped across the Gulf of Fonseca, which would call into question the basic reason for U.S. hostility toward Nicaragua.

[78] *Washington Post*, August 9, 1984, p. A31.

became important was the quality of aerial photography, the origin of U.S. M-16 rifles carried by the guerrillas, and other technical issues, all of which were related to Communist adventurism. Thus the administration could lose battle after battle, from the White Paper to the Tardencillas affair, and still emerge victorious because success, in this case, was defined as shifting the focus of debate from poverty to Communism.

In discussions of El Salvador the administration was successful in minimizing and then simply ignoring any discussion of the domestic sources of the insurgency in the country. Little more than a month after assuming office, Secretary of State Haig identified Cuba as "the platform, the instigator, and the operative leadership behind the situation in El Salvador."[79] Throughout his two-year tenure as Secretary of State, Mr. Haig continued to interpret the instability in El Salvador exclusively as a Cuban-Soviet intervention. "First and foremost," he told Congress, "our problem with El Salvador is external intervention in the internal affairs of a sovereign state in the hemisphere—nothing more, nothing less. That is the essential problem we are dealing with."[80] The remarks by Haig's successor, George Shultz, were identical.[81]

The administration identified Nicaragua as "the preserve of a small Cuban-advised elite of Marxist-Leninists,"[82] and charged that "Nicaragua is being exploited as a base for the export of subversion and armed intervention throughout Central America."[83] But Nicaragua's role was more than that of a passive conduit, asserted Assistant Secretary of State Enders. "We are concerned about Nicaragua's continuing large-scale support for the guerrillas

[79] *Foreign Assistance Authorization for Fiscal Year 1982*, p. 3, *supra* n. 45. This occasion also marked the Reagan administration's initial public threats directed at Cuba. Mr. Haig told the Senate Foreign Relations Committee that efforts to counter Communist moves in Latin America with military assistance had had "a distinguished lack of success in the past." Now, he asserted, "the problem has to be dealt with in a more forthright way, at its source" (p. 13). When the Secretary of State was later asked to identify whether the source was Cuba or the Soviet Union, he responded that "with respect to El Salvador, first and foremost, Cuba." *Foreign Assistance and Related Programs Appropriations, Fiscal Year 1982*, pt. 1, p. 29, *supra* n. 47.

[80] *Presidential Certification of El Salvador*, p. 136, *supra* n. 20.

[81] Perhaps Mr. Shultz's most detailed public statement is in U.S. Congress, Senate, Committee on Appropriations, *Foreign Assistance and Related Programs Appropriations, Fiscal Year 1984*, 98th Cong., 1st Sess., 1983, pt. 1, pp. 1-82.

[82] Enders, "Building the Peace in Central America," p. 4.

[83] Secretary of State Haig in U.S. Congress, Senate, Committee on Foreign Relations, *East-West Relations*, 97th Cong., 2d Sess., 1982, p. 7.

in El Salvador and its similar activities in other Central American countries. This, together with Nicaragua's extraordinary arms buildup and the large-scale presence of Cuban military advisers, is the fundamental cause of tension within the region."[84]

The Reagan administration was also successful in focusing policy debates upon what it believed was a linkage between instability in Nicaragua and El Salvador and a broader Soviet-Cuban effort to undermine U.S. power throughout the Caribbean region. As President Reagan noted, "a determined propaganda campaign has sought to mislead many in Europe and certainly many in the United States as to the true nature of the conflict in El Salvador. Very simply, guerrillas, armed and supported by and through Cuba, are attempting to impose a Marxist-Leninist dictatorship on the people of El Salvador as part of a larger imperialistic plan."[85] This plan involved subversion on a broad scale. "The Soviets have continued to finance, train, and staff a Cuban military establishment which has by now become a significant instrument of Soviet expansion," reported an alarmed Jeane Kirkpatrick.

> The first fruits of these efforts are the new governments of Grenada and Nicaragua. . . . El Salvador, having arrived now at the edge of anarchy, is threatened by progressively well-armed guerrillas whose fanaticism and violence remind some observers of Pol Pot. Meanwhile, the terrorism relied on by contemporary Leninists (and Castroites) to create a 'revolutionary situation' has reappeared in Guatemala. Slower but no less serious transformations are under way in Guyana, where ties to Castro have become extensive, tight, and complex, and in Martinique and Guadeloupe, where Castroite groups threaten existing governments.[86]

This was not perceived as a scattershot approach to subversion; rather, it was viewed as a carefully orchestrated sequential pattern. The Soviets had developed, in Alexander Haig's words, "a priority target list, a hit list, if you will, for the ultimate takeover of Central America. . . . What we are watching is a four-phase op-

[84] *Foreign Assistance Legislation for Fiscal Year 1983*, pt. 6, p. 9, *supra* n. 29. For similar statements, see Enders, "The Central American Challenge," pp. 8-10, and Enders' prepared remarks before the Senate Foreign Relations Subcommittee on Western Hemisphere Affairs, February 1, 1982, and March 14, 1983.

[85] *Weekly Compilation of Presidential Documents* 18 (March 1, 1982): 222.

[86] Jeane J. Kirkpatrick, "U.S. Security and Latin America," *Commentary* 71 (January 1981): 29.

eration. Phase one has been completed, the seizure of Nicaragua. Next is El Salvador, to be followed by Honduras and Guatemala. It is clear and explicit."[87] Other administration officials extended the hit list, arguing that once Nicaragua, El Salvador, Honduras, and Guatemala had been conquered, "through a cluster of Central American satellites the Soviet Union would be able to direct insurgencies against Mexico and Panama."[88] The Latin American dominoes were falling. "Nicaragua is already gone," Senator Helms told an audience in 1983. "Cuba's already gone. If we sit out, Guatemala, Honduras, Costa Rica—bam, bam, bam."[89]

IN THE space of a few short years in the late 1970s and early 1980s, dramatic changes occurred in United States policy toward Central America. These changes were based upon policy makers' differing perceptions of the causes and consequences of the instability that flared throughout the region. To some officials the instability was caused by poverty and deprivation; other officials believed the instability was caused by Communist adventurism. Some were less certain about the relationship among poverty, Communism, and instability. There were equally strong disagreements and no small amount of uncertainty over the consequences of instability for U.S. security.

The evidence policy makers gathered to support their differing perceptions was often quite convincing—so convincing, in fact, that most officials were unable to understand how others could hold a different position. Were their opponents playing politics—using instability in Central America as a convenient vehicle to attack the administration? Were their opponents so blind or so ignorant that they could not see the truth when it was placed before their eyes?

The answer to both questions is probably yes. Policy makers regularly play politics with any convenient issue. Senator Lodge's position notwithstanding, nowadays politics never stops at the

[87] *Foreign Assistance Legislation for Fiscal Year 1982*, pt. 1, p. 194, *supra* n. 5.

[88] Ernest Graves, "U.S. Policy toward Central America and the Caribbean," in *The Central American Crisis: Policy Perspectives*, ed. Abraham F. Lowenthal and Samuel F. Wells, Jr., Working Paper 119, Latin American Program, The Wilson Center, Washington, D.C., 1982, p. 19.

[89] Address to Rotary International, Asheville, N.C., April 30, 1983. For similar comments see Enders, "The Central American Challenge," p. 10; *Impact of Cuban-Soviet Ties*, p. 46, *supra* n. 11; *U.S. Policy toward El Salvador*, p. 210, *supra* n. 58; and W. Scott Thompson, "Choosing to Win," *Foreign Policy*, no. 43 (Summer 1981), pp. 81-82.

water's edge; if something is to be gained by attacking an administration's policy toward Latin America, someone will do it. And, like the rest of us, many policy makers are both blind and ignorant when it comes to issues involving instability in Latin America. Regardless of how much we might already know, there is always more to learn about Latin America. Yet all of us are reluctant to accept new knowledge if it contains evidence that might make us uncomfortable with our current beliefs. Understandably, policy makers who have staked out a public position on instability in Latin America are particularly vulnerable to this type of self-deception.

But these shortcomings did not cause the contemporary disputes over United States policy toward Latin America. The reluctance to learn or to accept new evidence, along with the temptation to use Latin America for partisan political advantage, are only minor problems. The basic problem is that there is a profound difference among policy makers in their beliefs about what instability in Latin America means in terms of world politics. It is to the analysis of these beliefs that we now turn.

PART I

The Causes of Instability

· 2 ·

POVERTY

A SUBSTANTIAL proportion—probably a majority—of policy makers believes that poverty is the cause of instability in Latin America. For over thirty years a focus upon poverty has been the hallmark of liberal interpretations of instability in Latin America. Representative Gerry Studds, for example, would be expected to attribute "the primary cause of violence in El Salvador" to "human misery," just as Senators Leahy and Pell would be expected to conclude that "the large social and economic disparities which have dominated Salvadoran society bred this revolution."[1] There is no surprise in these statements; indeed, their absence would be surprising.

What is more interesting, therefore, is the strong agreement voiced by moderates. It is one thing for liberal members of Congress to attribute instability to poverty; it is quite another to hear from a deputy assistant secretary of defense that "the fundamental security problems of El Salvador and Honduras arise from deep economic and social grievances," or to hear from an official of the Agency for International Development, a bastion of moderation during the Reagan administration, that "the unrest we see in the region today is due in very large measure to the inequitable development patterns of the past and, in a number of countries, the present."[2] Perhaps Margaret Daly Hayes has provided the most succinct summary of the beliefs of her fellow moderates: "Social science literature, revolutionary doctrine, and common sense argue persuasively that at the root of political instability are poverty, illiteracy, frustrated aspirations, unequal income distribution, and inadequate economic opportunities for people."[3]

[1] The statements occurred during congressional hearings, printed in U.S. Congress, House, Committee on Foreign Affairs, *Foreign Assistance Legislation for Fiscal Year 1982*, 97th Cong., 1st Sess., 1981, pt. 7, p. 247; U.S. Congress, Senate, Committees on Foreign Relations and on Appropriations, *El Salvador: The United States in the Midst of a Maelstrom*, 97th Cong., 2d Sess., 1982, p. 6.

[2] U.S. Congress, House, Committee on Appropriations, Subcommittee on Foreign Operations and Related Agencies, *Foreign Assistance and Related Programs Appropriations for 1981*, 96th Cong., 2d Sess., 1980, pt. 1, p. 336; *Foreign Assistance Legislation for Fiscal Year 1982*, pt. 7, p. 106, *supra* n. 1.

[3] Margaret Daly Hayes, "The Stakes in Central America and U.S. Policy Re-

These liberal and moderate policy makers not only share an understanding that poverty leads to instability; they also agree on *how* poverty eventually leads to instability. This explanation is presented schematically in Figure 2.1. Briefly stated, these officials believe that to be destabilizing, poverty must first await the structural changes that erode traditional restraints upon behavior. Then, when two additional factors—political mobilization and elite intransigence—are also present, the result is instability.

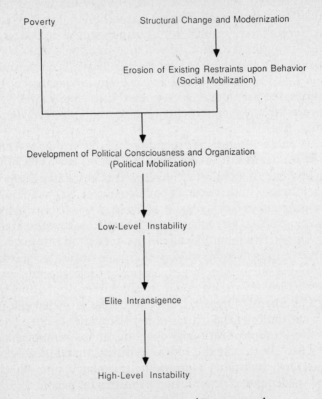

FIGURE 2.1. Perceptions of Poverty as the
Cause of Instability

sponses," *AEI Foreign Policy and Defense Review* 4 (1982): 18. At the time she wrote this article, Hayes was a staff member of the Senate Committee on Foreign Relations; she subsequently became the Washington representative of the Council of the Americas, a business lobby. For a similar statement by a typical congressional moderate, Representative Gus Yatron, see U.S. Congress, House, Committee on Foreign Affairs, Subcommittee on Inter-American Affairs, *U.S. Policy Options in El Salvador*, 97th Cong., 1st Sess., 1981, p. 156.

Readers who are familiar with Washington foreign policy making will recognize that few officials ever articulate in a clear and coherent manner the intellectual progression outlined in Figure 2.1. In my interviews no more than a handful of policy makers ever expressed the complete set of ideas that is described in the figure. In part that is because any such expression would require a substantial amount of time, and only a few policy makers are willing to grant an interviewer the time that would be necessary for a complete explanation. My experience is that to ask an unfamiliar official a question that would take an hour to begin to answer adequately—to ask, for example, "Do you believe that poverty leads to instability and, if so, how?"—is to guarantee a brief interview.

The lack of detailed responses also reflects in part the abstract nature of the question. Washington officials seem most comfortable when responding to concrete questions about who did what to whom, not to broader, more abstract questions about the motivations of social groups or the impact of structural changes upon group consciousness. Thus an interviewer is likely to receive a blank stare in response to the question, "What do you think Latin Americans believe are the results of the international political economy's emphasis on export agriculture?" But if one asks instead, "What do you think about the recent drop in world sugar prices?" or "What do you think will be the impact of the agrarian reform on coffee production in El Salvador?" the responses are often fairly detailed. They can be used, very carefully, to infer the respondent's beliefs about the general functioning of the international political economy.

Mostly, however, the lack of comprehensive responses resembling the pattern in Figure 2.1 stems from ignorance. It is not unfair to write that many U.S. policy makers do not know much about Latin America; they tend instead to be experts on U.S. foreign policy. There is, to be sure, great variation among policy makers. Some officials have spent years in Latin America; others cannot converse with a Latin American who does not speak English. Most have never heard of *comunidades eclesiales de base*; a few have spent years helping to create these grass-roots organizations in remote corners of Latin America. But on balance, many policy makers do not provide detailed responses to many questions because, given their limited knowledge of Latin America, they are unable to do so.

How, then, is it possible to assert that Figure 2.1 exists in the minds of U.S. policy makers if, for whatever reasons, the officials themselves do not express the beliefs? The answer—and here is a warning—is that Figure 2.1 is a product of inference. I sat down with the interview responses and other data from the group of policy makers that believes poverty is the fundamental cause of instability, and I pieced together a modal set of responses.

There are, of course, some obvious problems with this procedure, the most important of which is that it leads to an emphasis upon the beliefs of those respondents who talk the most. The many respondents who mentioned the revolution of rising expectations, who cited the development of new means of mass communication, and who noted the changing focus of traditional institutions such as the Church—they are all included. But it is easy to overlook the beliefs (or lack of beliefs) of the respondents who expressed nothing more than their view that poverty causes instability, despite the fact that in many cases these respondents are important policy makers. In creating Figure 2.1, I made a conscious effort to compensate for this tendency to over-specify—to avoid asserting the existence of beliefs that are not widely shared; moreover, in the pages that follow I note explicitly those instances where the analysis is based upon the responses of a relatively small number of policy makers.

It should be emphasized, however, that few respondents said nothing. Of those officials who expressed the belief that poverty causes instability, most were willing to elaborate on their views. And, perhaps because these policy makers share an understanding of the basic cause and effect relationship, their beliefs are remarkably similar. Respondents tend to differ primarily in the richness of the detail they are willing or able to communicate to a researcher. On four key beliefs, there is a near perfect consensus: 1) poverty is unjustifiably widespread in Latin America; 2) structural changes reduce the traditional passivity of the poor; 3) political mobilization heightens class tensions; and 4) elite intransigence makes instability inevitable. These beliefs, when combined, produced Figure 2.1; they are the topic of this chapter.

BELIEFS

Poverty Is Unjustifiably Widespread in Latin America

Poverty in Latin America strikes arriving visitors like the blow of a two-by-four between the eyes—the immediacy of human deprivation overwhelms the senses. Careful empirical analyses confirm this impression. One World Bank study concluded that "40 percent of households in Latin America live in poverty, meaning that they cannot purchase the minimum basket of goods required for the satisfaction of their basic needs, and . . . 20 percent of all households live in destitution, meaning that they lack the means of buying even the food that would provide them with a minimally adequate diet."[4]

What does it mean to live without a minimally adequate diet? It means, for example, that because from 10 percent to 50 percent of the women in Latin America have iron-deficiency anemia, their children are more likely to be born prematurely or small and therefore to have high mortality rates. Overall, malnutrition may account for as much as 57 percent of all infant deaths.[5] Children born to inadequately nourished mothers also tend to exhibit permanent neurological and mental dysfunction, for malnutrition during pregnancy and early childhood prevents the creation of neurons

TABLE 2.1
INCIDENCE OF POVERTY IN LATIN AMERICA
(percentage of households)

	Below Poverty Line			Below Destitution Line		
	Urban	Rural	National	Urban	Rural	National
All Latin America	26	62	40	10	34	19
Argentina	5	19	8	1	1	1
Brazil	35	73	49	15	42	25
Colombia	38	54	45	14	23	18
Costa Rica	15	39	24	5	7	6
Chile	12	25	17	3	11	6
Honduras	40	75	65	15	57	45
Mexico	20	49	34	6	18	12
Peru	28	68	50	8	39	25
Uruguay	10	—	—	4	—	—
Venezuela	20	36	25	6	19	10

SOURCE: Oscar Altimir, "The Extent of Poverty in Latin America," World Bank Staff Working Paper No. 522 (Washington, D.C.: World Bank, 1982), p. 82.

[4] Oscar Altimir, "The Extent of Poverty in Latin America," World Bank Staff Working Paper No. 522 (Washington, D.C.: World Bank, 1982), p. 78.
[5] Ruth R. Puffer, C. V. Serrano, and Ann Dillon, *The Inter-American Investigation of Mortality in Childhood* (Washington, D.C.: Pan American Union, 1971), pp. 2-6; World Bank, *The Assault on World Poverty: Problems of Rural Development, Education and Health* (Baltimore: The Johns Hopkins University Press, 1975), p. 359.

and neuronal connections. The result is irremedial mental retardation.[6]

If they survive the neonatal stage, a majority of preschool children in many Latin American countries will suffer from chronic protein-calorie malnutrition. Compared to well-fed children, these malnourished children contact more childhood diseases for longer periods of time. These diseases, in turn, place additional demands upon nutritional reserves. Gastrointestinal infections inhibit the absorption of nutrients, for example, thereby requiring an increase in consumption to maintain a given nutritional level. Fevers that accompany infections increase the metabolic rate, also increasing the need for additional food. But in the many cases where the required nutritional reserves do not exist, the effect is physically devastating: a cycle of "malnutrition, infection, severe malnutrition, recurrent infection and eventual death at an early age."[7]

There is now at least a small handful of analyses of poverty, in English, on every country in Latin America. On Guatemala, for example, the International Labor Organization reported in 1978 that an annual income of $320 was required for the satisfaction of basic needs but that 75 percent of the entire nation's population had an income of less than $215; in Guatemala's western highlands, where large-scale instability existed in the early 1980s, the average income was less than $140. A similar study by the World Bank reported that half of Guatemala's population was unable to satisfy minimum requirements for food and shelter, that one-third of the rural population was malnourished, that infant mortality was at least 79 per 1,000 (the figure for the United States is 11 per 1,000), and that over 80 percent of rural children under the age of five had inadequate diets.[8]

[6] Jean Mayer, "The Dimensions of Human Hunger," *Scientific American* 235 (September 1976): 41.

[7] Ibid., 41; World Bank, *Assault on World Poverty*, p. 360.

[8] World Bank, *Guatemala: Economic and Social Position and Prospects* (Washington, D.C.: World Bank, August 1978), pp. 18-19, 99-101; Shelton H. Davis and Julie Hodson, *Witness to Political Violence in Guatemala: The Suppression of a Rural Development Movement* (Boston: Oxfam America 1982), pp. 44-46; Jerry Stilkind, *Guatemala's Unstable and Uncertain Future* (Washington, D.C.: Center for Development Policy, 1982). For a general AID overview of poverty in Latin America in the early 1980s, see U.S. Congress, House, Committee on Foreign Affairs, *Foreign Assistance Legislation for Fiscal Year 1983*, 97th Cong., 2d Sess., 1982, pt. 6, pp. 42-106. See also International Labor Organization, *Employment, Growth and Basic Needs, A One-World Problem* (Geneva: ILO, 1976); World Bank, *Assault on World Poverty*. For a similar set of studies on El Salvador, consult the

While many policy makers are inevitably appalled by the poverty they perceive in much of Latin America, they would probably be less outraged if the poverty were not so obviously juxtaposed with ostentatious wealth. Compassion rather than anger was the most prominent reaction to famine in Africa during the 1980s, for example, because it is irrational to be outraged at the weather. But there is no drought in Latin America; most Latin American countries have more food than their citizens need. In the 1980s, only three have had less than 95 percent of requirements.[9] The nutritional problem in Latin America stems not from the lack of food but rather from its uneven distribution. While the poor starve, the rich eat brie.

Policy makers' perceptions of extreme inequality are grounded in solid empirical evidence. The best available data suggest that income is distributed extremely unequally in Latin America.[10] As Table 2.2 indicates, several Latin American countries are distributional nightmares from the standpoint of equity. The problem, moreover, seems to be getting worse. As Portes and others have demonstrated, "the relative share of income going to the poorest groups actually shrank" during the 1960 to 1975 period of rapid economic growth in Latin America.[11] Table 2.3 indicates that the poorest Latin Americans—the informal proletariat—"bore the brunt of rapid income concentration to the advantage of all other classes, particularly the top ones."[12] Simon Kuznets' theory that income concentration grows during the initial periods of economic growth and then levels off and falls with increased economic modernization has clearly not yet found empirical verification in Latin America. Economist David Felix concedes that

excellent study by Carmen Diana Deere and Martin Diskin, "Rural Poverty in El Salvador: Dimensions, Trends, and Causes," mimeographed (Geneva: International Labor Organization, January 1983).

[9] World Bank, *World Development Report 1985* (New York: Oxford University Press, 1985), pp. 220-221. The three are El Salvador and Peru (both 90 percent) and Haiti (84 percent).

[10] Insightful recent analyses are by David Felix, "Income Distribution and the Quality of Life in Latin America: Patterns, Trends, and Policy Implications," *Latin American Research Review* 18, no. 2 (1983): 3-33, and Enrique Iglesias, "Development and Equity: The Challenge of the 1980s," *CEPAL Review* 15 (December 1981): 7-46.

[11] Alejandro Portes, "Latin American Class Structures: Their Composition and Change during the Last Decades," *Latin American Research Review* 20, no. 3 (1985): 24; Iglesias, "Development and Equity"; Felix, "Income Distribution," p. 5.

[12] Portes, "Latin American Class Structures," p. 26.

TABLE 2.2
INCOME DISTRIBUTION IN SELECTED LATIN AMERICAN
COUNTRIES
(percentage share of household income)

	Percentile Groups of Households	
	Poorest 40%	Wealthiest 10%
Brazil	7.0	66.6
Peru	7.0	61.0
Panama	7.2	61.8
Mexico	9.2	57.7
Venezuela	10.3	54.0
Costa Rica	12.0	54.8
Chile	13.4	51.4
Argentina	14.1	50.3
El Salvador	15.5	47.3

SOURCE: World Bank, *World Development Report 1985* (New York: Oxford University Press, 1985), pp. 228-229.

TABLE 2.3
INCOME DISTRIBUTION IN LATIN AMERICA AND THE
UNITED STATES, 1960-1975
(by percentage)

	Share of Total Income	
Income Strata	1960	1975
Latin America		
Richest 10%	46.6	47.3
20% below richest 10%	26.1	26.9
30% below richest 10%	35.4	36.0
Poorest 60%	18.0	16.7
Poorest 40%	8.7	7.7
United States		
Richest 10%	28.6	28.3
20% below richest 10%	26.7	26.9
30% below richest 10%	36.7	36.9
Poorest 60%	34.8	34.8
Poorest 40%	17.0	17.2

SOURCE: Alejandro Portes, "Latin American Class Structures: Their Composition and Change during the Last Decades," *Latin American Research Review* 20, no. 3 (1985): 25.

further economic growth may "transform the linear trend of rising inequality into a parabolic Kuznets curve, but I see little basis for expecting this to happen within the next couple of decades from market forces operating under prevailing institutions and economic policies."[13]

One cannot help but sense the moral outrage among many policy makers when describing this poverty and inequality as they perceive it in places like El Salvador. During one informal interview with a small group of lobbyists and activists in Washington, two of whom had worked in the Carter administration, I apparently reacted too coolly as they cited a list of dismal social statistics. Sensing my eminent educability, they referred me to a recently released movie entitled *Witness to War: An American Doctor in El Salvador*. Seeing the movie does help to understand how Latin American poverty creates a sense of injustice among many policy makers. In one scene, for example, a ragged Salvadoran peasant stands in a field with a large house in the background and, looking sadly into the camera, tells why he chose rebellion: "I worked on the hacienda over there, and I would have to feed the dogs bowls of meat or bowls of milk every morning, and I could never put those on the table for my own children. When my children were ill, they died with a nod of sympathy from the landlord. But when those dogs were ill, I took them to the veterinarian in Suchitoto."[14] This type of image is both heart-rending and outraging. The juxtaposition of dead children and pampered dogs is designed to overwhelm the viewer, to evoke a visceral sympathy for rebellion against the fundamental unfairness of a degenerate social structure. It is remarkably effective.

But emotional films do not facilitate dispassionate analysis. Are Latin American peasants driven to rebellion because of poverty and injustice, or is it that many U.S. policy makers *think* Latin American peasants are driven to rebellion because they themselves would revolt if forced to live like Latin American peasants? Quite obviously these officials are attributing to the poor of Latin America not only their own standards of injustice but also their own reaction to the injustice. But no one will ever convince them

---

[13] Felix, "Income Distribution," p. 26. For further discussion of this subject, see several chapters, especially that of William Cline in Robert Ferber, ed., *Consumption and Income Distribution in Latin America* (Washington, D.C.: OAS, 1980).

[14] A print version of the film's story lacks the emotional immediacy of the film but is otherwise a faithful reproduction. Charles Clements, *Witness to War: An American Doctor in El Salvador* (New York: Bantam Books, 1984).

that each society develops its own standards of injustice, or that the standards of a middle-class U.S. citizen might be irrelevant to an interpretation of injustice in Latin America. When an intrepid interviewer made such a suggestion, one respondent shot back: "You don't have to be a U.S. citizen to know that it's wrong to starve children so that a few can live like Arab shieks. That's wrong here; that's wrong in Guatemala; that's just plain wrong."

In other words, many policy makers believe that instability in Latin America is the manifestation of a broadly human phenomenon: rebellion growing out of a "natural" opposition to a situation that is fundamentally unfair. These officials argue that Latin Americans are naturally influenced by an attitude Barrington Moore has identified as universal: "There are indications of a widespread feeling that people, even the most humble members of society, ought to have enough resources or facilities to do their job in the social order, and that there is something morally wrong or even outrageous when these resources are unavailable."[15] Moore concludes that human beings in general (including Latin American peasants and U.S. policy makers) act in response to the belief "that personal and private retention *without use* of resources that are in short supply and needed by others is somehow immoral and a violation of the higher rights of the community."[16]

Ironically, Jeane Kirkpatrick has most effectively summarized the impact of this belief on U.S. citizens and their policy makers: "The extremes of wealth and poverty characteristic of traditional societies . . . offends us," she wrote in 1979. "Moreover, the relative lack of concern of rich, comfortable rulers for the poverty, ignorance, and disease of 'their' people is likely to be interpreted by Americans as moral dereliction pure and simple. The truth is that Americans can hardly bear such societies and such rulers."[17] These perceptions of stark poverty and an uncaring selfish elite is

---

[15] Barrington Moore, Jr., *Injustice: The Social Bases of Obedience and Revolt* (White Plains, N.Y.: M.E. Sharpe, 1978), p. 47. See also Manus I. Midlarsky, "Scarcity and Inequality: Prologue to the Onset of Mass Revolution," *Journal of Conflict Resolution* 26 (March 1982): 3-38.

[16] Moore, *Injustice*, p. 38.

[17] Jeane J. Kirkpatrick, "Dictators and Double Standards," *Commentary* 68 (November 1979): 42. The irony lies in the fact that while Ambassador Kirkpatrick is able to identify this sense of injustice as a core belief among U.S. citizens, one searches in vain through her many speeches and writings for an indication that she, too, might believe that the extremes of poverty and wealth in Latin America are unjust. Indeed, as the tone of the quotation suggests, Ambassador Kirkpatrick seems to be contemptuous of U.S. citizens who hold this belief.

what most inflames many policy makers. In some cases it is the driving force in their lives. Crusaders for the poor of Latin America, they have chosen Washington as the field of battle.

Some officials make a simple, direct linkage between the human suffering caused by poverty and distributional injustice to instability. They believe that the poor would rather fight than starve. Others recognize, however, that throughout history people have regularly starved quietly without dissent; the most common response to poverty has been resignation. How, then, are poverty and injustice converted into instability? The answer for many policy makers is that the postwar process of structural modernization has made poverty increasingly unbearable in contemporary Latin America.

Structural Changes Reduce the Traditional Passivity of the Poor

"Structural change" is an imprecise term that conveys a variety of meanings. Most policy makers approach the subject by focusing upon the postwar spread of ideas that has been accelerated by the technological revolutions in transportation and communication. This focus resembles Marion Levy's "solvent" approach to modernization, in which citizens of less-developed countries, "plagued with problems of stability," adopt selected ideas that were first developed in industrialized societies.[18] They import these ideas not because they are coerced by invading armies but because they believe that the ideas will help them to be "better off."

These imported ideas subvert the status quo. Some of the ideas are tangible. The importation of North Atlantic techniques of production, for example, has ripple effects throughout a society's social structure. It increases the degree of specialization of organization, decreases self-sufficiency, creates new media of exchange (cash markets), causes demographic shifts (urbanization), and alters traditional family relationships. Other imported ideas are attitudinal rather than physical. For example, Oscar Altimir of the

[18] Marion J. Levy, Jr., *Modernization and the Structure of Society: A Setting for International Affairs*, 2 vols. (Princeton, N.J.: Princeton University Press, 1966), 2: 741-764. For what is still one of the most thoughtful discussions of the transfer of "European" concepts of modernity to the Third World, see A. Irving Hallowell, "Sociopsychological Aspects of Acculturation," in *The Science of Man in the World Crises*, ed. Ralph Linton (New York: Columbia University Press, 1945), pp. 171-200.

Economic Commission for Latin America and the Caribbean (ECLAC) notes the "invasion" of consumerism: "This predominant lifestyle, to which the great majority of the population of Latin America aspires, . . . is the lifestyle projected from the industrialized countries and which, in Latin America, is transmitted through consumer marketing, the mass media and the values of the leading groups."[19] Whatever their form, the important point here is that many policy makers believe these imported ideas have an ability to erode traditional social structures.

Because these structural changes are largely based upon increased economic interaction, international trade and foreign investment are often cited as crucial vehicles of instability. To many analysts, foreign market forces "distort" local economies to the detriment of the poor. Walter LaFeber, for example, argues that in El Salvador "the ruling families channeled Alliance aid into their own industrial enterprises or used it to buy up more land, on which they grew crops for export rather than food for their countrymen."[20]

As LaFeber's comment suggests, a focus upon the predominantly agricultural countries of the Caribbean basin will center attention upon the conversion of agricultural land from production for domestic needs into production for foreign consumption. In the case of El Salvador, writes William Durham,

> large areas of the Pacific lowlands were converted to cotton plantations in the 1950's and early 1960's as large landholders sought to take advantage of favorable market conditions. The economic value of land in this region increased rapidly, and this "greatly reduced the land area available to *campesinos* [peasants] whether they were *colonos* [hacienda laborers], sharecroppers, or renters," or indeed squatters. . . . Sugar production for export was another part of the diversification phase. The total land area planted annually in sugarcane increased from 10,000 hectares in the 1930's to 28,000 hectares by 1971, and export production increased nearly 1,000 percent in the same period. As the agricultural censuses reveal, much of this expanded production took place on large holdings (over 50 hectares), where, once again, we may infer that tenant farmers were displaced. The expanded production of lesser

[19] Altimir, "The Extent of Poverty in Latin America," pp. 76-77.
[20] Walter LaFeber, "Inevitable Revolutions," *Atlantic Monthly* 249 (June 1982): 80.

commercial crops (tobacco and sesame in particular) has probably also contributed to the increasing concentration of land.[21]

It is not necessary that this transformation be implemented by agents of an uncaring dictatorship or a rapacious oligarchy, although the case at hand—El Salvador—has had more than its fair share of both groups. Most policy makers who discuss this issue perceive the "culprits" to be the impersonal forces unleashed by the importation of capital-intensive production techniques and, especially, by foreign demand for inexpensive labor or the products of tropical agriculture. Obviously, they say, it is not wrong for U.S. citizens to consume more sugar, bananas, and coffee; but as production for U.S. consumers has increased, it has unfortunately led to a decrease in the production of agricultural staples.[22]

Two consequences are often cited as contributing to instability. First, there is less food: "Food is scarce not because the land is incapable of producing enough for the resident population, but rather because large areas have been underutilized or dedicated to the production of export crops."[23] Second, there are absolutely fundamental changes in the rural social structure. As Portes notes, "a vast peasantry dedicated exclusively to subsistence production is a phenomenon of the past in most of Latin America. In those

[21] William H. Durham, *Scarcity and Survival in Central America: Ecological Origins of the Soccer War* (Stanford: Stanford University Press, 1979), p. 44. LaFeber also argues that "between 1960 and 1970 thousands of Salvadorans were evicted from land so that more sugar could be grown" for export. LaFeber, "Inevitable Revolutions," p. 80. See also David Browning, "Agrarian Reform in El Salvador," *Journal of Latin American Studies* 15 (November 1983): 405. For a detailed analysis of this process, see Ridgway Satterthwaite, "Campesino Agriculture and Hacienda Modernization in Coastal El Salvador: 1949 to 1969," Ph.D. diss., University of Wisconsin, 1971, especially p. 280 where the author observes that agricultural modernization has meant that peasants "are sinking into a relatively worse poverty than before." For a general summary of the effects of economic modernization upon the standard of living of the poor, see Paul Streeten, "From Growth to Basic Needs," *Finance and Development* 16 (September 1979), especially p. 29.

[22] On this process in the specific case of coffee in Guatemala, see Sanford A. Mosk, "The Coffee Economy of Guatemala, 1850-1918: Development and Signs of Instability," *Inter-American Economic Affairs* 9 (Winter 1955), especially pp. 13-15; David McCreery, "Debt Servitude in Rural Guatemala, 1876-1936," *Hispanic American Historical Review* 63 (November 1983): 735-759; David McCreery, "Development and the State in Reform Guatemala, 1871-1885," papers in International Studies, Latin America Series No. 10 (Athens: Ohio University Center for International Studies, 1983.

[23] Durham, *Scarcity and Survival in Central America*, p. 54.

countries where rural subsistence enclaves persist, they are rapidly being undermined."[24] What has occurred in El Salvador, for example, is that

> the population so displaced from its traditional means of livelihood has been obliged to occupy a disadvantaged and dependent position within the new agrarian structure. Those employed or resident on private estates depend on exploitative conditions of wage employment, cash rentals, or *colono* or sharecropper status. For those *descampesinados* obliged to join the increasing number of migrant landless workers, currently estimated at over one-half of the rural population, dependence is in the form of a competitive search for temporary seasonal employment on private farms. It is this displacement of population with no compensating measures to provide for their welfare which has so adversely affected rural society and in particular the village community.[25]

In general, discussions of this type are not very frequent in Washington. Readers may have noted that the quotations used in the preceding paragraphs are all from writings by social scientists, not from interviews with policy makers, most of whom lack the personal background needed to discuss the impact of changing land use patterns upon the poor in Latin America. What these policy makers perceive is overwhelming poverty. They are not always sure how it came to exist, but most are convinced that the postwar evolution of international trade and investment is somehow involved in the process. Thus while I have yet to meet a Washington official who has read William Durham's book, when interview questions use Durham's ideas to probe beliefs about the role of

[24] Portes, "Latin American Class Structures," p. 16.

[25] Browning, "Agrarian Reform in El Salvador," p. 405. A large number of more general studies has confirmed Durham's findings in El Salvador: agricultural modernization apparently causes increased class polarization and inequality. For a general overview of this issue, see Keith Griffin, *The Political Economy of Agrarian Change: An Essay on the Green Revolution* (Cambridge, Mass.: Harvard University Press, 1974), especially pp. 60-82. Migdal's study of the effect of export agriculture in increasing the level of peasant political activity seems the closest academic parallel to the beliefs of many U.S. policy makers. See Joel S. Migdal, *Peasants, Politics, and Revolution: Pressures toward Political and Social Change in the Third World* (Princeton, N.J.: Princeton University Press, 1974), pp. 14-22. For a critical evaluation of the perceptions by Midgal and others, see Theda Skocpol, "What Makes Peasants Revolutionary?" *Comparative Politics* 14 (April 1982): 351-375.

changes in the structure of production for international trade, many policy makers embrace the ideas because they help articulate their understanding of the relationship between poverty and structural change, on the one hand, and growing instability, on the other.

Basically, what many policy makers believe is that foreign trade and investment promote economic modernization, and this modernization, in turn, undermines traditional social structures. Then, rather than discuss changes in land use or trade patterns, these policy makers turn instead to the more visible effects of economic modernization, particularly population growth, urbanization, literacy, and media exposure—the types of changes that for many years have been subsumed under Karl Deutsch's category of social mobilization.[26] Everyone recognizes that these transformations create an environment in which Latin America's poor can begin to assess their poverty from new perspectives. Economic modernization includes the revolution in electronics, for example, which means that radios can now communicate political messages to mass audiences. Economic modernization also includes the concentration of people in urban centers, which facilitates political recruitment. As new labor unions, new parishes, and new cooperatives are formed, the door is opened to attitudinal changes that lead, eventually, to instability.

At the same time that foreign trade and investment has encouraged economic modernization, other foreign forces have contributed to structural change in Latin America. Of greatest importance to most policy makers are foreign governments, particularly the United States and the Soviet Union. For example, the Peace Corps is often cited as "subversive" of the fatalistic acceptance of poverty in Latin America. The picture being perceived in Washington is of bright, energetic, and reasonably well-scrubbed U.S. volunteers entering remote Latin American villages, gaining the confidence of the impoverished population, and eventually succeeding in the twin struggles against an uncaring government and centuries of ingrained passivity. Success may be slow in coming, but in the end the effect of these intrusive Americans is to impregnate Latin America with a new consciousness of poverty. A similar role is played by the U.S. aid program. U.S. aid funds build schools, health clinics, and roads, and these open up new vistas for

[26] Karl W. Deutsch, "Social Mobilization and Political Development," *American Political Science Review* 55 (September 1961), pp. 493-514.

Latin America's poor. Meanwhile, prodding AID development specialists force reluctant Latin American bureaucrats to expand the role of government in addressing the problems of poverty. In both cases, the U.S. aid program destabilizes traditional social structures. Many policy makers view this destabilization as the most positive bilateral contribution the United States can make to a brighter future for Latin America.

The Soviet Union is also perceived as increasingly active in Latin America. Some of this activity is thought to be simple subversion—a topic for the following chapter—but there are other forms of Soviet involvement that, like the various U.S. aid programs, are believed to encourage new attitudes about poverty. Radio broadcasts from the Soviet Union, Eastern Europe, and Cuba—over four hundred hours per week in the 1980s—and Spanish and Portuguese language editions of Soviet periodicals, along with most other forms of cultural interaction, are generally perceived as fairly unimportant. One exforeign service officer, for example, writes that "the list of Party, parliamentary, trade, educational, scientific, and cultural delegations to Latin America on visits of two or three weeks is long and impressive until one considers how thinly they are spread over more than twenty countries."[27] The one Soviet program that worries almost all policy makers is the program of educational exchanges. With 7,600 Latin American students studying in the Soviet Union in 1983, and about 3,000 additional students in Eastern European universities, the training of students is generally perceived as the Soviet Union's most important impact on Latin America.[28]

In summary, a substantial group of policy makers believes that for several decades a number of forces, most of them foreign in origin, have been attacking the foundation of Latin America's traditional social structure. Modernization is occurring, and one aspect of this modernization is a new consciousness among many Latin Americans that their poverty need no longer be accepted with the

[27] Cole Blasier, *The Giant's Rival: The USSR and Latin America* (Pittsburgh: University of Pittsburgh Press, 1983), p. 11.

[28] These students are not perceived by Blasier or by contemporary officials as brainwashed ("very few could easily be misled about the nature of Soviet society"); instead, "most are sincerely grateful to the USSR for the opportunity to receive a university or professional school education, an opportunity they would probably never have had at home." Ibid., p. 12. For examples of two policy makers' worries about Soviet education of Latin American students, see *Foreign Assistance Legislation for Fiscal Year 1983*, pt. 8, p. 101 (Representative Michael Barnes), and p. 103 (Representative Clement Zablocki), supra n. 8.

fatalistic passivity of an earlier era. That old-fashioned term, the revolution of rising expectations, is alive and well in the minds of many Washington officials.

Political Mobilization Heightens Class Tensions

Political mobilization is the bridge between awareness and action. It is the process whereby people who have been mentally transformed by the revolution of rising expectations begin to express themselves politically. This, argue many policy makers, is what is occurring today in Latin America. One class—the poor—is pushing its way into the political system.

Political mobilization can occur in an almost infinite variety of ways. But when this mobilization involves large numbers of people at the same time there is always an idea that serves as a focus and an institution that serves as a coordinator of the transformation. The idea in contemporary Latin America is poverty, and the principal institutions are the new generation of grass-roots organizations. In the 1960s and 1970s, many of the region's more traditional political institutions—particularly trade unions and center-left political parties—came under increasingly effective attack by repressive governments. The result was the creation of new institutions to serve as political vehicles to address the problems of poverty. Most of these new organizations, from cooperatives to neighborhood self-help committees, are in some way connected to the Catholic Church.

There is no agreement in Washington (or anywhere else) on the date when the Church began its complex set of changes in social doctrine that led, eventually, to the transformation of religion in Latin America. Perhaps 1961 is a reasonable starting point, for it was in that year that Pope John XXIII issued his papal encyclical *Mater et Magistra*. Writing in the first-person plural, Pope John sent a jolt of electricity coursing through the somnolent hierarchy:

> We are filled with an overwhelming sadness when we contemplate the sorry spectacle of millions of workers in many lands and entire continents condemned through the inadequacy of their wages to live with their families in utterly subhuman conditions. . . . In some of these lands the enormous wealth, the unbridled luxury, of the privileged few stands in violent, offensive contrast to the utter poverty of the vast ma-

87

jority. . . . Economic progress must be accompanied by a corresponding social progress, so that all classes of citizens can participate in the increased productivity. The utmost vigilance and effort is needed to ensure that social inequalities, so far from increasing, are reduced to a minimum.[29]

Two years later, a second encyclical, *Pacem in Terris*, told governments it was time to "give considerable care and thought to the question of social as well as economic progress, and to the development of essential services."[30] To those governments that might not know what was meant by essential services, Pope John provided a detailed list. In the process he identified the Church with a concept of human rights that has since come to be known as the right to the fulfillment of basic needs:

Man has the right to live. He has the right to bodily integrity and to the means necessary for the proper development of life, particularly food, clothing, shelter, medical care, rest, and, finally, the necessary social services. In consequence, he has the right to be looked after in the event of ill-health, disability stemming from his work; widowhood, old age; enforced unemployment; or whenever through no fault of his own he is deprived of the means of livelihood.[31]

Pacem in Terris then proceded to question the legitimacy of governments that failed to promote these rights: "Any government which refused to recognize human rights or acted in violation of them, would not only fail in its duty; its decrees would be wholly lacking in binding force."[32]

Five years later these messages revolutionized the pastoral mission of the Church in Latin America. Meeting in Medellín, Latin America's bishops began a process of reorienting the Church toward what later came to be known as "a preferential option for the poor."[33] This led to the creation throughout Latin America of

[29] Claudia Carlen, ed., *The Papal Encyclicals, 1958-1981*, 5 vols. (Wilmington, N.C.: McGrath Publishing Company, 1981), 5: 67.

[30] Ibid., p. 114.

[31] Ibid., p. 108.

[32] Ibid., pp. 113-114.

[33] The term "preferential option for the poor" is from the final document of the Third General Conference of Latin American Bishops, which convened in Puebla, Mexico, in 1979: "We affirm the need for conversion of the whole Church to a preferential option for the poor, an option aimed at their integral liberation." For a sensitive and highly detailed analysis of the changing Church in Central America, see

Christian grass-roots communities (*comunidades eclesiales de base*, or CEBs), the organizations that Latin American Catholics now use to translate the new social doctrine into everyday experience. Alongside the expansion of CEBs came an explosion in the number of lay workers to promote the new concepts of Catholic social justice.

These new organizations spread like wildfire through poor Latin American communities. In El Salvador, for example, "starting in the early 1970s, more and more priests and nuns began trying to stir the country's deeply impoverished peasantry, first by forming *comunidades de base* from which lay preachers—known locally as Delegates of the Word—emerged as community leaders, then by encouraging the landless peasants to campaign for an agrarian reform."[34]

Many of these religious workers were foreigners. The most obvious example was in Nicaragua's remote Zelaya province, where U.S. Catholic Capuchins and Protestant Moravians had been active for decades building chapels, schools, and providing such diverse social services as prenatal care and agricultural extension services. The seeds sown by these activists encouraged the creation of a new consciousness among the rural poor. They became socially mobilized, as journalist Penny Lernoux reported after visiting the Capuchin parishes:

> The Zelaya Christian communities were local bookstrap operations which encouraged the people to elect representatives to be trained in Capuchin seminars as teachers, midwives, artisans, agronomists, and religious leaders. None of these people are experts in their fields. Many of the primary schoolteachers, for example, are themselves barely able to read and write. But because they form part of the community, and therefore have a stake in its progress, they are not about to abandon their jobs, as all the previous urban-educated teachers, agronomists, and doctors did. So eager for knowledge are these people that they will hike for three and four days through the rain forests and mountains in order to attend a lo-

---

Phillip Berryman, *The Religious Roots of Rebellion: Christians in Central American Revolutions* (Maryknoll, N.Y.: Orbis Books, 1984). For a specific example of how the changing doctrine affected the Church's educational activities, see Charles J. Beirne, S.J., "Jesuit Education for Justice: The Colegio in El Salvador, 1968-1984," *Harvard Educational Review* 55 (February 1985): 1-19.

[34] *New York Review of Books*, May 28, 1981.

cal seminar. The Capuchins created a whole library of pamphlets, posters, and books written in the language of the Zelaya peasants as teaching tools, all with the same message: love your neighbor, know your legal rights and be proud of your Indian peasant heritage.

But then Lernoux observed that this social mobilization seemed to lead, almost inexorably, to political action. First,

the people constructed the cornerstone of Zelaya's Christian communities, a chain of rustic chapels, each with a communal dining hall. The purpose of these halls is to encourage the people to pray, meet, and eat together once a week.... As a result of these meetings, Zelaya's Christian communities soon spawned a whole series of other organizations, including parent-teacher associations in charge of the rural schools and agricultural clubs. In the most advanced Christian communities ... these associations and clubs have served to unify the people against the National Guard and the large cattle ranchers. If any member is falsely accused by the military or the ranchers, the community unites behind him. Moreover, the people know their legal rights and are not afraid to talk back to the National Guard.[35]

Until a few years ago the overwhelming proportion of U.S. policy makers had missed this transformation entirely. Still today specifics about the changing role of the Church in Latin America are little understood in Washington; indeed, one of the most perceptive students of the Latin American Church, Michael Dodson, argued as late as 1984 that the grass-roots organizations are "perhaps the greatest misunderstanding U.S. policy makers have of Central America today."[36]

Yet Washington officials know something is occurring because the knowledge has been thrust upon them by a relentless decade

[35] Penny Lernoux, "Notes on a Revolutionary Church: Human Rights in Latin America," mimeographed report to the Alicia Patterson Foundation, February 1978, pp. 89-91.

[36] Michael Dodson, "Democratic Ideals and Contemporary Central American Politics," paper presented at the Annual Meeting of the International Studies Association, Atlanta, Georgia, March 1984, p. 15. For a history of one such grass-roots Church effort in the Salvadoran town of Aguilares, see Phillip Berryman, "El Salvador: From Evangelization to Insurrection," in *Religion and Political Conflict in Latin America*, ed. Daniel H. Levine (Chapel Hill: University of North Carolina Press, 1986), pp. 58-78.

of lobbying by church groups. No member of Congress, for example, has been exempt. In 1981, the chairman of the Senate Foreign Relations Committee, Charles Percy, told his colleagues that

> we have some extraordinarily fine, well-meaning church people across our country—I certainly have them in Illinois— who are concerned about this problem. Their concern is very deep. . . . Just a few weeks ago a group had a sit down in my Illinois office and refused to leave until I assured them, by telephone from Washington, that I would oppose all military assistance to El Salvador. Well, I made it clear to them when I talked to them that I had no such intention. They made it clear that they had no such intention of leaving my office until I gave that assurance. So, regretfully, at closing time, the U.S. marshals moved these fine people out of the office. Now, you don't like to do that.[37]

This action by church people followed years and years of active lobbying by ecumenical delegations—letter after letter, visit after visit to the Senator's offices in Illinois and Washington. Most letters and virtually every visit contained a message from a missionary or a lay worker similar to the communications of Lernoux.

In other words, while it is true that some Washington officials remain unaware of the new pastoral emphases of the past two decades, their number is declining rapidly. The majority today tends to be better informed. Some officials have experience working with the Church in Latin America; virtually all the rest have been the target of church-based lobbying. On balance, most contemporary policy makers know something of the Church's role in fostering instability.

Although some policy makers complain about "leftist" priests who stir up trouble in Latin America, Church-encouraged political mobilization is generally perceived as pragmatic, nonideological, bread-and-butter politics. To Ambassador Robert White, for example, in El Salvador "the guerrilla groups, the revolutionary groups, almost without exception began as associations of teachers, associations of labor unions, campesino unions, or parish organizations which were organized for the definite purpose of getting a schoolhouse up on the market road."[38] Along with other

[37] U.S. Congress, Senate, Committee on Foreign Relations, *Foreign Assistance Authorization for Fiscal Year 1982*, 97th Cong., 1st Sess., 1981, p. 440.
[38] U.S. Congress, House, Committee on Foreign Affairs, Subcommittee on Inter-

policy makers, White perceives a process in which grass-roots organizations seek to alleviate the most immediate and distressing features of poverty. There is no role here for the abstract forces of ideology; this is politics at its most basic, immediate level.

It should be emphasized that many policy makers believe this instability is more than an often-violent negotiation process between the noble poor who deserve more and the ignoble rich who have too much. Many officials have a larger view—an ideology—that explains instability as part of the march of history. "As we revolted against paying taxes," writes retired diplomat John Paton Davies, "so if it takes a revolution to make 'oligarchs' pay their taxes and divide up those big plantations, let the revolutions rip."[39] This belief—that when the circumstances of life are perceived as unjust, people will come together to do something about it—is broadly held by foreign policy officials. From this perspective the twentieth century in Latin America has been characterized by a continuous renegotiation of the social contract between rich and poor, powerful and powerless, in which the rich and powerful have been forced by newly mobilized groups to surrender part of their wealth and power in order to maintain peace.[40] As Paton's remark suggests, few policy makers verbalize their perceptions in this abstract fashion. Instead, they talk about concrete demands for land reform or for better schools by people who were, until just yesterday, satisfied with what they had. Once deprived citizens organize, instability is the inevitable result of the negotiation process to determine whether there will be taxation without representation, land reform, or a new school; it is evidence of the continuous piecemeal renegotiation of the social contract.

In summary, the intellectual glue linking poverty to instability is the belief that structural changes have led to the expansion of the population that is politically relevant. The most salient political characteristics of these new participants are (1) their poverty, (2) their contact with modernity, and (3) their organization into increasingly effective pressure groups. As new groups of impoverished citizens have crowded into limited political space, the jos-

---

American Affairs, *Presidential Certification on El Salvador*, vol. 1, 97th Cong., 2d Sess., 1982, p. 228.

[39] John Paton Davies, *Foreign and Other Affairs* (New York: W.W. Norton, 1964), pp. 31-32. "The catch is," Paton continues, "that the tidy, constructive ends we wish to see achieved are not predictably served by violent, erratic means over which control is easily lost, quite possibly to our adversaries."

[40] Moore, *Injustice*, p. 18.

tling for position has produced instability. This low-level instability is perceived as neither pathological nor exceptional; on the contrary, to many policy makers instability represents a logical, expected outcome of modernization.

But there is instability and then there is *real* instability, and it is the latter that captures the attention of U.S. policy makers. Only the few officials responsible for a given Latin American country are aware of that country's chronic low-level instability, an instability that characterizes all modern political systems. Strikes, demonstrations, or even an occasional assassination are simply not seen by most Washington officials. It is only when instability threatens to get out of hand—to become uncontrollable by the existing government—that it becomes a focus of attention in the United States.

Elite Intransigence Makes Instability Inevitable

Many policy makers believe that low-level instability is converted into high-level instability in Latin America when political elites refuse to institute timely reforms. What happens, they say, is that eventually the poor rebel when the government continually refuses to build the school.

Everyone in Washington perceives significant resource limits on the ability of governments to address the demands of Latin America's newly mobilized poor. To illustrate this perception policy makers often compare the region's economy to a pie. In Latin America, they say, the political pie is growing like topsy, while the economic pie is small and, when population growth is considered, it is not growing very rapidly; at times it may even be shrinking. Were it growing at a more rapid pace, then more of the demands of the newly mobilized poor could be met, for the financial benefits of growth could be channeled disproportionately in their direction. Until such growth occurs, however, the resources to meet the needs of the poor must necessarily be obtained by the redistribution of existing wealth.

But except under revolutionary circumstances, redistribution is painfully slow at best; at worst it is a politician's promise whose fulfillment becomes less likely with each passing day.[41] As previously noted, recent periods of economic prosperity in Latin Amer-

[41] William Ascher, *Scheming for the Poor: The Politics of Redistribution in Latin America* (Cambridge, Mass.: Harvard University Press, 1984).

ica have resulted in income concentration rather than income distribution. Despair, discouragement, and frustration are the words many policy makers use to describe the effect of unfulfilled promises on Latin America's poor. Not surprisingly, they say, many among the poor eventually perceive promises of reform to be nothing more than a trick. As their impatience grows, the poor take steps to accelerate the reform process, to destabilize an unresponsive political system.

To many officials, elite intransigence, typically accompanied by an increase in repression, is guaranteed to increase instability. Some policy makers cite John Kennedy's tired dictum—"those who make reform impossible make revolution inevitable"—as if it were a law of nature.[42] In the late 1970s and 1980s, for example, the unbending selfishness of El Salvador's oligarchy became the frequent keystone of explanations for the civil war in that country. To former Undersecretary of State William D. Rogers, instability in El Salvador was "tied to centuries of discrimination and injustice by greedy oligarchs, a passionate disrespect of civil and political rights of its citizens by the Army and a rigid bureaucratic system that was not capable of gradual change."[43] Similarly, former NSC staffer Robert Pastor argued that the cause of the Salvadoran crisis was that "the regime's leadership—the oligarchy—refused either to adapt existing institutions to these changes or to permit the political emergence of new groups. They thought—and continue to think—that stability could be imposed forcefully by insisting on the status quo."[44] As was often the case, Ambassador Robert White expressed this view most forcefully: "The Salvadorans on whom the success of the Reagan formula depends are rotten to the core. Nothing we can do can instill morale into a Salvadoran military officer corps which has earned the contempt of the civi-

[42] What today seems trite, however, is not necessarily incorrect. Dress up Kennedy's argument in the garb of modern social science and you have Skocpol's more recent thesis: a crucial factor in creating revolutions is the ability of elites to keep the state from accommodating pressures for reform. Theda Skocpol, *States and Social Revolutions: A Comparative Analysis of France, Russia, and China* (London: Cambridge University Press, 1979). For an analysis and critique, see Jerome L. Himmelstein and Michael S. Kimmel, "States and Revolutions: The Implications and Limits of Skocpol's Structural Model," *American Journal of Sociology* 86 (March 1981): 1,145-1,154.

[43] William D. Rogers and Jeffrey A. Meyers, "The Reagan Administration and Latin America: An Uneasy Beginning," *Caribbean Review* 11 (Spring 1982): 17.

[44] Robert A. Pastor, "Three Perspectives on El Salvador," *SAIS Review* 1 (1981): 39.

lized world by its routine practice of torture and assassination. Nothing we can do can prevent the economic collapse of the country as the rich and powerful systematically export the wealth of El Salvador into their foreign bank accounts."[45] Many policy makers are outraged by the thought that Latin American oligarchs would prefer to buy condominiums for themselves in Florida than schools for the poor in their own country.

Tensions increase as new political demands from the poor confront the stone wall of elite intransigence. In 1981, former Assistant Secretary of State Vaky captured this perception well when he wrote that "unjust situations are unstable over the long run. Demands for reform and change can be suppressed for a while, but the likely consequence is to push demands into violence and to make the change more disrupting when it occurs."[46] Civil disobedience increases as newly politicized groups battle for public policies that will fulfill their needs. Land seizures, strikes, demonstrations, "terrorist" attacks, and guerrilla warfare become increasingly common, while repression by paramilitary and government security forces becomes increasingly brutal. Low-level instability becomes high-level instability.

Now Latin America has the attention of U.S. policy makers, who begin to concentrate upon the developing conflict in much the same way that geologists flock to the sides of a volcano as signs of activity flare. Standing on the rim and looking down into the crater, one group of officials shares an identical understanding of the cause of the "confusion, effervescence, and revolutionary agitation" that roils ominously below them: the poor are mad as hell, and they're not going to take it anymore.

Contrary Beliefs

Not everyone agrees. Virtually all policy makers recognize the existence of poverty in Latin America, and they often comment on it when discussing the causes of instability in the region. But the comments of some policy makers are made almost in passing, as an afterthought or an aside, subordinate to whatever major observation is being made. Note, for example, how quickly Roger Fontaine, the first Latin Americanist on the Reagan National Security

---

[45] *Presidential Certification on El Salvador*, 1: 135, *supra* n. 38.

[46] Viron P. Vaky, "Hemispheric Relations: 'Everything Is Part of Everything Else,' " *Foreign Affairs* 59 (1981): 641.

Council staff, touched on the issue of poverty during a discussion of instability:

> Few would suggest that the Cubans have manufactured the crisis in Central America out of whole cloth. On the contrary, the turmoil in the region has its origins in the historical inequities between the social classes that have generated political and economic frustrations among large numbers of citizens. Nevertheless, the Carter administration has found it convenient to minimize the degree to which Cuba and local Marxist guerrillas have heightened and manipulated these frustrations into increasingly destabilizing acts designed to promote Marxist revolution.[47]

Similarly, when speaking of the Caribbean, the State Department's John Bushnell acknowledged the importance of poverty in causing instability but only in the context of a more important cause, Cuban adventurism: "These island states find themselves critically strained by stagnant agricultural sectors, the low output of industries, and unemployment rates of up to 35 percent, all contributing to discontent and political instability which invite Cuban adventurism. . . . The crucial problem is unemployment, particularly of youths just entering the labor force—the groups most susceptible to Cuban-inspired exploitation."[48] Many officials make this type of statement; it amounts to a pro forma observation that poverty exists in Latin America. In the 1980s, Colonel Samuel Dickens of the American Legion spoke of poverty only to make his point that Communism "is going to exploit any weakness"; similarly, Myles Frechette, a career official, noted "Cuba's readiness to exploit instability and injustice" but never elaborated on the region's injustice; instead, he focused upon Cuba.[49]

As these examples suggest, many policy makers focus upon Communist adventurism, the topic of the next chapter, when they discuss instability. Buried in their discussions is an acknowledgment that poverty causes instability, but it is the minimum possible recognition. Some officials would not even mention poverty as a cause of instability were they not occasionally obliged to do

[47] Roger Fontaine, Cleto DiGiovanni, Jr., and Alexander Kruger, "Castro's Specter," *Washington Quarterly* 3 (Autumn 1980): 16.

[48] *Foreign Assistance Legislation for Fiscal Year 1982*, pt. 7, pp. 71-72, supra n. 1.

[49] *Foreign Assistance and Related Programs Appropriations for 1981*, pt. 1, p. 360, supra n. 2; Myles R. R. Frechette, "Letter to the Editor," *Foreign Policy*, no. 48 (Fall 1982), p. 178.

so. Secretary of State Haig, for example, was prodded by a liberal
senator:

> SENATOR MATHIAS: I think we would not have problems
> with Communists in El Salvador if there were not basic social
> and economic problems that have preceded the arrival of the
> Communists on the scene.
> SECRETARY HAIG: I think that is absolutely right, Senator.[50]

Secretary Haig, who at the time of this dialogue was seeking funds
for the U.S. foreign aid program, was on his best behavior, eager to
agree with everyone who had a vote on the aid bill. But Mr. Haig
never spoke publicly at his own initiative about poverty as the
cause of instability in Latin America. When he chose the topic, he
talked about Communism.

In the early 1980s, no one could avoid mention of the destabilizing effect of the worldwide recession that struck Latin America
particularly hard. In unveiling his Caribbean Basin Initiative in
1982, for example, President Reagan spoke of "temporary economic suffering" linked to the recent rise in the price of petroleum
and the decline in prices for the products of tropical agriculture:

> In 1977, 1 barrel of oil was worth 5 pounds of coffee or 155
> pounds of sugar. Well, to buy that same barrel of oil today,
> these small countries must provide 5 times as much coffee—
> nearly 26 pounds—or almost twice as much sugar—283
> pounds. This economic disaster is consuming our neighbors'
> money, reserves, and credit, forcing thousands of people to
> leave for other countries, for the United States, often illegally,
> and shaking even the most established democracies.[51]

Similarly, Secretary of State Shultz told senators that the world
recession and inflation of the late 1970s and early 1980s had resulted in "a rise in unemployment and underemployment which
is of truly major proportions—25% to 40% in many countries.
Added to the evils of inflation, spiraling foreign debt, and major
balance-of-payments problems, it amounts to an almost classic
recipe for social discontent and loss of confidence in the future."[52]

The existence of temporary economic troubles was eventually
accepted by nearly all policy makers as a cause of at least some of
the instability in Latin America during the late 1970s and early

[50] *Foreign Assistance Authorization for Fiscal Year 1982*, pp. 26-27, *supra* n. 37.
[51] *Weekly Compilation of Presidential Documents* 18 (March 1, 1982): 219.
[52] Prepared statement before the Senate Committee on Finance, August 2, 1982.

1980s. Deputy Assistant Secretary of State Bushnell perceived social unrest as a result of "increasingly burdensome energy costs" that began to rise in 1973,[53] Assistant Secretary Enders noted how "economic crisis has *recently* subjected fragile institutions to additional stresses,"[54] and Assistant Secretary Motley observed that *"during the last several years,* per capita gross domestic product has fallen by 35% in El Salvador, 23% in Costa Rica, 14% in Guatemala, and 12% in Honduras. In 4 years, El Salvador has lost 15 years of economic development."[55]

The conflict among policy makers, however, is not over the impact of occasional recessions—everyone agrees that recessions are destabilizing—but over the question of whether *structural* poverty causes instability. As we saw in the preceding chapter's discussion of reform in El Salvador, to many officials the answer is no. "Revolutions are not born out of resentments of landless peasants but (as Plato understood) in the bosom of the ruling elite," wrote Jeane Kirkpatrick. "Revolutions in our times are born in the middle class and carried out by sons of the middle class who have become skilled in the use of propaganda, organization and violence."[56] Why do middle-class youth encourage revolutionary instability? The answer of many policy makers, typified by Senator Jesse Helms, is that these middle-class youngsters mistakenly believe there is structural poverty: "The terrorism that is now tearing apart the social fabric of El Salvador is the product of middle-class intellectuals who have substituted ideology for reality. The belief that revolution arises from 'oppression' of the masses is a romantic concept that almost never bears the test of examination."[57]

Additional evidence of the rejection of the linkage between structural poverty and instability is to be found in Congress, where a number of members regularly vote against any major U.S. aid program, not because they are selfish but because they believe

[53] *Foreign Assistance Authorization for Fiscal Year 1982*, p. 419, *supra* n. 37.

[54] Prepared statement before the Senate Committee on the Judiciary, Subcommittee on Security and Terrorism, March 12, 1982. Emphasis added.

[55] Prepared statement before the House Committee on Appropriations, Subcommittee on Foreign Operations, March 27, 1984. Emphasis added.

[56] Jeane J. Kirkpatrick, "U.S. Security and Latin America," in *Rift and Revolution: The Central American Imbroglio*, ed. Howard J. Wiarda (Washington, D.C.: American Enterprise Institute, 1984), p. 356. This chapter is a slightly longer version of the magazine article of the same title.

[57] U.S. Congress, Senate, Committee on Foreign Relations, *Nomination of Robert E. White*, 96th Cong., 2d Sess., 1980, p. 7.

the aid program is inefficient and unnecessary.[58] Even when they have agreed to support economic aid to help with "stabilization" and "recovery" from difficult but transitory economic problems, they have done so only reluctantly. Thus while the Reagan administration made large economic aid requests in response to the recommendations of the Kissinger Commission, many members of Congress clearly perceived the aid as a necessary evil—necessary for (1) helping Central America weather the recession and, (2) obtaining support for the administration's Central American policy in general and its military aid request in particular.

Further evidence of this perception is provided by the *type* of economic aid requested by the executive branch. As Table 2.4 indicates, for FY1986 the Reagan administration asked Congress for $831 million in economic aid for Central America. Of that amount, $110 million (or 13 percent) was in the form of food, which is primarily a form of relief and only secondarily development assistance.[59] The largest amount, $509 million (or 61 per-

TABLE 2.4
ECONOMIC AID REQUEST FOR CENTRAL AMERICA, FY1986
(thousands of dollars)

	Development Assistance	Food for Peace (PL480)	Economic Support Fund	Total
Belize	6,800	0	4,000	10,800
Costa Rica	14,350	23,000	150,000	187,350
El Salvador	89,800	50,837	210,000	350,637
Guatemala	33,000	18,983	25,000	76,983
Honduras	45,000	17,956	80,000	142,956
Panama	22,600	0	40,000	62,600
Total	211,550	110,776	509,000	831,326

SOURCE: U.S. Department of State, Bureau of Public Affairs.

[58] There are many reasons why members of Congress vote against economic aid bills, and it is typically impossible to identify any one reason as the most prominent. Members who believe aid is unnecessary generally believe also that graft prevents aid from reaching the intended recipients, that governments should not interfere with free markets in allocating resources, and that countries receiving aid are often ungrateful (and sometimes hostile) to the United States.

[59] It is true, however, that counterpart funds that a government receives from the sale of donated or concessional U.S. food are often invested in development projects, but this money can also be used for other nondevelopmental purposes. For an example of AID's losing battle to use counterpart funds for development in El Salvador in the 1980s, see U.S. General Accounting Office, "Providing Effective Eco-

cent), was from the Economic Support Fund (ESF), a dollar cash transfer that recipients use to pay for foreign purchases.[60] ESF is best described as an interstate subsidy. It is not a loan but a gift of cash. The typical procedure is for the Agency for International Development and the recipient government first to agree upon the use to which ESF money will be put. Then AID either writes a check or directly deposits the appropriate amount in the recipient government's central bank account in the United States.[61] Once the money is in the hands of the recipient government, importers apply for the dollars from the central bank, paying for them in local currency, which the central bank then uses for such domestic operating expenses as civil service salaries.

In most cases ESF simply keeps troubled economies afloat. In FY1983, for example, the Reagan administration requested $326 million in ESF "in support of the immediate U.S. objective of assisting stabilization and recovery programs in the Caribbean Basin." Of that amount, the United States gave El Salvador $120 million. The government of El Salvador then sold these dollars for 300 million *colones*, which were used, in turn, to cover about 16 percent of the government's operating budget.[62] The financial ef-

---

nomic Assistance to El Salvador and Honduras: A Formidable Task," Report No. GAO/NSIAD-85-82 (Washington, D.C.: GAO July 3, 1985), p. 20.

[60] Today's ESF began as a form of military aid. When first created during the Mutual Security Act period (1951-1961), ESF was called Defense Support, and it was designed specifically to pay for the import bills of recipients who, for one reason or another, were devoting their own resources to the purchase of arms. Israel was (and remains) the principal recipient of ESF, with $1.2 billion of the $3.8 billion appropriation for FY1985. During the Kennedy years, the program's name was changed from Defense Support to the less bellicose title of Security Supporting Assistance. In 1971, Security Supporting Assistance was specifically moved by Congress from Part I of the Foreign Assistance Act of 1961—the part that governs development assistance—to Part II of the same act—the part that governs military aid. In 1978, Security Supporting Assistance was renamed today's Economic Support Fund. For the move of ESF from economic aid to military aid, see Section 202 of PL92-226; for the renaming of Security Supporting Assistance to ESF, see Section 10(a) of PL95-384.

[61] The recipient, in turn, provides AID with periodic reports on how the money has been spent. For the nuts and bolts of ESF procedures, see U.S. Congress, House, Committee on Foreign Affairs, *Foreign Assistance Legislation for Fiscal Year 1985*, 98th Cong., 2d Sess., 1984, pt. 8, pp. 6-8; *Foreign Assistance Legislation for Fiscal Year 1983*, pt. 6, *supra* n. 8; *Foreign Assistance Legislation for Fiscal Year 1982*, pt. 1, *supra* n. 1; and U.S. Congress, Senate, Committee on Appropriations, *Foreign Assistance and Related Programs Appropriations, Fiscal Year 1982*, 97th Cong., 1st Sess., 1981, pt. 1.

[62] The dollars were sold to Salvadoran citizens who could demonstrate the need for hard currency to pay for imports. U.S. Agency for International Development,

fect of Food for Peace (Title I) is about the same, with the principal difference being that, rather than give dollars, the United States loans dollars on concessional terms that the recipient government uses to buy food from private U.S. suppliers at prevailing market prices. The government then sells the food to domestic purchasers—typically wholesalers—at local market prices and uses the proceeds to finance other expenses.

Policy makers who view Latin American poverty as structural do not like to use ESF.[63] They prefer to provide economic aid through the traditional Development Assistance budget administered by the Agency for International Development. This aid finances development projects such as health clinics, schools, water purification plants, and market roads. Many officials were disappointed that only 25 percent (or $211 million) of the FY1986 administration economic aid request for Central America was in Development Assistance; they were further discouraged to discover later that even this figure was artificially inflated in the case of aid to El Salvador. The General Accounting Office (GAO) reported that "in El Salvador, development assistance projects have emphasized public works construction, employment, and commodity procurement needed to improve economic conditions in the short term. AID has funded relatively few activities to improve El Salvador's institutional capability to manage development activities."[64]

Congressional Presentation, Fiscal Year 1983 (Washington, D.C.: AID, 1982), p. 14; U.S. General Accounting Office, "Providing Effective Economic Assistance to El Salvador and Honduras," p. 12. For a general discussion of the use of ESF funds by Latin American recipients, see *Foreign Assistance Legislation for Fiscal Year 1983*, pt. 6, pp. 7 and especially 34-110, *supra* n. 8. An example of the use of ESF for purposes other than keeping shaky economies afloat was the 1979 amendments to the military bases agreement with the Philippines. In return for a continued U.S. military presence at Clark Air Force Base and Subic Bay Naval Base, the United States agreed to provide the Marcos government with $200 million in ESF during a five-year period, FY1980-FY1984, for development projects. See U.S. Congress, Senate, Committee on Appropriations, *Foreign Assistance and Related Programs Appropriations, Fiscal Year 1983*, 97th Cong., 2d Sess., 1982, pt. 1, pp. 666-667. In another example, in agreeing to provide ESF to Guatemala in the mid-1980s, Congress provided that the aid be "for development activities . . . aimed directly at improving the lives of the poor in that country, especially the indigenous population in the highlands." Section 101, PL98-473.

[63] See, for example, the comments by Representative Michael Barnes and by Larry Minear, secretary of the Interreligious Task Force on U.S. Food Policy, in *Foreign Assistance Legislation for Fiscal Year 1983*, pt. 2, pp. 64, 76-78, and pt. 8, p. 100.

[64] U.S. General Accounting Office, "Providing Effective Economic Assistance to El Salvador and Honduras," p. ii.

It is important not to let the details of the U.S. aid program obscure the principal point to be made. Those policy makers who believe that unrest in Latin America results from temporary economic setbacks, not from basic structural poverty, consider ESF and Food for Peace the appropriate aid vehicles. They understand that these are band-aid programs, used to prevent an infection of instability while the cure—international economic recovery—is given time to work. The recovery will trigger a resurgence in demand for Latin America's export products, halt the recession, and end any instability that is caused by economic factors.

These policy makers also disagree with the belief that structural changes lead to social and political mobilization. Most of the time their disagreement is manifested by silence. In their comments about instability many officials rarely discuss anything other than Communist adventurism: the conversion of land into the production of export crops, changing Catholic social doctrine, Peace Corps volunteers, urbanization, and other alleged agents of change are simply not mentioned.

On the rare occasions when these policy makers indicate a belief that foreign actors encourage instability in ways other than subversion, they almost always perceive a sinister force, a nefarious plot. As noted previously, many officials are concerned, for example, about scholarships for Latin American students to study in the Soviet Union.[65] The best example of this perception, how-

[65] The Reagan administration was particularly concerned about these scholarships. See, for example, the prepared statement by Deputy Assistant Secretary of State James H. Michel before the House Foreign Affairs Subcommittee on Inter-American Affairs, February 28, 1985. Michel's statement relied heavily upon data from U.S. General Accounting Office, "U.S. and Soviet Bloc Training of Latin American and Caribbean Students: Considerations in Developing Future U.S. Programs," Report No. GAO/NSIAD-84-109 (Washington, D.C.: GAO, August 16, 1984). The administration was somewhat careless in using the GAO study; indeed, it only cited those parts of the GAO analysis that supported its own policy position, while failing to indicate that there were other parts of the study that contradicted the view offered by the administration. The GAO, for example, stated that "the United States has always enjoyed a large lead over Soviet bloc countries in the total number of developing country students enrolled in their respective universities" (p. 12). But the Reagan administration only provided data on *federal government* scholarships, choosing to ignore the fact that the U.S. system of higher education is based upon state-level and private financing. The administration also ignored the GAO's conclusion that Latin America continues to be a low priority for Soviet educational exchanges: 10 percent of total Soviet scholarships went to Latin America in 1982 versus 11 percent in 1972 (p. 13). In addition to these distortions, Deputy Assistant Secretary Michel ignored his own predecessor's observation that the

ever, is the interpretation given to the occasions when the Catholic Church becomes involved in a major episode of instability. When that occurs, many policy makers perceive the Church as subversive. While no one suggests that the Church's doctrine is subversive, lower-level Church officials are regularly accused of encouraging instability for ulterior political motives.

An excellent example is the interpretation given to the rape and murder of four U.S. churchwomen in El Salvador in December 1980. Visiting the shallow gravesite where the bodies were dumped, Ambassador Robert White appeared visibly shaken.[66] To White and others who believe that poverty causes instability, the women were helping alleviate poverty in El Salvador. Having dedicated their lives to working with the poor, they had become martyrs of a just cause.

Several members of the incoming Reagan administration suggested, in contrast, that the churchwomen had received their just desserts. In Jeane Kirkpatrick's view, "the nuns were not just nuns. The nuns were political activists ... on behalf of the Frente." The nominee to be Assistant Secretary of State for Human Rights and Humanitarian Affairs, Ernest Lefever, similarly perceived the churchwomen as leftist political activists who fostered instability. Although Secretary of State Haig noted that "the facts on this are not clear enough for anyone to draw a definitive conclusion," he also observed that the churchwomen might have been asking for trouble: "I would like to suggest to you that some of the investigations would lead one to believe that perhaps the vehicle that the nuns were riding in may have tried to run a roadblock or may have accidentally been perceived to have been doing so, and there may have been an exchange of fire."[67]

Following a storm of protest, these officials reconsidered their original statements. Ambassador Kirkpatrick first denied having

---

United States was educating large numbers of Latin Americans: in 1982, Deputy Assistant Secretary Stephen Bosworth told Congress that in 1980/1981 there were 30,188 students from Latin America and the Caribbean studying in U.S. universities (that is, many times the number studying in the Soviet Union) and that the number was growing rapidly. See *Foreign Assistance Legislation for Fiscal Year 1983*, pt. 8, p. 107, *supra* n. 8.

[66] A picture of White at the gravesite was published in *Newsweek*, December 22, 1980, p. 53. He is reported to have said, "This time they won't get away with it. They just won't." *Washington Post*, December 5, 1980, p. 1.

[67] The comments are all reprinted in U.S. Congress, Senate, Committee on Foreign Relations, *The Situation in El Salvador*, 97th Cong., 1st Sess., 1981, pp. 197, 204. Mr. Haig did not comment on the fact that the women were also raped.

made her statement, but after the reporter for the Tampa *Tribune* produced the verbatim notes upon which his original article (December 25, 1980) was based, Kirkpatrick issued an explanation that was, in fact, an apology. Secretary of State Haig tried to make a joke out of the entire matter. An incredulous senator asked, "Are you suggesting . . . the nuns may have run through a roadblock?" Haig responded: "Oh, not at all. No, not at all. My heavens. . . . The dear nuns who raised me in my parochial schooling would forever isolate me from their affections and respect."[68]

Administration critics were not laughing. They were fully aware that the churchwomen were encouraging instability in an indirect fashion because they were contributing to the political mobilization of the poor. Daniel Levine's assessment of the churchwomen captures this belief perfectly: "Although their work is not political in conventional partisan terms, it cannot help being political in a larger sense, because it necessarily comes up against issues of power and privilege. Service of this kind, with its pervasive stress on solidarity and sharing, identifies with people whose lives are shaped by oppressive structures. Serving them brings one quickly up against the limits and the violence of threatened elites."[69] As Ambassador White noted, "there is no doubt who killed the nuns. They were killed by government security forces. That is the absolute fact."[70] Why were they killed? In the view of many policy makers the churchwomen were targeted because they were helping the poor in El Salvador to reform a system that is unjust. The churchwomen were subversive in a positive sense.

Underlying the dramatically different interpretation by other

[68] *Foreign Assistance Authorization for Fiscal Year 1982*, p. 36, *supra* n. 37. For Kirkpatrick's retraction, see *The Situation in El Salvador*, pp. 235-236, *supra* n. 67. For examples of the general furor, see U.S. Congress, House, Committee on Foreign Affairs, Subcommittee on Inter-American Affairs, *U.S. Policy toward El Salvador*, 97th Cong., 1st Sess., 1981, p. 8, and U.S. Congress, House, Committee on Banking, Finance, and Urban Affairs, Subcommittee on International Development Institutions and Finance, *Human Rights and U.S. Policy in the Multilateral Development Banks*, 97th Cong., 1st Sess., 1981, p. 426.

[69] Daniel H. Levine, "Whose Heart Could Be So Staunch?" *Christianity and Crisis*, July 22, 1985, p. 312. For a fuller appreciation of the positive appraisal of this subversion, see the biography of one of the churchwomen murdered in El Salvador: Ana Carrigan, *Salvador Witness: The Life and Calling of Jean Donovan* (New York: Simon and Schuster, 1984).

[70] Robert White, "There Is No Military Solution in El Salvador," *The Center Magazine* 14 (July-August 1981): 10.

officials is their profoundly different perception of poverty in Latin America. They viewed the women as destabilizing a social order that, to use Senator Helm's words, "has done a good job in creating wealth and distributing it broadly." While El Salvador may have been victimized by the world economic recession of the late 1970s and early 1980s, its socioeconomic structure is perceived as fundamentally just and sound. Anyone who would seek basic structural changes in that system deserves, therefore, to be called a subversive in the negative sense of the word.

The differing perceptions of this incident encapsulate perfectly the first answer we can give to the question of why there is profound disagreement over an appropriate policy toward Latin America: there are deeply differing beliefs about the role of poverty in causing instability. But this is only part of the answer. There are equally strong disagreements over the role of another factor—communism—in causing instability. It is to a consideration of that factor that we now turn.

· 3 ·

COMMUNISM

SINCE the end of World War II, the foreign policy of the United States has been dedicated to the containment of Communism. The United States pursues this policy because of the belief that Communism is a threat to our peace and security.

U.S. citizens differ in their understanding of the precise nature of the threat. Many conceive of the threat in terms of an economic model: communism vs. capitalism. To most U.S. policy makers, however, communism as an economic system is no threat; indeed, it is perceived as such an objective failure that no one would willingly choose it over the capitalist ("free market" is the term used in Washington) mode of production. Other U.S. citizens conceive of the threat in terms of political ideology: communism vs. pluralism or, more commonly, dictatorship vs. democracy. Here again, however, few U.S. officials are worried by the competition. Not many people are so ignorant, they say, as to choose "democratic" centralism over liberal democracy.

To U.S. policy makers, the problem is more complex than the containment of the idea of communism. That, perhaps, is because communism now seems more complex. In an earlier era when communism appeared to be monolithic, the threat was simpler. The 1964 edition of Thomas Bailey's standard textbook, *A Diplomatic History of the American People*, interpreted the Chinese Revolution as "undeniably the most staggering blow yet suffered by the free world in its life-and-death struggle against Communism. With a half billion or so Chinese now in the Communist camp, a frightening shift had occurred in the world balance of power. The Moscow Communists had started from scratch in 1917. Now, thirty-two years later, they controlled about one-fourth of the world's land and over one-third of its population."[1] By 1982, in contrast, the Reagan administration's Undersecretary of State for Security Assistance, James Buckley, was asking Congress to give military aid to the Chinese Communists. "The fact is," he said, that

[1] Thomas A. Bailey, *A Diplomatic History of the American People*, 7th ed. (New York: Appleton-Century-Crofts, 1964), p. 785.

106

the PRC has gone its own way for years and years, quite separate from the Soviet bloc. It is a country that we are now seeking a closer security cooperation with. The entire relationship between our two countries has dramatically changed in the last half-dozen years. . . . The Chinese tie down huge numbers of Soviet military, and we believe that an enhancement of their capacity to continue that tie down is to our benefit as well.[2]

As Mr. Buckley suggests, today most U.S. policy makers perceive communism as a threat only when it is coupled in some way with the armed might of a superpower, the Soviet Union. In fact, said Secretary of State Shultz, "our relations with China and Yugoslavia show that we are prepared for constructive relations with Communist countries regardless of ideological differences. Yet, as a general principle in the postwar world, the United States has and does oppose Communist expansionism, most particularly as practiced by the Soviet Union and its surrogates."[3] To speak of the containment of communism, then, is to focus upon the threat posed by the Soviet Union.

Having read in the *Communist Manifesto* that "Communists everywhere support every revolutionary movement against the existing social and political order of things," many U.S. policy makers have been particularly wary of Soviet expansion into those areas of the world where the social and political order of things has been favorable to the United States. Latin America is one such area. As a representative of the Conservative Caucus told Congress in 1981, "there is no place in America for Socialism or Communism. Only by their expulsion will there be peace and freedom in the Western Hemisphere."[4]

Before beginning the discussion of policy makers' beliefs about Communism, a note of explanation is necessary: the structural similarity between this chapter on Communism and the previous chapter on poverty is more apparent than real. There is an imbalance in the presentation of evidence—more quotations indicating

[2] U.S. Congress, House, Committee on Foreign Affairs, *Foreign Assistance Legislation for Fiscal Year 1983*, 97th Cong., 2d Sess., 1982, pt. 2, pp. 45-46.

[3] Address to the National Committee on American Foreign Policy, New York, October 2, 1985.

[4] U.S. Congress, House, Committee on Foreign Affairs, Subcommittee on Inter-American Affairs, *U.S. Policy toward El Salvador*, 97th Cong., 1st Sess., 1981, p. 313.

officials' beliefs and perceptions and fewer data on the expansion of Communism. This imbalance might appear as a bias. That is not my intention. The problem here is with the data, not with its interpretation. When a policy maker says, "Peasants revolt because they are poor," a researcher may not be able to confirm the causal relationship, but at least there are substantial data to confirm the existence of both peasants and poverty. On the other hand, when a policy maker says, "The Soviets expand into Latin America because they are driven by an evil ideology," a researcher looking for data faces a much more difficult task. First, the presence of the Soviets (or their surrogates), unlike the presence of peasants, is not so obvious in Latin America. Everyone who has spent time in Latin America has observed that the region is full of peasants; Soviets and Cubans, in contrast, are seen less frequently. Second, because poverty can be seen, researchers feel qualified to confirm its existence. But who knows what evil lurks in the minds of men? In the preceding chapter there was a table on the distribution of income; in this chapter there is no table on the distribution of evil. I have no data to offer in support or opposition to the assertion that the Soviets are evil. To be sure, the Soviets have done some very nasty things; so has the United States. And, to be fair, the Soviets have also done some nice things, although probably not as many nice things as the United States. How it all adds up depends upon who is doing the addition. The purpose of this chapter is not to assess the nature of Soviet Communism but rather to describe and analyze U.S. policy makers' beliefs about Communism as the cause of instability in Latin America.

In a broader sense this chapter begins a discussion of Latin America's place in the postwar struggle between East and West, a discussion that continues in the chapters that follow. In addition, this chapter also introduces an analysis of Latin America's geopolitical relationship with the United States, a relationship that worried U.S. officials long before the Russian Revolution or, for that matter, the birth of Karl Marx. The best approach to these broad topics, however, is to focus first upon a narrower concern: U.S. policy makers' beliefs about Communism as a cause of instability in Latin America.

Unlike the preceding chapter, where beliefs about poverty could be presented in tabular form as a perceptual progression (from poverty and structural change to social mobilization to political mobilization to elite intransigence to instability), policy makers' beliefs about Communism are highly interdependent but not sequential. There is, instead, a constellation of beliefs that, when taken together, explain why Communism causes instability in Latin America. They begin with the belief that has dominated postwar U.S. foreign policy.

BELIEFS

Soviet Communism is Evil and Expansionist

At the 1982 unveiling of his Caribbean Basin Initiative, President Reagan began the discussion of instability by citing the region's declining terms of trade in the midst of a global economic recession. Then, quite suddenly, there was a shift in focus, and the world price of oil and the demand for sugar and coffee became irrelevant to an explanation of instability. The president had begun to discuss Communism.

It was more than a change in subject; it was a change in mood. When the word "Communism" is inserted into the context of a discussion of instability in Latin America, the effect on many policy makers is not unlike that of turning up the speed of a model train. These officials may be chugging along, talking leisurely about the adverse impact of declining world prices for sugar or yawning their way through a discussion of Latin political culture. Then someone mentions Communism and the speed picks up. Words come faster, delivered with much more conviction. Adjectives become more colorful; conditional verbs are replaced by straightforward declarations. The discussion hurtles along. Latin America slips off the side track and onto the main line.

And what a dangerous line it is. The 1982 Department of Defense "Assessment of the Global Military Situation" begins with an assertion of the fundamental premise of postwar U.S. foreign policy: "The Soviet Union poses a greater danger to the American people than any other foreign power in our history. Only the Soviet Union has the power to inflict tens of millions of casualties on our population. Only the Soviet Union has massive and modern conventional and nuclear forces deployed, directly confronting our friends and allies in Europe and Asia. Only the Soviet Union has the forces and geographic proximity to threaten the free world's major source of energy."[5]

[5] U.S. Department of Defense, *Annual Report to the Congress, Caspar W. Weinberger, Secretary of Defense, Fiscal Year 1983* (Washington, D.C.: GPO, 1982), p. II-3.

In a sense this statement is endorsed by all policy makers. There is a generalized perception that the USSR could destroy the United States. The Soviets obviously constitute a credible threat. Moscow's strength dictates that the United States focus its foreign policy upon deterring the use of Soviet power. No one questions the simple historical observation that the survival of any nation requires a prudent regard for the physical strength of any potential adversary.

To many U.S. policy makers, however, the Soviet Union is, in addition to being powerful, possessed by an ideological curse—Communism. The result is demonic. To Eugene Rostow, the Reagan administration's first Director of the Arms Control and Disarmament Agency, the USSR resembles Darth Vader's evil Empire, "seeking not to preserve but to destroy the state system that was organized under the Charter of the United Nations in 1945, and to replace it with an imperial system dominated by its will."[6] As for past Soviet behavior, President Reagan remarked in 1982 that "the record is clear. Nowhere in its whole sordid history have the promises of communism been redeemed. Everywhere it has exploited and aggravated temporary economic suffering to seize power and then institutionalize economic deprivation and depress human rights."[7] A year later Mr. Reagan characterized Soviet Communism as "the focus of evil in the modern world."[8]

As these statements suggest, to many policy makers the East-West conflict is no mere political dispute; it is an issue of fundamental morality. At his first press conference President Reagan asserted that the Soviets "reserve unto themselves the right to commit any crime, to lie, to cheat"; on the other hand, "we operate on a different set of standards."[9] Two years later Mr. Reagan argued that "Soviet leaders have openly and publicly declared that the only morality they recognize is that which will further their cause, which is world revolution."[10] To Secretary of State Shultz, "the fundamental difference between East and West is not in economic or social policy, though those policies differ radically, but in the moral principles on which they are based. It is the difference

---

[6] Eugene V. Rostow, "Re-Arm America," *Foreign Policy*, no. 39 (Summer 1980), p. 8.

[7] *Weekly Compilation of Presidential Documents* 18 (March 1, 1982): 221.

[8] *Weekly Compilation of Presidential Documents* 19 (March 14, 1983): 369.

[9] *Public Papers of the Presidents of the United States, Ronald Reagan, 1981* (Washington, D.C.: GPO, 1982), p. 57.

[10] *Weekly Compilation of Presidential Documents* 19 (March 14, 1983): 368.

between tyranny and freedom—an age-old struggle in which the United States never could, and cannot today, remain neutral."[11] This belief in the immorality of Soviet Communism and the morality of the United States has had an impact upon some officials' perceptions of Latin America. In the view of Representative Le-Boutillier, "there are black and white issues and the black and white in this case is that the Soviets and the Cubans are wrong, and we are right. Castro and the Russians go around butchering people. We don't."[12]

Added to the belief that the Soviet Union is inherently evil is a closely related belief that Communism is driven by the dictates of its ideology to relentless expansion. Today this belief is often dismissed as extremist, uncharacteristic of mainstream opinion. Take, for example, the following analysis from the Heritage Foundation:

> Prime Minister Margaret Thatcher is presumably under pressure to completely emancipate Belize. Should this occur and Guatemala invade, only the Marxists would benefit: West Indian revolutionaries are being readied in Grenada and, moreover, Guatemala would be condemned as an aggressor and the insurgents would receive worldwide support and sympathy. The last barrier between the castroites and *Reforma* fields in Southern Mexico would have been breached. The question is one of timing. The Sandinista Simon Bolivar Brigade is ready for action. The Soviets have sufficient airlift capacity to move the 3,000 Russian combat troops in Cuba to any area in the Caribbean. Debate over the date rages in Moscow, Havana and Managua.[13]

I showed this quotation to dozens of policy makers in the early and mid-1980s. Those who believe that poverty causes instability almost invariably rolled their eyes or shook their heads in disbelief. Some challenged the facts—no one had ever heard of the Sandinista Simón Bolívar Brigade, for example—but most remarked about the utter implausibility of Soviet-triggered West Indian revolutionaries invading Guatemala or Soviet soldiers storming the beaches of Belize. To these policy makers, only analysts who have

[11] Address to the Creve Coeur Club of Illinois, Peoria, Illinois, February 22, 1984.
[12] U.S. Congress, House, Committee on Foreign Affairs, *Foreign Assistance Legislation for Fiscal Year 1982*, 97th Cong., 1st Sess., 1981, pt. 1, p. 188.
[13] "Soviet Penetration of the Caribbean," mimeographed, Heritage Foundation National Security Record No. 22, June 1980, p. 3.

taken leave of their senses would believe that "the debate over the date rages in Moscow, Havana and Managua."

In fact, however, only minor differences in rhetoric separate the John Birch Society and Heritage Foundation analysts from an extremely large group of U.S. foreign policy officials who share the fundamental belief in restless Soviet expansionism. "Let's not delude ourselves," Mr. Reagan told an audience on the campaign trail in 1980, "the Soviet Union underlies all the unrest that is going on. If they weren't engaged in this game of dominoes, there wouldn't be any hot spots in the world."[14] Similarly, after defining terrorism as the destabilization of U.S. allies, Secretary of State Haig charged the Soviets with "the major responsibility today for the proliferation and hemorrhaging, if you will, of international terrorism as we have come to know it."[15]

Why do the Soviets behave this way? The answer from many policy makers is that Moscow is motivated by an ideology that requires expansion. "We know that the ideology of the Soviet leaders does not permit them to leave any western weakness unprobed, any vacuum unfilled," President Reagan told a Los Angeles audience in 1983.[16] On another occasion he noted that Communists predict that their ideology will achieve "eventual domination of all peoples on the earth."[17] In 1984, Secretary of State Shultz reiterated this view of expansionism as a metaphysical imperative of the Soviet character: "The Soviet Union, most importantly and uniquely, is driven not only by Russian history and Soviet state interest but also by what remains of its revolutionary ideology to spread its system by force."[18] "Growth," wrote one of the administration's early advisers on Latin America, "is an organic psychological compulsion of the Soviet state."[19] Like an untreated cancer, the Soviet menace is compelled by its nature to attempt to spread until it destroys Western Civilization.

It is difficult to overemphasize the extent to which this view of Soviet Communism as an evil, expansionary force has pervaded many officials' perceptions of instability in Latin America. Presi-

[14] *Wall Street Journal*, June 3, 1980, p. 1.
[15] *Foreign Assistance Legislation for Fiscal Year 1982*, pt. 1, pp. 192-193, *supra* n. 12.
[16] *Weekly Compilation of Presidential Documents* 19 (April 4, 1983): 485.
[17] *Weekly Compilation of Presidential Documents* 19 (March 14, 1983): 369.
[18] Address to the Creve Coeur Club of Illinois, Peoria, Illinois, February 22, 1984.
[19] Pedro A. Sanjuan, "Why We Don't Have a Latin America Policy," *Washington Quarterly* 3 (Autumn 1980): 32.

dent Reagan's ambassador-at-large and U.N. Ambassador, Vernon Walters, wrote in his memoirs that the United States was correct to applaud the Brazilian military's ouster of the democratic Goulart government in 1964 in order to forestall a Communist takeover. "Perhaps there have been some excessive shows of zeal under the present regime," Walters admitted, but "they are very small alongside of what would have happened if Brazil had gone Communist. We would not have isolated cases of police brutality such as occur in many countries. We would have had another Gulag archipelago."[20] To Walters, as instability grew, the choices were narrowed to two: the military or the Communists, and regardless of how brutal the generals who seized power may have become—and they were viciously brutal[21]—they were by far the lesser of the two evils.

With this approach General Walters presaged the dilemma posed a decade later by Jeane Kirkpatrick: "No problem of American foreign policy is more urgent than that of formulating a morally and strategically acceptable, and politically realistic, program for dealing with non-democratic governments who are threatened by Soviet sponsored subversion."[22] Ambassador Kirkpatrick then proceeded to generalize the position adopted by General Walters: given the extraordinarily evil nature of Communism and its relentless expansionary dynamic, she argued, sometimes the United States must ally itself with unsavory Third World dictators in order to protect its own security.

The broad popularity of Kirkpatrick's analysis is a direct result of the belief that the Soviet Union is, as Mr. Reagan said, "the focus of evil in the modern world." In the work of analyst Caesar Sereseres, a Reagan administration appointee and a long-time defender of the Guatemalan military, "the elaborately organized and internationally linked guerrilla movement" in Guatemala is juxtaposed with "the politically innovative Ríos Montt military regime."[23] Given these two choices, many policy makers opted for

[20] Vernon A. Walters, *Silent Missions* (Garden City, N.Y.: Doubleday, 1978), p. 389.

[21] For the details of this brutality, see the remarkable chronicle edited by Joan Dassin, *Torture in Brazil: A Report by the Archdiocese of São Paulo* (New York: Random House, 1986).

[22] Jeane J. Kirkpatrick, "Dictators and Double Standards," *Commentary* 68 (November 1979): 34.

[23] U.S. Congress, House, Committee on Foreign Affairs, Subcommittee on Western Hemisphere Affairs, *United States Policy toward Guatemala*, 98th Cong., 1st Sess., 1983, p. 59. Sereseres served with his former RAND colleague, Luigi Einaudi,

the latter group in Guatemala, however brutal it may have been, because the former was "internationally linked" (codewords for "Communist") and therefore worse by definition. Another analyst with close ties to the Reagan administration, Howard Wiarda, noted of El Salvador that "it is difficult not just morally but also politically, let us recognize, to support any regime that rapes and murders nuns and others. Yet it is a regime we feel we cannot abandon for fear of a Fidelista-style takeover."[24] In this way many in Washington have come to accept Kirkpatrick's distinction between the authoritarianism of an existing non-Communist Latin American government and the totalitarianism of Soviet-sponsored Communist insurgents:

> Generally speaking, traditional autocrats tolerate social inequities, brutality, and poverty while revolutionary autocracies create them. Traditional autocracies leave in place exiting allocations of wealth, power, status, and other resources which in most traditional societies favor an affluent few and maintain masses in poverty. But they worship traditional gods and observe traditional taboos. They do not disturb the habitual rhythms of work and leisure, habitual places of residence, habitual patterns of family and personal relations. Because the miseries of traditional life are familiar, they are bearable to ordinary people who, growing up in the society, learn to cope, as children born to untouchables in India acquire the skills and attitudes necessary for survival in the miserable roles they are destined to fill. Such societies create no refugees. Precisely the opposite is true of revolutionary Communist regimes. They create refugees by the millions because they claim jurisdiction over the whole life of the society and make demands for change that so violate internalized values and habits that inhabitants flee by the tens of thousands in the remarkable expectation that their attitudes, values, and goals will "fit" better in a foreign country than in their native land.[25]

Kirkpatrick's point is that "a realistic policy" toward Latin America and other areas of the Third World "will need to face the un-

---

on the policy planning staff of the State Department's Bureau of Inter-American Affairs.

[24] Howard J. Wiarda, *In Search of Policy: The United States and Latin America* (Washington, D.C.: American Enterprise Institute, 1984), p. 48.

[25] Kirkpatrick, "Dictators and Double Standards," p. 44.

pleasant fact that, if victorious, violent insurgency headed by Marxist revolutionaries is unlikely to lead to anything but totalitarian tyranny."[26]

Early in the Reagan administration, critic Tom Farer, the former president of the Inter-American Commission on Human Rights, wrote that "while the eccentricity of Kirkpatrick's account may raise doubts about her competence for public service, what matters more is the effect her account is likely to have on policy makers who confuse it with reality."[27] Evaluations of reality aside, it is clear that the authoritarian/totalitarian distinction exerts a strong hold on a large number of policy makers. In the 1980s, many officials argued that while the government of El Salvador had its shortcomings—even quite serious shortcomings—it was morally preferable to a government led by the Communist insurgents. The extreme left would "establish a rigid grip," asserted Undersecretary of State Stoessel, and "close off all but one narrow path for the future development of the Salvadoran people."[28] Every so often I get a letter from someone in the States asking me to examine my conscience," remarked Ambassador Deane Hinton. "And do you know what I tell them? I tell them the alternative would be worse."[29] "I'll tell you this," said Jesse Helms: "If we sit back and do nothing or if we do the wrong thing, El Salvador will fall within 30 to 60 days. Here we are saying, 'We can't do anything with that government, they're not perfect.' Well, my friends, neither is our government. But at least it's anti-Communist."[30]

[26] Ibid., p. 45. For a critique of the totalitarian-authoritarian dichotomy, see Michael Walzer, "On Failed Totalitarianism," *Dissent* 30 (Summer 1983): 297-306. Walzer argues that "the line that marks off these two is hard to draw, and to insist upon its central importance doesn't serve any useful political or moral purpose. More accurately, that insistence serves a repugnant purpose: it provides an apologia for authoritarian politics" (pp. 305-306). See also Henry Steele Commager, "Outmoded Assumptions," *Atlantic Monthly* 249 (March 1982): 12-22 and especially page 16, where Commager argues that Kirkpatrick's terms do nothing more than classify governments by "whether they are authoritarian on our side or not."

[27] Tom J. Farer, "Reagan's Latin America," *New York Review of Books*, March 19, 1981, p. 10.

[28] U.S. Congress, Senate, Committee on Foreign Relations, *The Situation in El Salvador*, 97th Cong., 1st Sess., 1981, p. 7.

[29] Interview with Ambassador Deane Hinton, San Salvador, January 9, 1983.

[30] Address to Rotary International, Asheville, North Carolina, April 30, 1983. In general, it is difficult to overestimate the popularity of the authoritarian/totalitarian distinction during the Reagan administration. Within the administration it was particularly popular with Secretary of State Shultz, with NSC staffer Constantine Menges, and with ARA Special Advisor Gorden Sumner. For Shultz, see two

The United States Is Vulnerable to Attack from Latin America

In 1980, Representative David Obey asked the crucial question about the growing instability in the Caribbean region: "Why are we worried . . . ? They don't have any weapons to threaten us. There are lots of small countries, as you indicated. Outside of Mexico, there are not very many people. Why should we worry about them?" The response from the State Department's John Bushnell was that "the thing that tends to worry most of your constituents and most of the American people is that they have learned enough geography to know that these places are pretty close to us."[31] Most policy makers agree with Bushnell. Geography is an extraordinarily potent weapon to use in any argument about Latin America because it is so obviously important to anyone who looks at a map. Perhaps that is why the first page of the Kissinger Commission report cited geographic proximity as a fundamental reason for U.S. concern over instability: "Central America is our near neighbor. Because of this, it critically involves our own security interests."[32]

---

speeches: "Struggle for Democracy in Latin America," mimeographed, speech to the Dallas World Affairs Council and Chamber of Commerce, Dallas, April 15, 1983, and "America and the Struggle for Freedom," mimeographed, speech to the Commonwealth Club of California, San Francisco, February 22, 1985; for Menges, see Constantine C. Menges, "The United States and Latin America in the 1980s," in *The National Interests of the United States in Foreign Policy*, ed. Prosser Gifford (Washington, D.C.: University Press of America, 1982), p. 58; for Sumner, see U.S. Congress, House, Committee on International Relations, Subcommittee on Inter-American Affairs, *Arms Trade in the Western Hemisphere*, 95th Cong., 2d Sess., 1978, pp. 69, 307.

Earlier policy makers also endorsed the distinction. With Argentina's virulently anti-Communist Videla dictatorship in mind, Henry Kissinger argued for the need to "maintain the moral distinction between aggressive totalitarianism and other governments that, with all their imperfections, are trying to resist foreign pressures or subversion and that thereby help preserve the balance of power in behalf of all free peoples." Richard Nixon noted that "we don't like dictatorships, but we must recognize the difference between Communist dictatorships and non-Communist dictatorships." The Kissinger comment was reported in the *Washington Post*, September 25, 1977, p. C3; Nixon made his comment in a speech to the American Newspaper Publishers Association, San Francisco, April 12, 1986, and it was reported in the *Washington Post*, April 22, 1986, p. A12.

The totalitarian/authoritarian distinction even found its way into college foreign policy textbooks. See John Spanier, *American Foreign Policy since World War II*, 9th ed. (New York: Holt, Rinehart and Winston, 1982), p. 299.

[31] U.S. Congress, House, Committee on Appropriations, Subcommittee on Foreign Operations and Related Agencies, *Foreign Assistance and Related Programs Appropriations for 1981*, 96th Cong., 2d Sess., 1980, pt. 2, p. 362.

[32] U.S. National Bipartisan Commission on Central America (Kissinger Com-

The geopolitical role of Latin America in world politics is complex and multifacted. The five chapters that follow are required to provide a comprehensive answer to Representative Obey's query. Here, however, Mr. Bushnell's response is more than adequate, for it encapsulates perfectly a second belief held by many policy makers, one that gives special significance to Latin America as a potential focus of Communist expansionism: geographic proximity dictates not only that the United States should fear instability in Latin America but also that the Soviet Union should encourage it. Indeed, since World War II the fear of instability in Latin America has been based upon the seemingly logical assumption that the Soviet Union would covet any opportunity to encourage disorder in America's "backyard."

In the 1980s, this belief—that geographic proximity dictates Latin America's strategic role in Communist plans for expansion—dominated the perceptions of many officials. Secretary of State Shultz told members of the Senate Finance Committee that "you only need to glance at a map to see that it is indeed our third border. If this area should be dominated by regimes hostile to us or if it becomes the scene of prolonged social upheavals, the impact on our own economy and society would, indeed, be of major proportions."[33] Most recent administrations have justified U.S. military aid on the basis of Latin America's *location* rather than, for example, its contents.[34] Occasionally a skeptic will remark that Latin America's nearby location is increasingly irrelevant in our era of intercontinental missiles that can destroy Washington from launching platforms on the other side of the globe. The frequent response is that geography is important in a subjective sense: people believe it is important, so it is. Sereseres argues, for example, that "geographic proximity remains a critical psychological and, therefore, political-military fact of life."[35] Stated differently, even if geography is in fact of decreasing importance because of changing weapons technology, many officials believe it remains crucial

---

mission), "Report of the National Bipartisan Commission on Central America," mimeographed [Washington, D.C.: Department of State(?), January 1984].

[33] Prepared remarks before the Senate Finance Committee, August 2, 1982.

[34] The State Department has argued, for example, that Guatemala is important because of its "geopolitical centrality." *United States Policy toward Guatemala*, p. 23, *supra* n. 23. See also the justification for military aid to various Latin American countries in U.S. Department of State, *Congressional Presentation, Security Assistance Programs, FY1983* (Washington, D.C.: Department of State, 1982).

[35] Caesar D. Sereseres, "Inter-American Security Relations: The Future of U.S. Military Diplomacy in the Hemisphere," *Parameters* 7 (1977): 55.

because in our strategic thinking we are hooked, psychologically, on its centrality.

The belief that U.S. security requires geographic isolation—the exclusion of other powers from the hemisphere—is as old as U.S. foreign policy. As James Monroe asserted in his famous message to Congress:

> The citizens of the United States cherish sentiments the most friendly in favor of the liberty and happiness of their fellow men on that side of the Atlantic. . . . We owe it, therefore, to candor and to the amicable relations existing between the United States and those powers, to declare that we should consider any attempt on their part to extend their system to any portion of this hemisphere as dangerous to our peace and safety.

This doctrine struck a responsive chord among U.S. citizens, in part because it formally codified several geography-related principles that had become popular in the early years of U.S. independence: the two hemispheres policy, the no-transfer resolution, and nonentanglement.[36] Since then the Monroe Doctrine has become an integral component of the American foreign policy psyche. For the sake of its peace and safety the United States has sought the exclusion from the Western Hemisphere of all rival powers. The original focus upon incursions by the Holy Alliance in Latin America and Imperial Russia in the Pacific Northwest have, with time, given way to a focus upon European debt collectors, upon German imperialists, and upon Japanese fishing interests.

The Japanese case exemplifies the broad security concerns that underlie the Monroe Doctrine. It involved the 1912 attempt by an American company to sell its rights to use Mexico's Magdalena Bay to a Japanese concern. Upon learning of the potential transfer, Senator Henry Cabot Lodge obtained the unanimous support of his colleagues for the following resolution:

> When any harbor or other place in the American continents is so situated that the occupation thereof for naval or military purposes might threaten the communications or the safety of

[36] Ernest May argues that domestic political considerations outweighed strategic considerations in the formulation of the doctrine: "in the instance of the Monroe Doctrine, the positions adopted by American policymakers seems to me to be best explained as functions of their domestic ambitions." Ernest R. May, *The Making of the Monroe Doctrine* (Cambridge, Mass.: Harvard University Press, 1975), p. 255.

the United States, the Government of the United States could not see without grave concern the possession of such harbor or other place by any corporation or association which has such a relation to another Government, not American, as to give that Government practical power of control for national purposes.[37]

This, the Lodge Corollary to the Monroe Doctrine, was based upon "the principle that every nation has a right to protect its own safety, and that if it feels that the possession by a foreign power, for military or naval purposes, of any given harbor or place is prejudicial to its safety, it is its duty as well as its right to interfere. . . . The Monroe Doctrine was, of course, an extension of our own interests of this underlying principle—the right of every nation to provide for its own safety."[38]

To many policy makers the Monroe Doctrine excludes from this hemisphere not only nations but ideologies as well. But unlike the more recent containment policy or any of the containment-related postwar doctrines—Truman, Eisenhower, Nixon, Carter—the Monroe Doctrine rests firmly upon a geopolitical foundation. Its provisions therefore transcend ideology but also encompass any "foreign" form of government, including Communism. Indeed, the original doctrine was announced in terms of a completely different European system of government—monarchy—and was used as a justification for excluding Fascism before its focus settled, in our time, upon Communism.[39]

By the time of the tenth Inter-American Conference at Caracas in 1954, the Monroe Doctrine was defined in terms of anti-Communism. At that meeting John Foster Dulles "multilateralized" the Monroe Doctrine by gaining acceptance of his Declaration of Solidarity for the Preservation of the Political Integrity of the American States against International Communist Intervention. The heart of Dulles's resolution asserted that "the domination or control of the political institutions of any American state by the international Communist movement, extending to this hemisphere the political system of an extra-continental power, would constitute a threat to the sovereignty and political independence of the American States, endangering the peace of America, and

[37] *Congressional Record*, vol. 48, pt. 10, p. 10,045.
[38] Ibid. The Lodge Corollary is also known as the Magdalena Bay Resolution.
[39] Jorge I. Domínguez, *U.S. Interests and Policies in the Caribbean and Central America* (Washington, D.C.: American Enterprise Institute, 1982), p. 21.

would call for a meeting of consultation to consider the adoption of appropriate action in accordance with existing treaties." In accepting this declaration the OAS was perceived by Secretary Dulles to have "made a momentous declaration of principle. In effect, it makes as the international policy of this hemisphere a portion of the Monroe Doctrine which has largely been forgotten and which relates to the extension to this hemisphere of the political system of despotic European powers."[40]

In stark contrast to Dulles's use of the Monroe Doctrine to attack Communism, in 1962 the Kennedy administration agreed not to seek to overthrow the Castro government in return for the Soviet Union's agreement to withdraw its missiles from Cuba. Angered by this agreement, a group of prominent citizens led by Eddie Rickenbacker formed the Committee for the Monroe Doctrine to protest the Kennedy administration's agreement to let Cuba "remain as a Communist colony."[41] The committee argued that, unlike the foreign-dominated Mexican monarchy a century earlier, Cuban Communism represented more than a symbolic breach of the Monroe Doctrine; as an ally of the Soviet Union, Cuba was perceived as an immediate, objective threat to the peace and safety of the United States.

Unable to reverse the Kennedy-Khrushchev agreement, many U.S. policy makers have dedicated themselves since 1962 to ensuring that a second ideological breach does not occur. A high point in this activity arose in 1965, when a group of key representatives engineered passage by the House of an extraordinarily broad resolution stating that "subversive forces known as international communism, operating secretly and openly, directly and indirectly, threaten the sovereignty and political independence of all the Western Hemisphere nations." Using as a justification "the principles of the Monroe Doctrine," the resolution declared that the United States "could go so far as to resort to armed force . . . to forestall or combat intervention, domination, control, and colo-

[40] *Department of State Bulletin* 30 (March 22, 1954): 429. For the full text of the resolution, see Organization of American States, *Conferencias internacionales americanas, segundo suplemento, 1945-1954* (Washington, D.C.: Departamento Jurídico, Unión Panamericana, 1956), Resolution 93, pp. 363-364. It should be noted that the Caracas meeting occurred at a time when U.S. power was supreme in the OAS. "If the United States wanted to badly enough," complained one Latin American delegate, "it could have a resolution passed declaring two and two are five." *New York Times*, March 8, 1954, p. 1.

[41] *New York Times*, October 31, 1962, p. 19.

nization in whatever form, by the subversive forces known as international communism and its agencies in the Western Hemisphere."[42]

It seems reasonable to assume that the Monroe Doctrine has endured because it expresses an underlying principle of considerable significance: prudent people keep any potential enemy as far away as possible. This is not a principle invented by the Reagan administration in the 1980s. Earlier in the twentieth century Secretary of State Knox asserted that because Latin America is so close to the United States the Monroe Doctrine is "essential to our peace, prosperity, and national safety."[43] "The safety of the United States demands that American territory shall remain American," said Senator Elihu Root in 1914. "The principle which underlies the Monroe Doctrine," he continued, is "the right of every sovereign state to protect itself by preventing a condition of affairs in which it will be too late to protect itself."[44] During the tenure of Secretary of State Hughes, the Doctrine was conceived as "an assertion of the principle of national security."[45] Even after Undersecretary of State J. Rueben Clark's reinterpretation made the doctrine more palatable to Latin Americans, it retained this core meaning. The 1928 Clark Memorandum continued the interpretation of the doctrine as "an expression of the recognized right of self-defense."[46]

The fears many policy makers have of instability in Latin America must be understood not only as anti-Communism but as a logical adaptation of the Monroe Doctrine to the Cold War conditions that have existed since World War II. Most U.S. citizens have learned about the Monroe Doctrine as part of their socialization into the U.S. political system. They have learned that one reason the United States has been so successful in international relations is that it has excluded rivals from its backyard. Thus the threat of

[42] The vote to approve H.Res. 560 was 312 to 52. See *Congressional Record*, September 20, 1965, p. 24,347, for the complete resolution and pp. 24,347-24,364 for the fascinating debate, which demonstrated some interesting cracks in the Cold War consensus regarding Latin America.

[43] J. Ruben Clark, *Memorandum on the Monroe Doctrine* (Washington, D.C.: GPO, 1930), p. 175.

[44] Elihu Root, "The Real Monroe Doctrine," *Proceedings of the American Society of International Law*, 1914, pp. 12-13.

[45] Charles Evan Hughes, "Observations on the Monroe Doctrine," *American Journal of International Law* 17 (1923): 615.

[46] Clark, *Memorandum on the Monroe Doctrine*; William Everett Kane, *Civil Strife in Latin America: A Legal History of U.S. Involvement* (Baltimore: The Johns Hopkins University Press, 1972), p. 119.

Soviet incursions in the hemisphere is much more than a challenge to the postwar policy of containment; it is a challenge to the nation's two-hundred-year-old heritage of geographic isolation, a unique heritage that allows policy makers the inestimable luxury of not having to defend U.S. borders.

Latin America Is Vulnerable to Soviet Subversion

The two beliefs discussed to this point make Latin America a high-stakes focus of superpower rivalry. An expansionist state, the Soviet Union, motivated by an insidious ideology, Communism, is believed to be attempting to render inoperative the most time-honored axiom of U.S. national security policy: a southern border free from occupation by hostile forces. This bleak picture is darkened further by two additional beliefs, both of which increase the vulnerability of Latin America to the challenges posed by the Soviet Communism. The first of these is the belief that because Latin American political culture is inherently unstable, it is an easy prey for Soviet conquest.

The use of political culture to explain instability in Latin America is hardly the invention of the current generation of U.S. policy makers; rather it simply conserves in the present a long-standing, widely popular perception that Latin America has an "unfortunate" political heritage of instability. This view has been repeated so often for so long by so many respected leaders that it is often accepted as a verity.

It all began, perhaps, with John Quincy Adams. As Secretary of State (1817-1825) during much of the struggle for independence in Spanish America, Adams set the initial tone of U.S. policy toward republican Latin America. Recalling an 1821 conversation with Henry Clay about Latin Americans, Adams wrote in his diary: "So far as they were contending for independence, I wished well to their cause. But," he continued,

I had seen and yet see no prospect that they would establish free or liberal institutions of government. They are not likely to promote the spirit either of freedom or order by their example. They have not the first elements of good or free government. Arbitrary power, military and ecclesiastical, was stamped upon their education, upon their habits, and upon all their institutions. Civil dissension was infused into all their seminal principles. War and mutual destruction was in every member of their organization, moral, political, and physical. I had little expectation of any beneficial result to this country from any future connection with them, political or commercial. We should derive no improvement to our own institutions by any communion with theirs.[47]

Adams' early perception of Latin American political culture has persisted into the Cold War era. By 1950, the father of containment, George Kennan, had sensed his growing inefficacy within Dean Acheson's Department of State and decided on early retirement. Casting about for something to do during the lame-duck period prior to actually leaving government service, Kennan accepted an invitation to speak to a group of U.S. ambassadors gathered at Rio de Janeiro. Since time was no constraint, Kennan used the opportunity offered by the Rio meeting to visit more of Latin America, beginning with a trip by train from Washington to Mexico City. "I found the journey anything but pleasant," he reported.

At Mexico City the altitude bothered me; the city made upon me a violent, explosive impression. I felt that it never slept at night (perhaps because I myself didn't). The sounds of its nocturnal activity struck me as disturbed, sultry, and menacing. . . . Caracas, jammed in among its bilious yellow mountains, appalled me with its screaming, honking traffic jams, its incredibly high prices, its feverish economy debauched by oil money, its mushroom growth of gleaming, private villas creeping up the sides of the surrounding mountains. . . . Rio, too, was repulsive to me with its noisy, wildly competitive traffic and its unbelievable contrasts between luxury and poverty. . . . In Sao Paulo it was still worse. Montevideo, Buenos Aires, and the remaining places were more relaxed, but they all inflicted upon me a curious sense of mingled apprehension and melancholy. . . . In Lima, I was depressed by the reflection that it had not rained in the place for twenty-nine years, and by the thought that some of the dirt had presumably been there, untouched, for all that time. And the brief stops in Central America—again with the undercurrent of violence palpa-

[47] Charles Francis Adams, ed., *Memoirs of John Quincy Adams, Comprising Portions of His Diary from 1795 to 1848*, 12 vols. (Philadelphia, Pa.: J.B. Lippincott, 1874-1875), 5: 325.

ble even in the reception halls of the little airports—did nothing to cheer me up.[48]

All in all, it was clearly not an enjoyable trip. This fact was reflected in Mr. Kennan's overall impression of Latin America: "It seems unlikely that there could be any other region of the earth in which nature and human behavior could have combined to produce a more unhappy and hopeless background for the conduct of human life than in Latin America."[49]

How did Latin America fall into this condition? Like many career foreign service officers, Mr. Kennan suggested that Latin America's political problems were the perverse manifestation of deep psychological drives: the "subconscious recognition of the failure of group effort," he wrote in his report, "finds its expression in an exaggerated self-centeredness and egotism—in a pathetic urge to create the illusion of desperate courage, supreme cleverness, and a limitless virility where the more constructive virtues are so conspicuously lacking." Latin Americans, he concluded, were plagued by a "diseased and swollen human ego."[50]

More recently, Jeane Kirkpatrick, a principal architect of the Reagan administration's policy toward Latin America, echoed the long-prevailing U.S. view of Latin America: "Violence or the threat of violence is an integral part of these political systems—a fact which is obscured by our way of describing military 'interventions' in Latin political systems as if the system were normally peaceable. Coups, demonstrations, political strikes, plots, and counterplots are, in fact, the norm."[51] The civil war in El Salvador

[48] George F. Kennan, *Memoirs, 1925-1950* (Boston: Little, Brown, 1967), pp. 476-479.

[49] The point in quoting Mr. Kennan is to provide an example of how senior U.S. policy makers (Mr. Kennan was then Counselor of the Department of State) have used political culture to explain instability in Latin America. The purpose is not to suggest that words written nearly four decades ago reflect the current thinking of their author. Indeed, by 1967 Mr. Kennan had come to prefer the Latin American ego, "spontaneous, uninhibited, and full-throated," to the "carefully masked and poisonously perverted forms it assumes among the Europeans and Anglo-Americans." Ibid., p. 483. Later, Mr. Kennan noted that his 1950 observations on Latin American political culture were "only those of an uninitiated and poorly qualified observer" and "have been overtaken by intervening events." Letter from George F. Kennan, September 6, 1985. Portions of Mr. Kennan's 1950 report are reprinted in his *Memoirs*, pp. 476-484; the entire document may be found in the Kennan Collection at the Seely Mudd Manuscript Library, Princeton University.

[50] Kennan, *Memoirs, pp. 481-482.*

[51] Jeane J. Kirkpatrick, "U.S. Security and Latin America," *Commentary* 71 (January 1981): 34.

during the late 1970s and the 1980s, then, was not like the U.S. civil war—a contest over principles—but rather a reflection of "El Salvador's political culture [which] emphasizes strength and *machismo* and all that implies about the nature of the world and the human traits necessary for survival and success."[52] To Kirkpatrick, Latin America is a Hobbesian nightmare where life, if not solitary, is poor, nasty, brutish, and short. In Kirkpatrick's Latin America, physical force, intimidation, and instability are the "natural" features of human existence.

Kirkpatrick's analysis is supported by a modest literature. Political scientist Glen Dealy has provided the most subtle interpretation, arguing that a "spirit of caudillaje . . . pervades those areas of Western Christendom that have remained monolithically Catholic in culture since the Renaissance." According to Dealy, the essence of this spirit is "a rational, culturewide ethos propelling men in an unending, insatiable quest for public influence."[53] Since the goal of each citizen in such a culture is to exercise power over others, it is a "pipe dream" for Washington policy makers to think of Latin America as a likely site for U.S.-style pluralism and peace; Latins would rather fight than negotiate. Dealy writes that "America's heavy reliance upon an image of pluralist government effectively places its national security interests on a political vision outside the range of Central American philosophical and pragmatic possibility."[54]

In the hands of less-sophisticated analysts the discussion of political culture as a cause of instability sometimes degenerates into simple mudslinging:

> Like Cuba historically, El Salvador has had a long tradition of political violence. The tradition stretches back long before the present crisis. The country has had more deaths per capita from political violence than any other country in Latin America; *machetismo*, or the butchering of one's personal and po-

[52] Jeane J. Kirkpatrick, "U.S. Security and Latin America," in *Rift and Revolution: The Central American Ibroglio*, ed. Howard J. Wiarda (Washington, D.C.: American Enterprise Institute, 1984), p. 352. This chapter is an expanded version of the article with the same title cited in the preceding footnote.

[53] Glen Caudill Dealy, *The Public Man: An Interpretation of Latin American and Other Catholic Countries* (Amherst: University of Massachusetts Press, 1977), p. 4.

[54] Glen C. Dealy, "Pipe Dreams: The Pluralistic Latins," *Foreign Policy*, no. 57 (Winter 1984-1985), p. 127.

litical foes, is a way of life. . . . The entire political culture . . . is based on the display and use of violence.[55]

The sources used to support these claims are typically impressionistic rather than systematic: Kennan's two-month trip around the hemisphere, Dealy's chaperoning of a group of Latin American students, or Wiarda's nonfootnotes.[56]

What is important, however, is not the accuracy or inaccuracy of these perceptions but the fact that the perceptions exist. Kirkpatrick, Wiarda, and Dealy represent the present generation's contribution to the "political culture" school of Latin American political analysis. They put an academic gloss on the less clearly articulated perceptions of U.S. policy makers who, for nearly two centuries, have perceived Latin America as a region of gunslinging guerrilleros and moral degenerates. Not wanting to expose themselves to charges of ethnocentrism, these analysts often remark that Latin American political culture is neither better nor worse than that of the United States—just different. Nonetheless, it always seems to be the Latin Americans who make a practice of "the butchering of one's personal and political foes." We vote; they butcher.

Over time, the category of "political culture" has become a collection of vaguely specified characteristics, all of which receive the contempt of Anglo-Saxons when found among Latin Americans. In her highly negative evaluation of El Salvador, Kirkpatrick characterizes Salvadorans as valuing "competition, courage, honor, shrewdness, assertiveness, a capacity for risk and recklessness, and a certain 'manly' disregard for safety."[57] Outside the Latin American context, however, these are often thought to be highly attractive traits. And, attractive or unattractive, none of

[55] Wiarda, *In Search of Policy*, pp. 48-49.

[56] Wiarda cites two sources for his assertion: Merle Kling, "Violence and Politics in Latin America," in *Sociological Review Monographs*, No. 11, ed. Paul Halmos (Keele: University of Keele, 1967), pp. 119-132, and William F. Stokes, "Violence as a Power Factor in Latin Ameican Politics," *Western Political Quarterly* 5 (September 1952): 445-468. Neither source has much to say about El Salvador, neither compares El Salvador and Cuba, and neither says anything about deaths per capita from violence. Stokes only mentions El Salvador briefly (p. 459), and Kling's only discussion of the country notes that the level of violence there is relatively *low*, the third lowest in Latin America, in fact (p. 123). In other words, Wiarda may or may not be correct in his evaluation of Salvadoran political culture, but the sources he cites do not support the statements he makes.

[57] Kirkpatrick, "U.S. Security and Latin America," in *Rift and Revolution*, p. 352.

these characteristics is unique to Latin America. Most of the terms one State Department official used to describe Latin American political culture—"anti-democratic, anti-social, anti-progress, anti-entrepreneurial, and, at least among the elite, anti-work"—could probably be used to describe the culture of the State Department as well.[58]

The diplomatic taboo against invidious cultural comparisons makes policy makers circumspect in expressing their perceptions of Latin America's inferior political culture. Occasionally there is a mild public statement as, for example, when Assistant Secretary of State Thomas Enders noted that "bloodshed, terrorism, and guerrilla warfare have been comparatively common there since premodern times," but public comments such as this are rare.[59] Disparaging remarks about "the Latin mentality" are heard frequently in private conversations, however. Many, many officials regularly assert in private that Latin America is plagued by an overabundance of "hot-headed demagogues" and "irresponsible leaders." "I went to school in the 1960s," said the State Department's desk officer for Grenada in mid-1984, "and so I've been through all the imperialism and dependency crap. That isn't what screws up Latin America. What screws up Latin America is the Latin Americans. And they'll *always* screw it up, because *they're* screwed up."[60] In this environment, argue many policy makers, low-level instability is inevitable. This is exactly what Senator Henry Cabot Lodge thought of the Cubans some decades ago. "Some people," he wrote in 1906, are "less capable of self government than others."[61]

From this perspective, political life in Latin America is an ongoing effort to cope with the chronic low-level instability generated by a cultural penchant for "coups, demonstrations, political strikes, plots, and counterplots." Latin Americans are successful most of the time: one group of colonels replaces another in the Casa de Gobierno, demonstrators force the reconsideration of a hike in bus fares, workers strike to stem a decline in real wages

[58] Lawrence E. Harrison, *Underdevelopment Is a State of Mind: The Latin American Case* (Lanham, Md.: University Press of America, 1985), p. 165.

[59] Thomas O. Enders, "The Central America Challenge," *AEI Foreign Policy and Defense Review* 4 (1982): 9.

[60] Interview with Douglas Rohn, Washington, D.C., July 11, 1984.

[61] Henry Cabot Lodge, ed., *Selections from the Correspondence of Theodore Roosevelt and Henry Cabot Lodge, 1884-1918*, 2 vols. (New York: Charles Scribner's Sons, 1925), 2: 233.

caused by an economic austerity program, an earthquake levels the capital city, a hurricane destroys the year's banana crop—all these are natural phenomena, business as usual in Latin America. Once in a while, however, the Soviet Union, acting through Cuba and local radicals, invests significant resources in an effort to exacerbate this endemic instability. It is then, say many policy makers, that Latin America's cultural shortcomings are manipulated to contribute to the high levels of instability that threaten U.S. security.

Latin American Radicals Threaten U.S. Security

Latin American radicals scare almost all U.S. policy makers. It is easy to understand why: first, these radicals almost invariably demonstrate an abiding hostility toward the United States; second, they typically express an allegiance to some form of Marxism and friendship toward the Soviet Union; and third, according to some officials, they are uncommonly deceitful. In the view of many U.S. policy makers these three features make Latin American radicalism a threat to U.S. security.

ANTI-AMERICAN

It is difficult to find a Latin American radical who likes the United States. Radicals tend to view the United States as a reactionary force that, in alliance with local elites, is determined to maintain Latin America in a condition of international subservience. That is why Régis Debray reported some years ago on Latin American radicals' belief that "the path of independence passes by way of the military and political destruction of the dominant class, organically linked to the United States by the comanagement of its interests." Because of their commitment to independence, Latin American radicals struggle against their domestic exploitative elites, whom "the United States, with a century-old cunning, uses . . . as a screen which attracts the bulk of popular discontent, and receives the most violent attacks."[62]

But what happens when popular discontent rises, when the attacks are successful, and the struggle is about to be won? Just as "the tanks burst into flames and the guns captured by the people are turned against the last bastions of reaction, there will appear

---

[62] Régis Debray, *Strategy for Revolution* (New York: Monthly Review Press, 1971), p. 76.

128

on the horizon the warships and troop-carrying planes of the United States—the last and most powerful line of defense of the exploiting classes in any country in the hemisphere."[63] The marines storm ashore, shoot the radicals, spray for mosquitos, and revitalize the local rum industry. Meanwhile, U.S. negotiators search out the most despicable humans in the country, often identified by their ability to speak English. Once the right individual has been selected from among the pool of qualified traitors, the marines reembark, leaving the country to rot for another generation.

It need not be said that this is hardly the way most policy makers conceptualize U.S. behavior in inter-American relations. They tend to view the United States as a benevolent big brother, helping out with a variety of problems that Latin Americans cannot handle by themselves. That is why the U.S. president often sends his wife winging southward to survey the damage and offer assistance in the recovery from any major natural disaster—earthquakes are a particular favorite of our First Ladies. Such trips capture perfectly the U.S. perception of its selfless relations with Latin America. No, Latin Americans need not reciprocate—imagine the reaction in Washington if the president of Ecuador were to offer to send his spouse to survey a hurricane's damage in Mississippi. Since the United States has such enormous power and wealth, we can take care of ourselves, Latin Americans cannot, and it is our duty to help them out. And that is exactly what we do: $14.9 billion in economic aid alone between 1946 and 1983.

Thus many U.S. policy makers are deeply affronted by the negative remarks of Latin American radicals. They believe that these radicals are unfair because, far from being grateful for U.S. aid, they spit in Uncle Sam's eye. Of the Sandinistas, President Reagan complained that "the words of their official anthem describe us, the United States, as the enemy of all mankind."[64] And sure enough, the offending words are right in the middle of Carlos Mejía Godoy's "Hymn of Sandinista Unity":

[63] José Cuello and Asdrubal Domínguez, "The Dominican Republic: Two Years After," *World Marxist Review*, March 1968, p. 39.

[64] *Weekly Compilation of Presidential Documents* 20 (May 14, 1984): 679. For additional examples of umbrage on this issue, see the statements by Senator Robert Kasten and by Otto Reich, who held several political appointments in the Reagan State Department, in U.S. Congress, Senate, Committee on Appropriations, *Foreign Assistance and Related Programs Appropriations, Fiscal Year 1983*, 97th Cong., 2d Sess., 1982, pt. 1, pp. 3, 23.

We struggle against the yankee,
Enemy of humanity.

Anyone listening to a group of Sandinistas sing this song will be impressed, moreover, by the extra volume and vigor given to these two lines. They sound as if they really mean it.

In their heart of hearts U.S. officials do not believe they deserve this criticism. Certainly Mr. Reagan believed he never did anything to merit such opprobrium. "As soon as I took office, we attempted to show friendship to the Sandinistas and provided economic aid to Nicaragua. But," he remarked with a sigh, "it did no good."[65] On another occasion Mr. Reagan complained that "the Government of Nicaragua has treated us as an enemy. It has rejected our repeated peace efforts."[66] Like the rest of us, the president did not want to be rejected; no one would want two and a half million Nicaraguans to believe that you are the enemy of humanity.

All U.S. policy makers want to know why Latin American radicals can be so mistaken. The answer accepted by many officials is that the United States is caught at a difficult moment of transition. Since at least the 1920s, they say, nationalism—the desire to assert the independence of one's national identity—has been the most powerful political force in Latin America. It has been the misfortune of the United States to become the target of this nationalism. An example, they say, is Nicaragua, occupied by U.S. Marines from 1912 to 1933 and then ruled by the *entreguista* Somoza governments with U.S. support from 1933 to 1979. As a consequence of this involvement, the United States became the target of Nicaraguan nationalists, just as the Soviet Union has become the target of nationalists in Poland and Afghanistan. There is nothing personal in this dislike; rather it is a side effect of nationalism. Xabier Gorostiaga, the Spanish Jesuit advisor to the Torrijos government in Panama and the Sandinista government in Nicaragua, expressed this explanation best: "The great problem for the United States in understanding Latin America is that it confuses anti-imperialism with anti-Americanism. Latin America is not anti-American, but it is anti-imperialist."[67]

Many policy makers accept this explanation of anti-American sentiment. Radicals, they say, are condemning a system of inter-

---

[65] *Weekly Compilation of Presidential Documents* 20 (May 14, 1984): 679.
[66] *Weekly Compilation of Presidential Documents* 19 (May 2, 1983): 610.
[67] Interview with Xabier Gorostiaga, Managua, Nicaragua, November 3, 1984.

action, not the United States. But why is it, ask other policy makers, that the Sandinista hymn does not say they are struggling against imperialism instead of the Yankee? These officials answer their own question: all the talk about nationalism and antiimperialism is just a smokescreen designed to disguise the fact that Latin American radicals are Communists.

PRO-SOVIET

When Great Britain was preparing to grant independence to British Guiana in the early 1960s, officials of the Kennedy administration were understandably jittery about the transition. Cuba had been a sobering experience. The focus of U.S. concern was Cheddi Jagen, whom Kennedy adviser Arthur Schlesinger, Jr., recognized as "the most popular leader in British Guiana." Unfortunately, Jagen was also "unquestionably some sort of Marxist."[68] Given this perception of Jagen's political orientation, U.S. officials approached Great Britain with a plan of proportional representation that would swing the preindependence election in favor of Jagen's opponent, Forbes Burnham. This accomplished, "British Guiana seemed to have passed safely out of the Communist orbit."[69]

The scenario described by Mr. Schlesinger is as old as the Cold War. It follows a set pattern: a popular Latin American radical announces an attraction to Marxism, then does or says something either (a) hostile toward the United States or, (b) friendly toward the Soviet Union. This is taken as evidence that the radical is not a harmless Marxist (à la Francois Mitterrand) but a dangerous Communist (à la Joseph Stalin). Given this perception, the United States then moves to protect its security by neutralizing the threatening radical. This pattern is now quite familiar to students of inter-American relations. It has been seen in Guatemala against Arbenz, in Cuba against Castro, in Brazil against Goulart, in Chile against Allende, in Grenada against Bishop, and in Nicaragua against the Sandinistas. As a regular feature of inter-American re-

[68] Professor Schlesinger concluded Jagen was a Communist because, first, "his wife, an American girl whom he had met while studying dentistry in Chicago, had once been a member of the Young Communist League"; second, "his party lived by the cliches of an impassioned, quasi-Marxist, anticolonialist socialism"; and third, during an appearance on NBC's "Meet the Press," Jagen had "resolutely declined to say anything critical of the Soviet Union." Arthur M. Schlesinger, Jr., *A Thousand Days: John F. Kennedy in the White House* (Boston: Houghton Mifflin, 1965), pp. 774-775.

[69] Ibid., pp. 778-779.

lations this pattern is not restricted to conservative U.S. administrations—witness the Carter administration's hostility toward the Bishop government in Grenada.

The pattern takes its initial form when U.S. policy makers assess the meaning of Marxism in Latin America. There is considerable evidence to demonstrate that Marxism is far from a rigid dogma; indeed, the range and diversity of Marxism in Latin America is simply remarkable.[70] Some Marxists are the incarnation of policy makers' worst nightmares—Leninists who propose the violent overthrow of a repressive bourgeois state and a prolonged dictatorship of the proletariat, during which a vanguard party oversees the transition to a noncapitalist mode of production and integration into the socialist family of nations. This is Jeane Kirkpatrick's "Pol Pot" left or Robert Pastor's "Leninist cadres."[71] Other Latin American radicals are fairly mild Marxists who seem to have evolved in much the same fashion as European democratic socialism. Arrayed between these two poles is every type of Marxist that one can imagine. Their only unifying characteristic is that they dislike the status quo in Latin America.[72]

Given this diversity, the term "Marxism" lacks specific meaning in Latin America. Thus when Vice President George Bush used a Nicaraguan postage stamp with a picture of Karl Marx as evidence that the Sandinistas were setting up a Marxist state, many observers asked, "What kind of Marxist state?"[73] No one seemed to know for sure. Even Nicaraguan Vice President Sergio Ramírez, one of the nine Sandinista *comandantes*, seemed vague: "We are marxists," he said, "but this doesn't mean that we follow a rigid model."[74] Another of the nine *comandantes*, Tomás Borge, told an interviewer that he was a Communist, but he seemed fairly un-

[70] See the excellent survey by Sheldon B. Liss, *Marxist Thought in Latin America* (Berkeley: University of California Press, 1984).

[71] Kirkpatrick, "U.S. Security and Latin America," p. 29; Robert A. Pastor, "Three Perspectives on El Salvador," *SAIS Review* 1 (Summer 1981): 43.

[72] My impression is that the majority of Latin American Marxists are what C. Wright Mills would call "plain" Marxists: people who "are generally agreed that Marx's work bears the trademark of the nineteenth-century society but that his general model and his ways of thinking are central to their own intellectual history and remain relevant to their attempts to grasp present-day social worlds." C. Wright Mills, *The Marxists* (New York: Dell Publishing Company, 1962), p. 98.

[73] Parenthetically, others asked why the stamp should be used as evidence of anything, since it was part of a series on famous people that included Pope John Paul II, George Washington, and Babe Ruth. *New York Times*, March 30, 1985, p. 5.

[74] Interview with Sergio Ramírez, Managua, Nicaragua, November 2, 1984.

certain about the direction of the revolution he was helping to lead.[75]

Most U.S. policy makers relied heavily upon Nicaraguan foreign policy to determine their perception of the nature of the regime's radicalism. Here the signs were ominous. Six weeks after the Sandinista victory, the Nicaraguan delegation to the Sixth Nonaligned Summit in Havana supported the Soviet-Cuban position on Cambodian representation. Then in early 1980 Nicaragua abstained on the U.N. General Assembly resolution condemning the Soviet invasion of Afghanistan, an abstention for which they would never be forgiven by many U.S. officials. Shortly thereafter four principal Sandinista leaders (Tomás Borge, Humberto Ortega, Moisés Hassán, and Henry Ruiz) flew to Moscow to sign an agreement for Soviet military assistance, and within a year the Sandinista government had signed aid agreements with several other Eastern Bloc countries. This was all that many members of the Republican party needed to know about the Sandinistas. In mid-1980, they inserted into their party platform a statement deploring "the Marxist Sandinista takeover of Nicaragua."

The rest is simple history. Regardless of what the Sandinistas were doing inside Nicaragua, the die was cast once the newly elected U.S. policy makers had perceived the Sandinistas as both anti-United States and pro-Soviet. As Secretary of State Shultz remarked in 1985, "we must oppose the Nicaraguan dictators not simply because they are Communists but because they are Communists who serve the interests of the Soviet Union and its Cuban client and who threaten peace in this hemisphere. . . . had they not become instruments of the Soviet global strategy, the United States would have had a less clear strategic interest in opposing them."[76] Once Nicaraguan leaders had been perceived as "instruments of Soviet global strategy," however, the question then became procedural: how to get rid of the Sandinistas. This was understood to be a difficult task, for many U.S. officials considered the Sandinistas, like all Latin American radicals, to be uncommonly cunning adversaries.

DECEITFUL

One of the most persistent problems facing U.S. policy makers is in distinguishing Communist radical revolutionaries from

---

[75] *Playboy*, September 1983, especially p. 60.

[76] Address to the National Committee on American Foreign Policy, New York, October 2, 1985.

other Latin Americans who are so disgusted with their existing government that they become, for a time, radical revolutionaries. The Eisenhower administration's Assistant Secretary of State for Inter-American Affairs, Roy Rubottom, captured this dilemma perfectly on the day Batista fled Cuba:

> This kind of revolution, of course, is made to order for the Communists. We know that they are bound to be doing everything they can to take advantage of it. But there is also widespread feeling, on the basis of all the information that we can gather, against Batista, per se, in Cuba, on the part of many people who are not Communists. We have thousands of them here in the United States and we have, for example, representatives who have come in to see some of our people from time to time representing civic organizations, the lawyers, the doctors, the teachers, and so on, who are, we are absolutely certain, not Communists at all.[77]

In the eyes of many U.S. policy makers Cuba became a perfect example of what happens when officials like Rubottom fail to make the necessary distinction between Communist and non-Communist revolutionaries: the Communists always win.

The reason why they win is that Latin American Communists possess an ability denied to their non-Communist countrymen: the ability to conquer by deceit. As a U.S.-dominated OAS committee reported in 1962:

> Lacking sufficient numerical strength and without Soviet and Chinese Communist military forces close enough to this hemisphere to give them support, the Communists are dangerous only to the extent that they usurp the strength of others; that is, the students, workers, rural workers, writers, and even political leaders. Since the great majority of the citizens of the Americas believe in the ideals of national independence and individual liberty, and reject intervention and dictatorship, the Communists can strengthen themselves, and even come to power, only through a program of deceit that assumes many and varied forms.[78]

[77] U.S. Congress, Senate, Committee on Foreign Relations, *Executive Sessions of the Senate Foreign Relations Committee (Historical Series)*, vol. 10, 85th Cong., 2d Sess., 1958 (made public November 1980), p. 774.

[78] U.S. Senate, Committee on the Judiciary, Subcommittee to Investigate the Administration of the Internal Security Act and Other Internal Security Laws, *Or-

This problem of the 1950s and 1960s remained a problem in the 1980s. In Nicaragua, reported Ambassador William Jorden (the man who told Somoza that Mark Schneider was "a horse's ass"), "superior organization and tighter discipline enabled the Communists to survive the trials of civil strife longer and with more organizational coherence than most of their rivals. Many of the others . . . were gradually forced into silent conformity, into exile, or into the welcoming embrace of the Sandinistas."[79] Thus by late 1980, the CIA's Nicaragua analyst, Ronald Seckinger, perceived the disappearance of pluralism; moderates had been forced out of power, he said, by "nine Communists" who were better organized.[80]

According to President Reagan, the Nicaraguan case was made doubly difficult because the Communists received sound advice from Cuba, the hemisphere's masters of deceit. Calling the Sandinistas to Havana,

> Castro cynically instructed them in the ways of successful Communist insurrection. He told them to tell the world they were fighting for political democracy, not communism. But most important, he instructed them to form a broad alliance with the genuinely democratic opposition to the Somoza regime. Castro explained that this would deceive Western public opinion, confuse potential critics, and make it difficult for the Western democracies to oppose the Nicaraguan revolution without causing great dissent at home.[81]

This deceit, said one official in the Reagan State Department, is the common approach of all Communists: "Only after the attainment and consolidation of power may the revolution begin to eat

---

ganization of American States Combined Reports on Communist Subversion, 89th Cong., 1st Sess., 1965, pp. 3-4. See also the report's section entitled "Deceit in Communist Methods," pp. 16-26. This Senate document is a combined printing of four unpublished reports by the OAS Special Consultative Committee on Security Against the Subversive Action of International Communism. Formed in 1962 at the urging of the United States, the committee continues to exist under the shortened title of Special Consultative Committee on Security. Although it is much less active than in the past, it remains military dominated and apparently dedicated to the task of illustrating the nature of suffocating orthodoxy.

[79] William J. Jorden, *Panama Odyssey* (Austin: University of Texas Press, 1984), p. 670.

[80] Remarks at a Department of State seminar, Washington, D.C., January 9, 1981.

[81] *Weekly Compilation of Presidential Documents* 20 (May 14, 1984): 677-678.

its own children."[82] In President Reagan's view what happened in Nicaragua was that "no sooner was victory achieved than a small clique ousted others who had been a part of the revolution from having any voice in government."[83] Secretary Shultz's perception was similar: "An extraordinary national coalition against Somoza has cracked and shriveled under the manipulation of a handful of ideologues, fortified by their Cuban and Soviet-bloc military advisors."[84]

The Reagan administration's opposition to negotiated power sharing in El Salvador was a direct outgrowth of the belief in the superior discipline of Communist revolutionaries. In 1981, a representative of the Conservative Caucus said that "it is neither in the best interest of El Salvador, the United States nor the cause of freedom to aid or abet negotiations which would lead to participation in a future government of El Salvador by the left-wing political elements. History shows clearly that such coalition governments usually end up as Marxist governments."[85] Caught in the crossfire between the Reagan administration and Latin American Communists was the hardy band of non-Communist Salvadoran revolutionaries whom many U.S. officials perceived as little more than patsies, weak leaders who could not withstand the onslaughts from disciplined Leninist cadres. Their frustration was obvious. "The problem is that you are asking for an answer for the future, to prove that something is not going to happen in the future," complained rebel leader Rubén Zamora to *New York Times* reporter Raymond Bonner. "I cannot give you proof about the future."[86]

CONTRARY BELIEFS

In the 1980s, many policy makers rejected President Reagan's assertion that "the Soviet Union underlies all the unrest that is

[82] The comment was made by Kenneth N. Skoug, Jr., during his address "The United States and Cuba" to the Face to Face Program of the Carnegie Endowment for International Peace, Washington, D.C., December 17, 1984 (mimeographed).

[83] *Weekly Compilation of Presidential Documents* 19 (May 2, 1983): 610.

[84] Address to the Dallas World Affairs Council and Chamber of Commerce, Dallas, Texas, April 15, 1983.

[85] *U.S. Policy toward El Salvador*, p. 312, *supra* n. 4.

[86] Raymond Bonner, *Weakness and Deceit: U.S. Policy and El Salvador* (New York: Times Books, 1984), p. 105.

going on," calling it "arrant nonsense"[87] or "towering igno-rance."[88] These officials rarely argued that the Soviet Union was disinterested in the instability that had developed in Central America; rather, their position was that Moscow's policies did not *cause* the unrest. Ambassador Robert White argued that "it is a grave, and if we continue it, fatal, error to believe that we are con-fronting primarily a case of Communist aggression in Central America. What we basically are confronting is an authentic revo-lution, born out of despair and discouragement because of a lack of economic opportunity and because of a distortion of the political process."[89]

Behind this differing assessment of the cause of instability in Central America are profoundly different beliefs about the nature of Soviet Communism. While many officials believe that the So-viets are both evil and expansionist, other policy makers are more inclined to perceive the Soviet Union as a state, like all states, with interests to promote and values to protect. While many of these interests and values are understood to conflict dramatically with those of the United States, on other issues the two superpow-ers are in agreement. Because there is some common ground, con-flicts are often thought to be amenable to peaceful resolution. This is particularly true of conflicts that involve Latin America, where (with the exception perhaps of Cuba) vital Soviet interests are not believed to be at stake. Negotiation is possible because the Soviets are not evil demons.

When these policy makers refer to the Soviet Union as "a state, like all states," they mean that there are certain basic security-re-lated goals that motivate the behavior of all states and that the So-viet Union should not be considered an exception simply because it is a Communist superpower. To these officials Soviet strength is primarily defensive and only secondarily expansionist. On the

[87] William D. Rogers and Jeffrey A. Meyers, "The Reagan Administration and Latin America: An Uneasy Beginning," *Caribbean Review* 11 (Spring 1982): 15.

[88] Robert White, "There Is No Military Solution in El Salvador," *The Center Magazine* 14 (July-August 1981): 8.

[89] *The Situation in El Salvador*, p. 160, *supra* n. 28. See also the additional state-ments by White and related assertions by representatives Jim Leach and Gerry Studds in U.S. Congress, House, Committee on Foreign Affairs, Subcommittee on Inter-American Affairs, *Presidential Certification on El Salvador*, vol. 1, 97th Cong., 2d Sess., 1982, pp. 65, 136; *Foreign Assistance Legislation for Fiscal Year 1982*, pt. 7, p. 247, *supra* n. 12. For an analysis of this position, see Walter LaFeber, "The Reagan Administration and Revolutions in Central America," *Political Sci-ence Quarterly* 99 (Spring 1984): 3-4.

other hand, the Soviets obviously merit special attention because (1) they are powerful and (2) they have organized their society (and advocate the organization of all societies) in ways that are inimical to the basic values of liberalism. That does not mean, however, that the Soviets stand poised to use their strength against the United States, ready to strike whenever the slightest opportunity presents itself. No doubt the Soviets would be delighted to exert substantial influence over, say, Belize, just as the United States would like to increase its influence in Poland. But to these officials the chances of a Soviet-Cuban invasion of Belize is as remote as a U.S. invasion of Poland.

While the Soviets would not use force to "steal" Latin America from the U.S. sphere of influence, they would certainly assist domestically generated insurgencies in the region. But many officials interpret this assistance as principled rather than as an example of Hitlerian malevolence. As Averell Harriman told President Carter, "You will have to remember that the Soviet philosophy is to support liberation movements wherever they occur, and they believe in this just as deeply as Americans believe in human rights."[90] In this view the Soviets are opportunists, ready to take advantage of instability to increase their influence, but certainly not the basic cause of instability in the Third World.

There is no better summary statement of these beliefs about Soviet Communism than President Carter's Notre Dame commencement address in 1977:

> Being confident of our own future, we are now free of that inordinate fear of communism which once led us to embrace any dictator who joined us in that fear. I'm glad that that's being changed. For too many years, we've been willing to adopt the flawed and erroneous principles and tactics of our adversaries, sometimes abandoning our own values for theirs. We've fought fire with fire, never thinking that fire is better quenched with water. This approach failed, with Vietnam the best example of its intellectual and moral poverty. But through failure, we have now found our way back to our own principles and values, and we have regained our lost confidence.
>
> By the measure of history, our nation's 200 years are very brief, and our rise to world eminence is briefer still. It dates

[90] Jimmy Carter, *Keeping Faith: Memoirs of a President* (New York: Bantam Books, 1982), p. 242.

from 1945, when Europe and the old international order lay in ruins. Before then America was largely on the periphery of world affairs, but since then we have inescapably been at the center of world affairs.

Our policy during this period was guided by two principles: a belief that Soviet expansion was almost inevitable but that it must be contained, and the corresponding belief in the importance of an almost exclusive alliance among non-Communist nations on both sides of the Atlantic. That system could not last forever unchanged. Historical trends have weakened its foundation.[91]

With the possible exception of the Soviet ambassador, virtually everyone in Washington believes that the Soviet system operates on "flawed and erroneous principles." Over time, however, many policy makers have come to believe that the Soviet Union is not the "focus of evil in the modern world," nor do they believe that Soviet expansionism "underlies all the unrest that is going on." As we have seen, these officials believe, in contrast, that unrest is a product of poverty and injustice.

[91] *Weekly Compilation of Presidential Documents* 13 (May 30, 1977): 774-775. Over the course of his presidency Mr. Carter gradually appeared to change his belief about Soviet expansionism. An interesting study of this evolution is Jerel A. Rosati, "The Impact of Beliefs on Behavior: The Foreign Policy of the Carter Administration," in *Foreign Policy Decision Making: Perception, Cognition, and Artificial Intelligence*, ed. Donald A. Sylvan and Steve Chan (New York: Praeger, 1984), pp. 158-191. The evolution culminated on New Year's Eve 1980, when the President told ABC News that "my opinion of the Russians has changed more drastically in the last week than even the previous two and a half years before that. It's only now dawning on the world the magnitude of the action that the Soviets undertook in invading Afghanistan." As he wrote his memoirs a year later, however, Mr. Carter appeared to revert to his Notre Dame language. There, in two separate passages, he remarked that he had asked the Senate to cease consideration of the SALT II treaty not because *his* opinion of the Russians had drastically changed but because of domestic political realities: the Soviet invasion of Afghanistan, he wrote, "wiped out any chance for a two-thirds vote of approval." Carter, *Keeping Faith*, pp. 264 and 265.

PART II

The Consequences of Instability: Strategic Access

· 4 ·

STRATEGIC RAW MATERIALS

Two important facts begin all discussions about the relationship between raw materials and national security: first, raw materials are not distributed equally among the nations of the world and, second, the locations where they are distributed are often not the locations where they are needed.

These two facts are the source of considerable concern in Washington, for over the course of the twentieth century the United States has developed into the world's largest importer of unprocessed raw materials. Until about 1910, U.S. supplies of most raw materials greatly exceeded domestic needs, but this surplus disappeared in the years immediately prior to World War I. Since 1920, the United States has become an increasingly large net importer of raw materials.[1] Moreover, because domestic supplies of some materials remain ample, aggregate figures on net imports tend to hide extraordinarily high import dependence upon a number of critically important raw materials.

The transition by the United States from being a net exporter to a net importer is due to a number of interacting factors: the general increase in domestic consumption of nearly all raw materials, the severe depletion of U.S. reserves of many minerals, the absence of any significant domestic sources of materials such as diamonds, chromite, and manganese, the discovery of foreign sources that can be exploited at lower cost than known U.S. reserves, and the development of demand for previously unnoticed and unwanted resources (such as columbium) for which there are inadequate domestic supplies. These changes in the relationship between international supply and U.S. demand have combined to make the United States heavily dependent upon foreign sources of raw materials. Some of the changes have been startling: the U.S. share of total world production of zinc, for example, has declined from 62 percent in 1920 to 9 percent today. At the turn of the century the United States produced about a third of the world's iron ore, copper, lead, and silver; today it produces about one-tenth. Overall,

[1] M. H. Govett, "Geographic Concentration of World Mineral Supplies, Production, and Consumption," in *World Mineral Supplies: Assessment and Perspective*, ed. G.J.S. Govett and M. H. Govett (Amsterdam: Elsevier, 1976), pp. 133-134.

the United States now imports 50 percent or more of its consumption of twenty-three metals and minerals and more than 90 percent of twelve of these metals and minerals.[2]

As the shift away from domestic sources has occurred, an ongoing debate has developed over the implications of reliance upon foreign suppliers. Since the time of Theodore Roosevelt this debate has always been initiated by policy makers and analysts who are looking for data to support their allegation that the United States faces a security crisis. Thus in 1980, for example, the Heritage Foundation published a study designed to demonstrate that "Russia, under the shadow of emerging nuclear superiority, is . . . tightening the noose on the oil and ore vital to the Free World's economies."[3] At about the same time the American Legion came to perceive the problem similarly: "Areas of the world once accessible to U.S. military forces are effectively controlled by Soviet or Communist governments placing the Soviets in a position to cut and choke off oil and limit our access to strategic materials. This oil, so vital to us and our allies, as well as access to strategic minerals essential to our survival as a free nation, is now placed in jeopardy."[4]

Only very rarely is a raw materials supply problem perceived as the result of some factor other than the nefarious intentions of a hostile state. The literature is thin on the subject of supply disruptions caused by nonhostile political instability, for example, despite the fact that the second best example of a supply disruption in recent years occurred for this reason: in the 1970s, unrest in Zaire led to cobalt supply disruptions, a 700 percent increase in wholesale prices, and a system of rationing by the major U.S. cobalt distributor.[5] The best recent example, of course, was the

[2] U.S. Congress, House, Committee on Banking, Finance, and Urban Affairs, Subcommittee on Economic Stabilization, *U.S. Economic Dependence on Six Imported Strategic Non-Fuel Minerals*, 97th Cong., 2d Sess., 1982, p. 3; U.S. Congress, House, Committee on Banking, Finance, and Urban Affairs, Subcommittee on Economic Stabilization, *A Congressional Handbook on U.S. Materials Import Dependency/Vulnerability*, 97th Cong., 1st Sess., 1981, p. 16; M. H. Govett, "The Geographic Concentration of World Mineral Supplies," *Resource Policy* 1 (December 1975): 363.

[3] "Soviet Penetration of the Caribbean," mimeographed, Heritage Foundation National Security Record No. 22, June 1980, p. 3.

[4] U.S. Congress, House, Committee on Foreign Affairs, Subcommittee on Inter-American Affairs, *U.S. Policy toward El Salvador*, 97th Cong., 1st Sess., 1981, p. 209.

[5] *A Congressional Handbook on U.S. Materials Import Dependency/Vulnera-*

OPEC price increase of the 1970s, which can safely be attributed not to the hostility of an adversary but to the temptation that afflicts most oligopolists—greed.

Once the alarm has been sounded and the debate begins, analysts tend to separate into two camps as they assess the allegations of a crisis. Some are perennial alarmists, while others argue in response that only the complete cessation of supplies of a small handful of materials would lead to significant security problems. The "crisis" side of the literature tends to be dominated by conservative policy makers who are untrained in the subject being discussed—the availability of raw materials—while the "noncrisis" side is dominated by similarly untrained moderates and liberals. This suggests that the subject is not being analyzed; it is probably more correct to say that it is being exploited.

Nowhere is this more true than in studies of Latin America's role in supplying the United States with strategic raw materials: a series of unsubstantiated claims by alarmists is the most prominent feature of the literature. To support his argument that the United States should be less timid about armed intervention in Latin America, one analyst has asserted that "undoubtedly, the greatest strategic significance of the region lies in its wealth of natural resources."[6] Similarly, former ambassador John Bartlow Martin, a journalist by profession, has argued that "in these days of shrinking natural resources, scarcities, cartels, and a determined effort by the Third World to bring about what it calls a new world economic order, we have suddenly become aware that we are critically dependent—at least for now—on the Caribbean."[7] The strategic value of South America is of even greater importance, assert three students of U.S. military power: "Perhaps the greatest significance of this continent is its wealth of natural resources."[8] In the midst of an ominously gloomy assessment of Central America, one former member of the Senate Foreign Relations Committee staff worried about Mexico because "Mexico is the second most important supplier of critical raw materials to the United

bility, p. 301, supra n. 2; Michael Shafer, "Mineral Myths," Foreign Policy, no. 47 (Summer 1982), p. 165.

[6] Gregory D. Foster, "On Selective Intervention," Strategic Review 11 (Fall 1983): 56.

[7] John Bartlow Martin, U.S. Policy in the Caribbean (Boulder, Colo.: Westview Press, 1978), pp. 6, 278.

[8] Trevor N. Dupuy, Grace P. Hayes, and John A. C. Andrews, The Almanac of World Military Power, 3rd ed. (New York: R.R. Bowker, 1974), p. 41.

States. It is the principal supplier of silver, zinc, gypsum, antimony, mercury, bismuth, selenium, barium, rhenium and lead to the United States."[9] But this assertion, like those of other alarmists, is only that: an assertion, unaccompanied by either data or analysis, yet clearly designed to make the reader worry.[10] Hayes never defines what is meant by "critical raw materials." Nor does she explain why there should be much concern about Mexican selenium, when only 7 percent of U.S. supplies come from Latin America, or lead, when the United States is a net exporter of that mineral, or bismuth, when its principal use in the United States—the manufacture of Pepto Bismol—is not related in any significant way to U.S. security.[11] Similarly, it may be true that "the Caribbean's exports of petroleum and its derivatives, and bauxite, aluminum, and nickel, contribute significantly to the economic security of the United States."[12] But it is also appropriate to ask what is meant by the term "economic security," why nickel is included in the list when only 6 percent of U.S. imports come from Latin America, and how much of the petroleum is actually produced in the Caribbean region and how much is simply transshipped in Caribbean ports?

No organization is more adept than the Department of Defense at the use of a potential raw materials crisis to buttress its posi-

[9] Margaret Daly Hayes, "The Stakes in Central America and U.S. Policy Responses," AEI Foreign Policy and Defense Review 4 (1982): 14.

[10] Hayes argues elsewhere that the availability of minerals from the Caribbean Basin countries is only a "convenience" and that "only in the event of a major global conflict would their access be critical." Margaret Daly Hayes, "United States Security Interests in Central America in Global Perspective," in Central America: International Dimensions of the Crisis, ed. Richard E. Feinberg (New York: Holmes and Meier, 1982), p. 91.

[11] Moreover, the nation's leading expert on bismuth notes that "bismuth can be replaced in most of its uses, and, if conditions develop that interrupt output, substitutes can be used." He also observes that the recent growth of demand in bismuth is owing to the luminescent characteristics it imparts to lip gloss and eye shadow; demand is therefore "subject to unpredictable changes in cosmetic fashions." James F. Carlin, "Bismuth," Mineral Commodity Profiles Series (Washington, D.C.: Bureau of Mines, 1979), pp. 8-9. For a similar analysis of demand for selenium, see U.S. Department of the Interior, Bureau of Mines, Minerals Yearbook 1980, 3 vols. (Washington, D.C.: GPO, 1981), 1: 931-933.

[12] Jorge I. Domínguez, U.S. Interests and Policies in the Caribbean and Central America (Washington, D.C.: American Enterprise Institute, 1982), p. 12. For a similar analysis from the Department of Defense, see U.S. Joint Chiefs of Staff, "United States Military Posture for FY1982," printed in U.S. Congress, House, Committee on Armed Services, Hearings on Military Posture and H.R. 2614 and H.R. 2970, 97th Cong., 1st Sess., 1981, pt. 1, p. 109.

tions regarding the strategic significance of Latin America. In 1981, General Ernest Graves, then director of the Defense Security Assistance Agency, the organization that administers U.S. military aid programs, told congressional committees considering his budget that "the United States and many of our allies have grown dependent on access to critical raw materials imported from Latin America—petroleum, natural gas, iron ore, and bauxite to name a few. . . . We can ill afford to run the risk of disruption of access to these critical raw materials."[13]

Overall, the more one probes behind these and similar comments, the more it appears that policy makers and analysts who perceive a crisis tend to differ from those who do not principally on the basis of which raw materials they select to analyze; that is, *the perception of a raw materials security crisis is largely dependent upon which raw materials are selected as data*. This means that definitions play a crucial role in the debate.

DEFINITIONS

Not all raw materials are strategic raw materials, and the discussion of Latin America's role in supplying strategic raw materials will not progress far if we accept the position of the chief staff officer of the Bureau of Mines, who argued in 1981 that "adequate supplies of virtually every known material are a strategic necessity."[14] Statements such as this are more than simply incorrect; they depreciate the importance of the raw materials that are truly essential to U.S. security. A far better approach is to begin by using the definition found in the Strategic and Critical Materials Stock Piling Revision Act of 1979 (PL96-41, Sec. 12), which limits the

[13] U.S. Congress, House, Committee on Foreign Affairs, *Foreign Assistance Legislation for Fiscal Year 1982*, 97th Cong., 1st Sess., 1981, pt. 7, p. 77. For identical or similar statements by other Reagan administration officials, see U.S. Congress, Senate, Committee on Foreign Relations, *Foreign Assistance Authorization for Fiscal Year 1982*, 97th Cong., 1st Sess., 1981, p. 426; U.S. Congress, House, Committee on Foreign Affairs, Subcommittee on Inter-American Affairs, *Presidential Certification on El Salvador*, vol. 1, 97th Cong., 2d Sess., 1982, p. 35; *Foreign Assistance Legislation for Fiscal Year 1982*, pt. 2, p. 117. For comments by senior U.S. military officers, see Paul B. Ryan, "Canal Diplomacy and U.S. Interests," *Proceedings of the U.S. Naval Institute* 103 (January 1977): 48, and Raymond A. Komorowski, "Latin America—An Assessment of U.S. Strategic Interests," *Proceedings of the U.S. Naval Institute* 99 (May 1973): 151.

[14] Quoted in *A Congressional Handbook on U.S. Materials Import Dependency/ Vulnerability*, p. 29, *supra* n. 2.

term "strategic and critical materials" to "materials that (a) would be needed to supply the military, industrial, and essential civilian needs of the United States during a national emergency, and (b) are not found or produced in the United States in sufficient quantities to meet such need."[15]

One problem with this definition is that it includes not only military needs but also "essential civilian needs," a term that is not defined by the law. In practice, the use of this vague term encourages an unrealistic guns-and-butter approach. It is unrealistic because, first, in a genuine national emergency of any significant duration (such as World War II), the term "essential" has always been redefined as the economy converted to wartime production. Second (and conversely), a protracted low-level conflict similar to the Vietnamese war would not affect the U.S. supply of essential raw materials unless it were to involve a small handful of areas, particularly southern Africa. In any event, the use of this definition will lead to a bloated list of about forty-five nonfuel minerals essential to an industrialized economy.[16] Of these, the United States is dependent upon foreign supplies for at least twenty-six.

A second problem with the definition in PL96-41 is that it encourages policy makers to equate the term "import dependence" with the narrower category of "security vulnerability." To import is not necessarily to be vulnerable, since a variety of factors may serve to reduce vulnerability. Among these are the ability to substitute one material for another, the size of domestic reserves, the location of foreign sources, the availability of alternative supplies (including higher-cost domestic sources), and the ability to do without and not damage U.S. military capabilities or essential civilian needs. In assessing security vulnerability it makes a difference, for example, that a stable, friendly, contiguous neighbor (Canada) supplies much of the U.S. need for imported nickel, while most of the U.S. cobalt supply is obtained from unstable, distant Zaire, whose friendship is subject to considerable uncertainty. Similarly, while platinum is on every list of strategic raw materials, the United States could cut its current consumption of platinum in half if automobile manufacturers temporarily ceased

[15] The phrase "during a national emergency" is defined (Sec. 3) as the first three years of a major conflict.

[16] U.S. General Accounting Office, "Actions Needed to Promote a Stable Supply of Strategic and Critical Minerals and Materials," mimeographed, report No. EMD-82-69 (Washington, D.C.: GAO, June 3, 1982), p. 5.

equipping cars with catalytic converters—a device that reduces air pollution—for the duration of a security crisis.

Taking these considerations into account, the General Accounting Office has concluded that the United States is vulnerable to a raw materials supply crisis with twelve of the forty-five nonfuel minerals: bauxite, chromium, cobalt, columbium, gold, manganese, nickel, platinum-group metals, tantalum, tin, titanium, and tungsten.[17] The GAO approach can be criticized, however, for failing to place sufficient emphasis upon domestic supplies (including supplies available through recycling) and upon the reliability of some foreign sources (that is, Canada). In order to avoid these problems the Congressional Research Service (CRS) has selected an approach that identifies nonfuel minerals that are (1) essential for the production of military equipment, (2) not found or found in only small amounts in the United States, and (3) for which substitutes are essentially unavailable.[18] Using this definition to compile its initial list of imported essential strategic raw materials, the CRS then pared it further by examining the reliability and diversity of sources of supply. The result was a list of eight nonfuel minerals—bauxite, chromium, cobalt, columbium, manganese, platinum-group metals, tantalum, and titanium—that are both essential to our military strength and subject to supply disruptions from foreign sources.[19] A ninth import—petroleum and petroleum products—can be added to create an acceptable working definition of essential strategic imported raw materials.

Changing Supply and Demand

The United States has developed an absolutely critical need for these nine raw materials, and it is therefore difficult to overestimate the national security consequences of major supply disruptions.

Aluminum made from *bauxite*, 94 percent of which is imported, has broad defense uses in military vehicles and ammunition. For example, a single Pratt & Whitney F-100 engine for the F-15 and F-16 fighters uses 720 pounds of aluminum.

[17] Ibid.

[18] The CRS report is printed in *U.S. Economic Dependence on Six Imported Strategic Non-Fuel Minerals*. See especially p. 2.

[19] *A Congressional Handbook on U.S. Materials Import Dependency/Vulnerability*, p. 130, *supra* n. 2.

Most *chromium* consumed in the United States is in the form of high carbon ferrochronium, from which stainless steel is made. Each F-100 engine contains 1,656 pounds of chromium. U.S. reserves of chromite are very limited in quantity and of poor quality.

Cobalt is a crucial—irreplaceable—component of alloys used in jet turbine blades and high-stress aircraft structural components such as landing gear and engine mounts. There are 910 pounds of cobalt in each F-100 engine. Cobalt is also essential for the production of missile controls, precision rollers, and recoil springs for tanks. "An uninterrupted supply of cobalt is essential if the U.S. defense industrial base is to function with any degree of efficiency," reports the CRS.[20] No cobalt has been mined in the United States since 1971.

Columbium is also essential to the manufacture of alloys used by the aerospace industry; for example, there are 171 pounds of columbium in each F-100 jet engine. The United States imports 100 percent of its columbium consumption.

Manganese is also an irreplaceable mineral used in making steel—there is no adequate substitute—for which the United States is totally dependent upon foreign suppliers. Manganese is one of the very few minerals (chromite is another) not found anywhere in the United States; even in an emergency there are no low-grade, high-cost domestic manganese deposits that could be exploited.

The *platinum-group metals* (platinum, palladium, iridium, osmium, rhodium, and ruthenium) generally occur together in nature, but nature, lamentably, did not favor the United States in their distribution. With only 1.3 percent of the world's known reserves, the United States must import over 85 percent of its needs. These needs are diverse, ranging from the nonessential (jewelry, automobile emission controls, and dental supplies) to the production of sensitive electronic equipment and the processing of petrochemicals.

Tantalum is another of the exotic raw materials that is both essential to U.S. security and for which no substitute is available. Specifically, there is no substitute at the present time for the use of tantalum in the control system capacitors of jet engines. Thus while there are only three pounds of tantalum in

20 *U.S. Economic Dependence on Six Imported Strategic Non-Fuel Minerals*, p. 35, *supra* n. 2.

each F-100 engine, those three pounds are simply essential. And the situation could become even more critical, for the new single-crystal jet engine blades contain about 10 percent tantalum. If, as expected, they become widely used in military aircraft, this single product alone will consume about half the current world production of tantalum.

Titanium is used in two principal forms: as nonessential titanium dioxide in paints, varnishes, and lacquers, and as essential titanium sponge metal in the aerospace industry. Of the titanium sponge consumed in the United States, more than 80 percent is used in aircraft production. The F-100 engine requires a prodigious amount—5,366 pounds, the single largest raw material input in the engine. New military weapons will require even larger quantities of titanium. Each B-1 bomber uses 125 tons of titanium sponge, for example, which is 0.5 percent of total U.S. consumption in 1980. Domestic reserves of titanium are about 1.4 percent of known world reserves; thus the United States regularly imports nearly all of the titanium it consumes.

While it is obviously true that a steady, adequate supply of these eight minerals—and imported petroleum, of course—are essential to U.S. security, there are a series of considerations that could reduce their importance. First, substitutes are sometimes available. About a third of U.S. chromium consumption could be exchanged for available substitutes, and about 40 percent of cobalt consumption is replaceable. But there are no adequate substitutes for some materials (manganese and tantalum are the outstanding examples), and for others such as chromium there are no substitutes for many applications.

Second, nearly all statistics on world reserves—their size and especially their location—are built upon shifting sands, for the discovery of new sources of many minerals has become a regular occurrence. The term "known reserves" represents the reserves that have been worth finding with existing technology at prevailing prices.[21] But both technology and prices change over time, and so do world reserves. With copper, for example, technological innovations have made it possible for the exploitable grade to drop from 5 percent in 1900 to about .005 percent today; that is, it has become profitable to exploit low-grade deposits. Moreover, U.S.

[21] For a discussion of this issue, see W. Beckerman, *In Defense of Economic Growth* (London: Jonathan Cape, 1974), p. 118.

reserves of copper have been directly related to price throughout the twentieth century. As prices have risen, so have reserves, from 18.5 million tons in 1930 to 93.0 million tons in 1980, and the largest increases in reserves have followed directly on the heels of large price increases. In 1960, when the price of copper was 36.5 cents per pound (in 1970 dollars), U.S. reserves totaled 32.5 million tons; by 1970, the price had risen to 58.2 cents per pound, and U.S. reserves had nearly tripled to 85.4 million tons.[22] There are similar data on other metals—world reserves of chromite rose by 675 percent from 1950 to 1970, for example—to indicate that "total world reserves" is hardly a fixed figure but rather is highly responsive to demand, prices, and technological change. Historically, reserves of nearly all essential raw materials have increased, and this trend may be accentuated by recent technological innovations that permit deep seabed mining. It has been estimated, for instance, that sea exploitation will increase world manganese, nickel, molybdenum, vanadium, and copper reserves by 150 percent and cobalt reserves almost infinitely.[23]

Third, much of U.S. consumption of the nine raw materials is unrelated to national defense or to essential domestic industries. Certainly bauxite is a critical raw material, but only 4 percent of all aluminum is consumed by defense industries. Similar figures characterize the usage of several of the other essential raw materials: defense needs account for 3 percent of total U.S. consumption of the platinum group metals, 5 percent of manganese, 7 percent of chromium, and 17 percent of cobalt.[24] In a national security emergency accompanied by supply shortages, luxury consumption such as aluminum soft drink containers and platinum

[22] *A Congressional Handbook on U.S. Materials Import Dependency/Vulnerability*, pp. 23-24, *supra* n. 2.

[23] D. S. Cronan, *Underwater Minerals* (London: Academic Press, 1980), pp. 84-95, 252-290; John Temple Swing, "Law of the Sea," *Bulletin of the Atomic Scientists* 39 (May 1983): 15; Shafer, "Mineral Myths," p. 159.

It is also important to note that among experts there is wide disagreement over the size of existing reserves in even the most intensely explored areas. For example, estimates of domestic U.S. petroleum reserves range from 587 billion barrels to 2,900 billion barrels. See W. P. Pratt and D. A. Brobst, "Mineral Resources: Potentials and Problems," mimeographed, U.S. Geological Survey Circular No. 698 (Washington, D.C.: U.S. Geological Survey, 1974), and Govett, "Geographic Concentration of World Mineral Supplies, Production, and Consumption," p. 107.

[24] *U.S. Economic Dependence on Six Imported Strategic Non-Fuel Minerals*, pp. 11, 20, 29, *supra* n. 2.

jewelry could be curtailed almost immediately, leaving existing supplies for defense uses.

Fourth, since 1939 the United States government has conducted a program of acquiring and, since 1946, of retaining stocks of strategic materials. The Strategic and Critical Materials Stock Piling Revision Act of 1979 is the current legislation governing this program. It consolidates a number of separate programs under an omnibus National Defense Stockpile, which is operated by the Federal Emergency Management Agency (FEMA) under the close supervision of Congress. FEMA's task is to develop and revise continually a list of strategic and critical materials and to maintain a stockpile for each material based upon projected usage during the first three years of an emergency of indefinite duration, assuming large-scale industrial mobilization. There currently are ninety-three stockpiled materials, ranging from the eight essential strategic nonfuel minerals to a number of other sometimes intriguing products—feathers and opium, for examples—that, while not essential, contribute to the "health, welfare, morale, and productivity of civilians during wartime."[25]

Fifth, because production of the nine essential raw materials is spread among several nations, it is realistic to think in terms of partial loss of supply rather than total loss. Expanded petroleum production in the 1970s suggests that even a repeat of the 1973/1974 OAPEC oil embargo would have a far less significant impact today. In September 1973, OPEC nations supplied 68 percent of U.S. crude oil imports; by 1984, this proportion had declined to 47 percent.[26] For each of the nine raw materials a large partial loss would be harmful, but it could be compensated in part by the reduction of nonessential consumption and the use of existing U.S. stockpiles.

Finally, demand might drop for some essential raw materials, just as U.S. requirements for zirconium and hafnium fell when the boom ended in the construction of nuclear reactors. Changing styles (the declining use of chrome-plated steel in automobiles) and changing technology (especially the development of substitutes) could alter the entire structure of international demand for many strategic raw materials. No one forecasts a repeat of the

---

[25] *A Congressional Handbook on U.S. Materials Import Dependency/Vulnerability*, p. 234, *supra* n. 2. For a complete list of the stockpiled materials, see pp. 236-241.

[26] U.S. Central Intelligence Agency, *International Energy Statistical Review* (Washington, D.C.: CIA, June 26, 1984), p. 4.

boom-and-bust history of natural rubber imports, which serves as the outstanding example of demand shifts caused by technological innovation, but lesser changes are clearly probable.

For all of these reasons the analysis of the impact of foreign supplies of raw materials should never be considered an exact science, and strident claims on either side of the issue should be greeted with a healthy skepticism. In the literature on this subject, non-committed analysts (that is, those who are not attempting to prove a point using raw materials as an example) tend to hold a position similar to that of the General Accounting Office, which has concluded that "the U.S. economy could indeed adapt to most supply disruptions, but . . . depending on the particular mineral or material involved, the time required for such adaptation could be lengthy, and some disruption could occur."[27]

Essential Strategic Raw Material Imports from Latin America

Latin America supplies the United States with about thirty mineral raw materials, plus petroleum products and a long list of non-mineral, nonfuel, and nonessential raw materials that includes everything from hides to alpaca yarn. With two prominent exceptions Latin America's role in supplying the nine essential strategic raw materials is generally modest but hardly insignificant. Of the eight nonfuel minerals, four (*chromium*, *cobalt*, the *platinum-group metals*, and *titanium*) are not imported from Latin America.[28] About 16 percent of U.S. imports of *manganese* comes from two Latin American countries, Brazil (10 percent) and Mexico (6 percent). About 17 percent of U.S. imports of *tantalum* come from Latin America, specifically from Brazil. Brazil is one of eighteen sources of U.S. imports of tantalum, second to Canada in importance, but closely rivaled by West Germany and Australia.[29] About a third of U.S. imports of *columbium* is also supplied by Brazil.[30]

[27] U.S. General Accounting Office, "Actions Needed to Promote a Stable Supply of Strategic and Critical Minerals and Materials," p. 7.

[28] U.S. Department of Commerce, Bureau of the Census, *U.S. Imports for Consumption and General Imports: TSUSA, Commodity by Country of Origin, Annual 1980* (Washington, D.C.: GPO, 1981), pp. 1-312, 1-315, 1-316.

[29] Ibid., p. 1-313.

[30] This figure and the figure for tantalum fluctuate dramatically from year to year, since U.S. imports are small (less than five million pounds of columbium, two million pounds of tantalum) and relatively minor changes in pounds can produce substantial changes in percentage terms. For various perspectives on this issue, see

Bauxite is one of the two essential strategic raw materials imported primarily from Latin America. Seventy-nine percent of U.S. imports come from the region, with Jamaica supplying the lion's share (52 percent), followed by Surinam (9 percent), Brazil (6 percent), Guyana (4 percent), the Dominican Republic (4 percent), and Haiti (2 percent). Brazil has very large reserves that have not been exploited.

The second major import from Latin America is petroleum. As Table 4.1 indicates, in 1985, more than one-third of U.S. imports of crude petroleum came from Latin America.[31] Nearly all of this was produced in two countries: Mexico (22.2 percent) and Venezuela (9.8 percent). The balance (about 5 percent) was supplied by Trinidad and Tobago, and Ecuador.[32] From the perspective of U.S. security interests, the dominant position of Mexico is particularly

TABLE 4.1
UNITED STATES IMPORTS OF CRUDE OIL

	World Total*	Latin America*	Percentage Latin America	Percentage Venezuela	Percentage Mexico
1947	97,532	97,146	99.6	77.4	5.7
1950	177,714	136,096	76.6	60.2	6.9
1955	285,421	156,326	54.8	49.3	1.9
1960	371,575	188,880	50.8	46.5	0.2
1965	452,040	175,615	38.8	34.9	0.6
1970	483,293	106,108	21.9	20.3	0.0
1975	1,498,181	234,597	15.7	9.6	1.7
1980	1,926,162	306,811	15.9	3.0	9.6
1985	1,173,738	438,310	37.3	9.8	22.2

* Thousands of Barrels
SOURCE: American Petroleum Institute, *Basic Petroleum Data Book: Petroleum Industry Statistics* (Washington, D.C.: American Petroleum Institute, May 1986), Section IX, Table 4.

---

ibid., p. 1-312; *A Congressional Handbook on U.S. Materials Import Dependency/Vulnerability*, pp. 130-144; U.S. Congress, House, Committee on Foreign Affairs, *Foreign Assistance Legislation for Fiscal Year 1981*, 96th Cong., 2d Sess., 1980, pt. 6, pp. 71-72.

[31] There is a substantial difference between the amount of U.S. petroleum imports *produced* in Latin America and the amount of U.S. petroleum imports produced elsewhere that *pass through* Latin America or, to be specific, through the Caribbean. The use of Latin American ports for refining and transshipment will be discussed in Chapter 6.

[32] American Petroleum Institute, *Basic Petroleum Data Book: Petroleum Industry Statistics* (Washington, D.C.: American Petroleum Institute, May 1986), Section IX, Table 4c.

important, for it is the only Latin American country with whom the United States shares a land border. Data on U.S. petroleum imports change rapidly, as reflected in the table; indeed, the first draft of this chapter, using 1981 data, placed imports from Latin America at 18 percent rather than the 1985 figure of 37.3 percent. This substantial change reflects both a recession-induced decline in total imports (a 16 percent drop between 1977 and 1983) that hit Middle Eastern producers hardest, and the explosive growth of Mexican production, which is ideally located across the Gulf of Mexico from the major cluster of U.S. petroleum refineries. In 1973, Mexican hydrocarbon reserves were estimated at fewer than 5,000 million barrels, and Mexico supplied 0.2 percent of U.S. crude petroleum imports. By 1986, estimates of these reserves had increased to 49,300 million barrels, a tenfold increase that made Mexico a major factor in the world petroleum market.[33]

Another fuel, natural gas, is also imported from Mexico. In 1984, the United States purchased 52 billion cubic feet from the country, about 6 percent of total U.S. imports. Nearly all the rest comes from Canada (90 percent of U.S. imports) and a smaller amount from Algeria.[34] Latin American reserves of natural gas amount to about 5 percent of known world reserves.[35]

WHAT can be said in summary about the U.S. supply of essential strategic raw materials from Latin America? First, one essential commodity, bauxite, is largely supplied by Latin America, particularly Jamaica. Second, Brazil supplies considerable amounts of columbium, manganese, tantalum, and bauxite, but in no case does Brazil supply more than a third of U.S. demand (columbium), and the proportion of bauxite, manganese, and tantalum is much

[33] George W. Grayson, *The Politics of Mexican Oil* (Pittsburgh: University of Pittsburgh Press, 1980), p. 239; U.S. Central Intelligence Agency, *International Energy Statistical Review*, p. 4.

Two Latin American countries are included among the twenty leading countries of the world in terms of petroleum reserves: as of 1986, Mexico was fourth with 7.0 percent of world reserves, and Venezuela was ninth with 3.7 percent. No other Latin American country has as much as 0.4 percent of world reserves. American Petroleum Institute, *Basic Petroleum Data Book*, Section II, Table 3c. For a contrasting analysis emphasizing the hope that Latin America's contribution to world reserves will increase, see Adolfo Moncivaiz, "U.S. National Security Concern for the 1980s: Oil Export Potential in Latin America," Ph.D. diss., University of Miami, 1981.

[34] American Petroleum Institute, *Basic Petroleum Data Book*, Section XIII, Table 6.

[35] Ibid., Section XIII, Table 7c.

less. Third, Mexico provides the United States with considerable petroleum and some natural gas and manganese. Fourth, Venezuela supplies the United States with some petroleum. In sum, the data indicate that U.S. dependence upon Latin America for essential strategic raw materials is fairly modest, and Latin America's supplies are extremely highly concentrated in Brazil, Jamaica, Mexico, and Venezuela.

Why is it, then, that so many U.S. policy makers accept the notion that Latin America is a principal source of essential raw materials? Part of the answer is probably historic. Prior to World War II, Hubert Herring, whose Latin American history text was required reading for more than one generation of U.S. students, summarized the conventional wisdom when he observed that "Latin America is cut to the order of aspiring empire builders. She has room, millions of empty acres, untouched forests, untapped mines. Her soil and her veins yield or can be made to yield every foodstuff, every produce, every mineral needed by an industrialized world."[36] At about that same time, Juan Trippe, who built Pan American World Airways' Latin American network, wrote that "in South America, as nowhere else on the globe, there are tremendous natural resources awaiting development."[37] These and similar assertions once led geopolitical analysts to consider Latin America "one of the greatest undeveloped reservoirs of raw materials in the world."[38]

So when in 1937 an influential German economist wrote to an American audience that "Germany must produce her raw materials on territory under her own management," U.S. policy makers understandably looked with concern at the vulnerable nations of Latin America.[39] As the Axis began its expansion, Secretary of State Cordell Hull warned that the United States faced the need to contain "lawless nations, hungry as wolves for vast territory with rich undeveloped natural resources such as South America possesses."[40]

[36] Hubert Herring, *Good Neighbors: Argentina, Brazil, Chile and Seventeen Other Countries* (New Haven: Yale University Press, 1941), p. 329.

[37] Juan T. Trippe, "The Business Future—Southward," *Survey Graphic* 30 (March 1941): 139.

[38] Raymond Leslie Buell, *Isolated America*, 2d ed. (New York: Alfred A. Knopf, 1940), p. 254.

[39] Hjalmar Schacht, "Germany's Colonial Demands," *Foreign Affairs* 15 (January 1937): 233.

[40] Cordell Hull, "Memorandum of Conversation with Luis Fernando Guachalla, Bolivian Minister to the United States, April 11, 1939." Department of State unpublished document 824.6363 St2/336. For general discussions of the issue of U.S.

Looking at the data of the time, concern over the security of Latin America's resources seemed highly appropriate. Immediately prior to World War II, Latin American nations were among the leading exporters of most of the ten most important strategic raw materials of the time. Of the ten, only two (coal and nickel) were not exported from Latin America. Of the remaining eight, Latin American exports were highly significant in world trade.[41] And, as luck would have it, the two minerals not exported from Latin America were available in abundance domestically (coal) or from neighboring Canada (nickel). It was highly rational, therefore, for U.S. policy makers to be concerned about the security of U.S. mineral supplies from Latin America.

Now, a long generation later, times have changed. First, the list of essential strategic raw materials that contained products such as lead has been superceded by one containing exotic minerals such as columbium and tantalum, most of whose current uses were unknown to our immediate ancestors. Over time the specific minerals considered essential have changed; in the process Latin America has lost its previous importance as an exporter of essential strategic raw materials.[42] Many U.S. policy makers have failed to recognize this change.

Second, since World War II there has been a significant shift in the pattern of raw materials production away from some traditional producing countries, particularly those in Latin America. In 1947, for example, 99.6 percent of all U.S. petroleum imports came from Latin America.[43] Today five developed countries—

responses to potential resource wars, see Alfred E. Eckes, Jr., *The United States and the Global Struggle for Minerals* (Austin: University of Texas Press, 1979), and David G. Haglund, " 'Grey Areas' and Raw Materials: Latin American Resources and International Politics in the Pre-World War II Years," *Inter-American Economic Affairs* 36 (Winter 1982): 23-51.

[41] Cuba was sixth in chromite. Chile was first, Mexico seventh, and Peru eighth in copper. Chile was eighth in iron ore. Mexico was first and Peru sixth in lead. Brazil was fifth and Cuba sixth in manganese. Venezuela was first, the Netherlands West Indies third, Colombia eighth, Mexico ninth, and Trinidad tenth in petroleum. Bolivia was second in tin. Mexico was second in zinc. The ten raw materials were selected in C. K. Leith, J. E. Furness, and Cleona Lewis, *World Minerals and World Peace* (Washington, D.C.: Brookings Institution, 1943), pp. 238-246. The information presented here combines the "crude" and "refined" categories for copper, lead, petroleum, tin, and zinc.

[42] Haglund, " 'Grey Areas' and Raw Materials," pp. 24-51.

[43] In Table 4.1 (see p. 155), the decline in Venezuela's proportion of U.S. imports is illustrative: it is linked primarily to the growth of other suppliers, not to a drop in Venezuelan production.

Australia, Canada, South Africa, the Soviet Union, and the United States—are the major world suppliers of most minerals. Third World countries are the dominant producers of only a small handful of raw materials: tin and tungsten in Southeast Asia, petroleum in the Middle East and Africa, cobalt in Zaire, and bauxite in the Caribbean region.[44] The production of all other important mineral raw materials is concentrated in developed countries.

The conclusion is obvious. As economist (and former Carter administration official) Richard Feinberg recently observed: "The image of a future of hydra-headed OPECs, of resource blackmail and even resource wars can inflate rhetoric, making posture-statements and politicians' speeches exciting and even frightening; but the campaign balloons burst once the underlying assumptions are examined."[45] The data indicate that even if Feinberg is wrong and a resource war erupts between the United States and a major adversary, it will not be fought over Latin America. As we shall see, there are several legitimate reasons for U.S. policy makers to be concerned about Latin America in the event of a security crisis. Access to essential strategic raw materials, however, is clearly not one of them.

[44] M. H. Govett and G.J.S. Govett, "The New Economic Order and World Mineral Production and Trade," *Resources Policy* 4 (December 1978): 230; Govett, "The Geographic Concentration of World Mineral Supplies," p. 357.

[45] Richard E. Feinberg, *The Intemperate Zone: The Third World Challenge to U.S. Foreign Policy* (New York: W.W. Norton, 1983), pp. 119-120.

· 5 ·

MILITARY BASES AND MILITARY SUPPORT

IN THEIR public pronouncements and especially in their statements before congressional committees, Washington policy makers often assert that Latin America provides the United States with two significant types of military assistance: access to military facilities and military support from Latin American armed forces. The purpose of this chapter is to analyze these two aspects of Latin America's role in U.S. policy makers' strategic planning. As in the preceding analysis of raw materials, the goal is not only to discuss the beliefs and perceptions of policy makers but also to assess the specific nature both of U.S. military bases in Latin America and of Latin American military support.

ACCESS TO MILITARY FACILITIES

For many years U.S. policy makers have engaged in a low-intensity debate over the importance of military bases in Latin America. The focus has typically been upon the Caribbean region, where existing U.S. bases have been characterized as "of major importance"[1] or, conversely, as "essentially irrelevant, and most likely inimical, to U.S. security interests."[2] But with occasional exceptions—Panama in the late 1970s and Honduras in the 1980s—the debate over these bases has aroused little interest and even less action. U.S. policy makers maintained the bases that had been acquired earlier and gave little thought to either additions or reductions.

These existing bases are all located in the Caribbean basin. In the northeast corner of the Caribbean is the Bahamian island of Andros, where the U.S. Navy maintains its Atlantic Undersea Test and Evaluation Center (AUTEC), a test facility for U.S. anti-submarine warfare (ASW) equipment. Andros sits alongside a one-hundred-mile-long trench, the Tongue of the Ocean, which serves as an ideal torpedo test range, close to home and easily protected

[1] Raymond A. Komorowski, "Latin America—An Assessment of U.S. Strategic Interests," *Proceedings of the U.S. Naval Institute* 99 (May 1973): 171.

[2] U.S. Congress, Senate, Committee on Foreign Relations, *United States Foreign Policy Objectives and Overseas Military Installations*, 96th Cong., 1st Sess., 1979, p. 198.

MAP 5.1. Latin America

from the prying eyes of any adversary. The Department of State characterizes AUTEC as "vital to our interests."[3]

The Bahamas also contain two shore facilities for the U.S. Navy's SONUS (Sound Surveillance System) antisubmarine program. SONUS is a worldwide network of fixed hydrophones that rest on the continental shelf, detect submarine movement, and send the information via satellite to the United States. Other Caribbean SONUS installations are located at Grand Turk, immediately south of the Bahamas in the Turks and Caicos, at Sabana Seca in Puerto Rico, and at Antigua in the Leeward Islands and Barbados in the Windwards.[4] When combined with two offshore SONUS arrays north of Puerto Rico, these shore facilities permit the United States to monitor all submarine traffic in the region—no submarine enters or leaves the Caribbean without the knowledge of the U.S. Navy. The significance of this knowledge will become apparent in the discussion of Caribbean sea lanes in Chapter 6; for now it is sufficient to note that because of AUTEC and its SONUS facilities, the Bahamas are the most important U.S. military installations in the Caribbean.[5]

In addition to its antisubmarine facilities, the United States once maintained a network of bases in the eastern Caribbean as part of its eastern missile test range. By the early 1980s, most of these had been abandoned or reduced in size and left to perform minor tasks such as oceanographic research. In addition, the United States maintains access to facilities in Barbados, Jamaica, St. Lucia, and Trinidad and Tobago.[6] The United States also con-

[3] U.S. Department of State, *Congressional Presentation, Security Assistance Programs, FY1983* (Washington, D.C.: Department of State, 1982), p. 419.

[4] A series of additional offshore SONUS arrays are positioned in the Gulf of Mexico. For a map that illustrates the SONUS system and underscores the significance of these Caribbean facilities, see Joel S. Wit, "Advances in Antisubmarine Warfare," *Scientific American* 244 (February 1981): 39-40.

[5] On this importance, see Robert D. Crassweller, *The Caribbean Community: Changing Societies and U.S. Policy* (New York: Praeger, 1972), pp. 401-402, and Vaughn A. Lewis, "The Bahamas in International Politics: Issues Arising for an Archipelago State," *Journal of Inter-American Studies and World Affairs* 16 (May 1974).

[6] The United States first gained access to these air and sea facilities in 1940, when Britain granted ninety-nine year base leases in exchange for fifty U.S. destroyers. This arrangement was modified by the U.S.-U.K. Mutual Defense Assistance Treaty of 1950 (TIAS 2017). Then, as Britain's Caribbean colonies gained independence beginning in the 1960s, they individually exercised their right to abrogate the ninety-nine-year leases. But the terms of the 1950 U.S.-U.K. treaty remain in effect in the four countries mentioned; they may be unilaterally abrogated

trols tiny (5 square kilometers) Navassa Island off the southwest tip of Haiti. Although claimed by Haiti, since 1916 the uninhabitated island has remained under the care of the U.S. Coast Guard, which maintains the island's only structure, a lighthouse.

The navy base at Roosevelt Roads and the navy gunnery range off Vieques are the principal military facilities in the U.S. Commonwealth of Puerto Rico. There are five other bases on the island, including the Air Force's Ramey Field (currently inactive) and the navy's communications base at Ponce.

And then, of course, there is Guantánamo, the oldest U.S. military base on foreign soil. The U.S. Navy base at Guantánamo Bay, Cuba, is the principal tangible product of the Platt Amendment, a document the United States required the Cubans to accept as a condition for the withdrawal of U.S. troops in 1903.[7] Article 7 of the Platt Amendment provided the United States with the right to establish "coaling or naval stations at certain specific points" on the Cuban coast. The United States originally sought four such facilities: at Guantánamo on the southeast coast, at Bahía Honda west of Havana on the northwest coast, at Cienfuegos, and at Bahía Nipe on the northeast coast. Agreement was reached with the Cubans on the use of the first two sites, and then in 1912 the United States relinquished its claim to Bahía Honda in return for more land (a total now of forty-five square miles) at Guantánamo. In 1934, when the U.S.-Cuban treaty containing the Platt Amendment was abrogated as part of FDR's Good Neighbor Policy, a new treaty was signed to permit the United States to maintain its base at Guantánamo in perpetuity. Article 3 of the 1934 treaty stipulates that the Platt Amendment's "agreement with regard to the

---

at any time. Thomas D. Anderson, *Geopolitics of the Caribbean: Ministates in a Wider World* (New York: Praeger, 1984), pp. 95-96, 111.

[7] The Platt Amendment is more accurately described as a treaty. In 1901, at the suggestion of Secretary of War Elihu Root, Senator Orville Platt, chairman of the Senate Committee on Relations with Cuba, submitted his amendment to the army appropriations bill for 1902. A prior amendment submitted by groups seeking U.S. withdrawal from Cuba had made the army's receipt of funds in 1902 dependent upon its prior withdrawal from Cuba, which the United States had ruled as a protectorate since defeating the Spanish in mid-1898. Secretary Root's goal was to ensure that U.S. withdrawal would not alter Cuba's status as a protectorate. Thus Article 3 of the Platt Amendment states that "Cuba consents that the United States may exercise the right to intervene," and Article 8 stipulates that "Cuba will embody the foregoing provisions in a permanent treaty with the United States." On June 12, 1901, by a majority of one, the Cuban constitutional convention accepted the Platt Amendment as a treaty and as an annex to the Cuban constitution.

naval station at Guantánamo shall continue in effect" for "so long as the United States of America shall not abandon the said naval station of Guantánamo or the two Governments shall not agree to a modification of its present limits."[8]

Early relations between U.S. officials at Guantánamo and the host government were not particularly difficult. Local Cubans worked on the base in various nonmilitary capacities, and the United States paid the Cuban government an annual rent of $4,085. The Cuban revolution changed these arrangements. As relations deteriorated in the early 1960s, fortifications were constructed by both sides. Fences went up around the eighteen-mile land perimeter—the Cubans have built a series of four fences, while the United States prefers less fence (a single fence topped with barbed wire) and more technology, specifically, one of the largest active minefields in the world.[9] Then in February 1964, the Cubans turned off the base's supply of fresh water in retaliation for the U.S. seizure of four Cuban fishing vessels. Washington's response was to install a desalinization plant, despite Fidel's offer in March 1964, to turn the water supply on once again. Today Guantánamo is an isolated outpost, surrounded on three sides by the land of a determined adversary and linked to the outside world only by sea and air. The U.S. Treasury continues to send Cuba an annual check for the rent, however, which the Castro government refuses to cash.[10]

Guantánamo is operated by the U.S. Navy. Base personnel include the staff of a shipyard (a new facility called a Shore Intermediate Maintenance Activity—SIMA—the only one of its kind overseas), a naval air station, a group performing electronic intelligence gathering functions, various support personnel, a few hundred marines for local defense, and, most important, a Fleet Training Group (FTG). The FTG trains entire ships' crews in all aspects of operations and maintenance. There are no significant major weapons on the base, and no ships are homeported at Guantánamo.[11]

[8] 48 Stat. 1682.

[9] The Cubans also have minefields on their side of the fences. Each day a few intrepid Cubans (88 in 1983) pass through a cattle chute at the northeast gate to work on the base, where they join 190 permanent Cuban exiles and 850 Jamaican workers who perform tasks ranging from skilled machinist to garbage collector. As the Cuban labor force ages, it is being replaced by U.S. and Jamaican citizens.

[10] In 1959, when the revolutionary government was a bit disorganized, Cuba cashed the check; none has been cashed since then.

[11] As of June 30, 1984, there were 2,302 U.S. military personnel stationed at Guantánamo: 1,821 Navy, 475 Marine Corps, 4 Army, and 2 Air Force. For general

From time to time Guantánamo serves as a site for military maneuvers, but these typically serve as a modern form of gunboat diplomacy rather than as an effort to sharpen military skills. After wrangling over the presence of a Soviet combat brigade in Cuba in 1979, for example, President Carter directed the navy to hold training exercises at Guantánamo as a demonstration of U.S. resolve. The navy responded by importing no fewer than ninety-three journalists and news photographers. Once they were positioned at strategic points above the beach, the marines rounded the bend in their landing craft, stormed ashore as the cameras whirred, secured an empty beach, dried off, ate lunch, and soon reembarked for the trip back to the United States.[12]

Some decades ago an occasional analyst could argue that Guantánamo provided a significant contribution to U.S. national security. One wrote in 1967 that Guantánamo "controls the Windward Passage to the Caribbean. It serves as an excellent base for anti-submarine warfare operations. It acts as a base for the protection of inter-American shipping. It can control the air routes from the east coast of the United States to the west coast of South America. It serves as a damaged ship haven. It has proven to be an excellent base for the training of ships and crews and it is the key point in the defense of the Panama Canal."[13] Even prior to World War II it is doubtful whether Guantánamo could ever fulfill all these roles, with a modest airstrip and one of nature's most easily blockaded ports. A recent evaluation is probably more accurate: "The installations at Guantánamo Bay are more symbolic than strategic; defense there against a determined ground assault is not possible. In the event of war, the base would serve mainly to provide a justification for retaliation."[14] Indeed, the base is particularly poorly situated for defense against a land attack. It is actually two bases, the

---

background on navy activity at Guantánamo, see Theodore K. Mason, *Across the Cactus Curtain: The Story of Guantánamo Bay* (New York: Dodd, Mead, 1984), and Gary Hopkins, "Guantanamo Bay: Big Mission in the Caribbean," *All Hands*, September 1983, pp. 20-30. For the Cuban government's view of the base, see Republic of Cuba, Ministry of Foreign Relations, Information Office, *Guantánamo: Yankee Naval Base of Crime and Provocation* (Havana: Instituto del Libro, 1970).

[12] The landing on October 17, 1979, by 1,800 marines from Camp Lejeune, North Carolina, was hampered by heavy rains, and what had been scheduled for dawn had to be delayed until 8:30 A.M. to provide better light for the TV cameras. After being informed of the navy's plans for the landing, one journalist called it a "media circus." *Washington Post*, October 6, 1979.

[13] Gary L. Maris, "International Law and Guantanamo," *Journal of Politics* 29 (May 1967), pp. 284-285.

[14] Anderson, *Geopolitics of the Caribbean*, p. 152.

naval air station on the leeward side of the bay and the naval base on the windward side, separated by two and a half miles of water and linked only by a ferry that cannot operate in rough weather. There is no bridge or tunnel between the two sides, and the connecting road is never used because it passes through Cuban territory.

The navy considered closing the facility in 1977, but in 1981 the Reagan administration decided to increase activities—the SIMA and an enlarged FTG are the principal additions—that today justify Guantánamo's continued operation. Despite this recent resurgence of activity, few Washington policy makers assert that Guantánamo is a necessary component of the U.S. military posture. In recognition of Guantánamo's vulnerability the navy now performs its major security tasks in the area from bases in the mainland United States (including Key West), from Panama, and from its Puerto Rican facility at Roosevelt Roads. Aside from its training function, which could be performed at a broad variety of other navy facilities,[15] most policy makers agree that the primary value of Guantánamo is that it serves as a profound irritant to the Cuban government.

Within the former Panama Canal Zone (at Quarry Heights on the Pacific side of the isthmus) is the headquarters of the United States Southern Command (SOUTHCOM), the organizational arm of the Department of Defense in Latin America.[16] SOUTHCOM has responsibility for the defense of the canal and for U.S. military representation throughout Latin America. This representation includes the supervision of military aid programs, the support of military assistance advisory groups (MILGPs, MAAGs, mobile training teams, etc.), and the coordination of inter-American training exercises.

[15] It should be noted, however, that unlike all U.S. mainland bases, Guantánamo has the special value of being located very close to deep water and away from major commercial shipping lanes. This permits the more efficient use of training time under the supervision of the FTG.

[16] SOUTHCOM traces its roots back to the first U.S. Marines who arrived in 1903 and stayed until 1914, when they were reassigned to the search for Pancho Villa in northern Mexico. The marines were replaced by an army coast artillery unit, which in 1917 was upgraded to the Panama Canal Department of the U.S. Army. In 1941, the department's name was changed to the Caribbean Command, and in 1963 the name was changed once more to the present Southern Command. For an informative article by the commander of SOUTHCOM in the early 1970s, see W. B. Rosson, "U.S. Southern Command in Latin America," *Commanders Digest* 16 (October 18, 1973): 2-11.

To fulfill its function of defending the Canal, which by treaty will cease in the year 2000, SOUTHCOM has stationed in Panama a modest force consisting of the 9,000-member 193d Infantry Brigade and a squadron of A-7 light attack jets. The 3d Battalion, 7th Special Forces Group (about 400 soldiers) is also based in Panama; its mission is to supply Mobile Training Teams (MTTs) to other Latin American countries, not to defend the Canal.

To fulfill its function as the U.S. military representative to Latin America, SOUTHCOM has consolidated its activities in the former Canal Zone, giving up many of the U.S. facilities but retaining control until noon, December 31, 1999, of about 40 percent of the former Canal Zone. At the time of the signing of the canal treaties on September 7, 1977, the United States and Panama also signed a complex Agreement on Certain Activities of the United States in the Republic of Panama, which governs the use of former military bases.[17] This agreement permits the United States to continue operating an electronic intelligence gathering facility on Galeta Island and three inter-American military networks that link each of the three major services to one another in most Latin American countries. (The networks are the Inter-American Military Network, the Inter-American Telecommunications System for the Air Force, and the Inter-American Naval Telecommunications Network.) The agreement also permits the United States to continue operating four schools for training Latin American military: the army's School of the Americas, the air force's Inter-American Air Force Academy, and two small navy schools—the Small Craft Instruction and Technical Team and the Inter-American Naval Telecommunications Network Training Facility. The Jungle Operations Training Center at Fort Sherman, on the west side of the Canal's Caribbean entrance, also continues to operate. It is the only U.S. jungle warfare training facility. Each year about 11,000 U.S. troops train there.

With the exception of the army's School of the Americas, permission to continue operating these facilities expires with the expiration of the Panama Canal Treaty on December 31, 1999, although there is nothing to prevent the United States and Panama from reaching an agreement to maintain U.S. facilities after that date. Because of the negative public opinion surrounding the

[17] For detailed information on the status of each U.S. military facility in the former Canal Zone, see U.S. Congress, Senate, Committee on Appropriations, Subcommittee on Military Construction, *Military Construction Appropriations for Fiscal Year 1979*, 95th Cong., 2d Sess., 1978, pp. 57-61.

School of the Americas, the agreement stipulates "that the authority of the United States to conduct schooling of Latin American military personnel in the United States Army School of the Americas shall expire five years after the entry into force of the Panama Canal Treaty [that is, October 1, 1984], unless the two Governments otherwise agree." Some U.S. policy makers believe the school is essential to U.S. security—"the importance of this school to U.S. interests in Latin America cannot be overstated," reported the House Armed Services Committee in 1981—and it was because of this belief that the Reagan administration initially sought to ensure its continued operation in Panama. Once the United States had constructed a new training facility at Puerto Castilla, Honduras, and had explored other options, however, the need to maintain the School of the Americas in Panama appeared to decrease, and in mid-1984 the U.S. and Panama announced it would be closed. Even before that date, the limited utility of the school had been demonstrated by the decision to train large groups of Salvadoran soldiers elsewhere.[18] The school reopened in January 1985, in a temporary home at Ft. Benning, Georgia. As of 1987, Pentagon officials were still seeking a permanent home for the facility.

In response to a perceived need to strengthen the U.S. military presence in Central America, early in the 1980s the Reagan administration developed unusual military facilities in Honduras that, while not officially designated as bases, were at least semi-permanent establishments of U.S. military personnel. In 1982, former Ambassador Robert White told Congress that "Honduras today is alive with American military uniforms, Green Berets on their way to the Salvadoran border, Air Force personnel manning the helicopters which can have no other purpose than to threaten Nicaragua, Army officers 'inspecting' the border of Nicaragua, a constant flow of military training teams from the Southern Command in Panama."[19] Two years later *New York Times* correspond-

[18] For a discussion of this issue, see U.S. Congress, House, Committee on Armed Services, *Report on the Inspection of Military Facilities in Panama and Bermuda*, 97th Cong., 1st Sess., 1981, p. 10; for a review of various perspectives on the School of the Americas, see Lars Schoultz, *Human Rights and United States Policy toward Latin America* (Princeton, N.J.: Princeton University Press, 1981), pp. 232-237.

[19] U.S. Congress, House, Committee on Foreign Affairs, Subcommittee on Inter-American Affairs, *Presidential Certification on El Salvador*, vol. 1, 97th Cong., 2d Sess., 1982, p. 127. In April 1982, the administration reported that there were sixty-one U.S. military personnel stationed in Honduras. Thousands more were there

ent Raymond Bonner reported that "the Reagan administration transformed Honduras from a banana republic where the United Fruit Company picked the country's presidents in the early twentieth century into a virtual military base ruled by Ambassador Negroponte and the Honduran minister of defense. . . . In contrast to El Salvador, where the American military personnel have maintained a low profile, in Honduras they swagger into Tegucigalpa bars, pistols on their hips, patches on their combat fatigues identifying them as 'jungle experts.' "[20]

Perhaps the most important of these Honduran military facilities is the thoroughly modernized air force base at Palmerola, but other significant installations have been constructed or enlarged throughout the country. One is a radar site atop Cerro la Mole, about twenty miles southeast of Tegucigalpa. Operated by fifty-seven U.S. Air Force technicians, in the early 1980s the facility's primary task was to monitor air traffic as part of the effort to control the alleged flow of arms from Nicaragua to guerrilla groups in El Salvador. For similar reasons the U.S. Air Force also opened a radar station on Tiger Island in the Gulf of Fonseca. In May 1984, U.S. military personnel at the station were supposed to play a vital role in a coordinated Salvadoran-U.S. effort to intercept arms flowing across the Gulf. As noted in Chapter 1, the operation was a total failure and resulted in a decision—later rescinded—to dismantle the station.

Whatever the case, the long-term purpose of these facilities has been unclear from the beginning. In early 1981, Secretary of State Haig noted ominously that "the air defense system we help a friendly state develop could one day serve as a pre-positioned shelter under which Western relief forces might move."[21] Other administration officials regularly denied any intention of establishing permanent bases in Honduras or anywhere else in Central America. In 1985, however, a leaked document from the U.S. Southern Command revealed that plans called for a military presence in Honduras for the balance of the decade, including the

---

during the nearly continuous military maneuvers held throughout the early and mid-1980s, however.

[20] Raymond Bonner, *Weakness and Deceit: U.S. Policy and El Salvador* (New York: Times Books, 1984), pp. 282-284. For a summary of the controversy concerning these facilities, see the report by congressional staffer Barry Sklar in *Congressional Record*, April 12, 1984, pp. S4416-S4418.

[21] U.S. Congress, House, Committee on Foreign Affairs, *Foreign Assistance Legislation for Fiscal Year 1982*, 97th Cong., 1st Sess., 1981, pt. 1, p. 154.

1,200 military personnel assigned to maintain U.S. facilities between military exercises.[22] These maneuvers, which involved as many as 10,000 U.S. troops, occurred almost continuously during the 1980s.

Honduras also became the site of the U.S.-operated Regional Military Training Center, known as the CREM, the Spanish acronym for *Centro Regional de Entrenamiento Militar*. The primary impetus for creation of the center was the need to train Salvadoran soldiers and the lack of an acceptable alternative location. As the large-scale buildup of Salvadoran forces began in 1981, U.S. advisors were confronted by (1) the refusal of the Panamanians to permit increased training at U.S. bases in the former Canal Zone, and (2) the informal agreement with Congress to limit to fifty-five the number of U.S. military personnel in El Salvador.[23] The first response—to train Salvadoran troops in the United States—was a stop-gap measure. In January 1982, the Ramón Belloso battalion was flown to Ft. Bragg, North Carolina, for four months of training, but in addition to being enormously costly this was an attraction for demonstrators (and the associated media coverage) protesting the administration's policies in Central America.

For a brief period in the early 1980s, the CREM was the U.S. Army's response to these problems. Built in 1981 by 126 U.S. Special Forces personnel, it sat near the site of an abandoned World War II airfield on the Honduran Caribbean near Puerto Castilla. By mid-1982, the CREM had begun the instruction of a 1,000-man Salvadoran battalion in fast-reaction techniques and four 350-man Salvadoran battalions in search-and-destroy tactics. By September 1984, the Puerto Castilla facility had completely trained seven Salvadoran battalions and provided a separate mortar course to three Salvadoran platoons—a total of 3,500 Salvadoran troops. The U.S. also used the CREM to train a smaller number of Honduran military and Costa Rican security forces.

From the beginning, however, the fragile Honduran political system dictated that the permanence of this U.S. facility would be subject to regular reappraisal by both U.S. and local officials. Washington's plans to continue training Salvadorans at the CREM suffered a setback in September 1984 when the Hondurans an-

[22] *Washington Post*, July 17, 1985, p. A5.

[23] Soon after his inauguration President Reagan increased the number of U.S. military personnel in El Salvador from twenty to fifty-five. Strong protests from a number of members of Congress led to a negotiated agreement not to increase the number further.

nounced a prohibition on the further training of foreign troops in their territory. This prompted a modest crisis within the Reagan administration and led to protracted negotiations, including a visit to Honduras by Vice President George Bush in March 1985. At the time of the Vice President's visit the Department of State was forced to reprogram $18.5 million intended for the CREM because of fears that the money could not be spent in Honduras before the end of the fiscal year.[24] During the negotiating period, one foreign service officer remarked in an interview, "What else can they do but agree? If we pull out, they're finished, and they know that without being told. Just to make sure, though, we had Bush tell them." In the end the Hondurans agreed to a continued U.S. military presence on their soil but not to a continuation of the training of Salvadoran soldiers. On June 19, 1985, the CREM graduated its last class—550 Honduran trainees—and closed its doors.

In summary, with the exception of these U.S. forces in Honduras whose role and permanence were still being defined in the mid-1980s, the United States maintains extremely modest military facilities in Latin America. In 1979, prior to the buildup in Central America, the estimated cost of maintaining U.S. military personnel in Latin America was $44 million. This was 0.5 percent of a worldwide total of $8.4 billion.[25]

There are probably two fundamental reasons why the Department of Defense spends only 1/200 of its overseas personnel budget to staff U.S. military facilities in Latin America. First, barring unusual circumstances, most Latin American governments do not want any foreign military forces on their soil. To Latin Americans the long history of U.S. armed intervention in the region militates against inviting the camel to stick its nose under the tent once again.

Honduras in the 1980s was one such unusual circumstance, one in which the Honduran military was simply bribed with extraordinary amounts of money provided in the "camel's nose" fashion that characterized U.S. military involvement throughout Central America in the 1980s. For FY1982, the Department of State asked for $10 million in Foreign Military Sales credits to "help Honduras protect itself against the dangers of foreign-supported insurgency

---

[24] The congressional deadline for notification of reprogramming is March 15. After that the money must be spent as authorized and appropriated or returned to the Treasury at the end of the fiscal year, September 30.

[25] *United States Foreign Policy Objectives and Overseas Military Installations,* p. 207, *supra* n. 2.

and help its efforts to stop the use of Honduran territory to support guerrillas in El Salvador."[26] At the time the amount seemed large, for it nearly tripled the $3.5 million FMS allocation in FY1980. Even before the end of FY1982, however, the $10 million came to be seen as insufficient, and the administration asked for more. Then as the United States expanded its use of Honduran territory for its campaigns against the Sandinistas in Nicaragua, the aid figures skyrocketed (see Table 5.1). By 1985, the administration's FY1986 military aid request for Honduras had risen to $88.2 mil-

TABLE 5.1
UNITED STATES ECONOMIC AND MILITARY ASSISTANCE TO HONDURAS
(millions of dollars)

	1978	1979	1980	1981	1982	1983	1984	1985[a]	1986[b]
Economic									
AID	12.4	20.7	45.8	25.7	31.1	31.2	39.5	41.5	45.0
Food for Peace	2.0	4.6	4.9	10.5	10.2	15.5	16.7	22.5	17.9
Total	14.4	25.3	50.7	36.2	41.3	47.7	56.2	64.0	62.9
Economic									
Support Fund	0	0	0	0	36.8	56.0	112.5	75.0	80.0
Military[c]									
FMS	2.5	2.0	3.5	8.4	19.0	9.0	0		0
MAP	0	*	*	0	11.0	38.5	77.5		87.0
IMET	.7	.3	.4	.5	1.3	.8	1.0		1.2
Total	3.2	2.3	3.9	8.9	31.3	48.3	78.5	62.5	88.2
Total	17.6	27.6	54.6	45.1	109.4	152.0	247.2	201.5	231.1

SOURCES: U.S. Department of State, Bureau of Public Affairs, *Foreign Assistance Program: FY1986 Budget and 1985 Supplemental Request*, Special Report No. 128 (Washington, D.C.: Department of State, May 1985), p. 14; Jonathan E. Sanford, "U.S. Foreign Assistance to Central America," mimeographed, Report No. 84-34 F (Washington, D.C.: Congressional Research Service, Library of Congress, March 2, 1984), p. 26; Larry Q. Nowels, "Central American and U.S. Foreign Assistance: Issues for Congress in 1984," Issue Brief No. IB84075 (Washington, D.C.: Congressional Research Service, Library of Congress, updated June 12, 1984), p. 12; U.S. Congress, House, Committee on Foreign Affairs, Subcommittee on Western Hemisphere Affairs, *U.S. Relations with Honduras—1985*, 99th Cong., 1st Sess., 1985, p. 75; U.S. Agency for International Development, *U.S. Overseas Loans and Grants and Assistance from International Organizations: Obligations and Loan Authorizations, July 1, 1945-September 30, 1983*, (Washington, D.C.: AID, 1984), p. 51.

NOTES: * = less than $100,000
[a] = preliminary figure
[b] = administration request
[c] = no inclusion of military support and training provided as a consequence of joint military exercises

[26] *Foreign Assistance Legislation for Fiscal Year 1982*, pt. 7, p. 72, *supra* n. 21.

lion—an increase of over 800 percent in four years—and Honduras had become, in the words of one official of the policy-planning staff in State's Bureau of Inter-American Affairs, "nuestra puta centroamericana"—our Central American whore. In Tegucigalpa in 1985, U.S. military personnel joked openly about the country and derided its independence—"Welcome to the USS Honduras," they would say to a newcomer, "the only U.S. aircraft carrier propelled by bananas." One can only imagine the response of the Hondurans.

Second—and again barring unusual circumstances—most policy makers agree that the United States does not need any additional military bases in Latin America. They are, in fact, prepared to reduce the number held at the present time. When eleven of the fourteen U.S. bases in Panama were closed in 1979, no U.S. national security official argued that U.S. interests had been compromised, and many argued the opposite—that U.S. security had increased by the process of decolonization. Similarly, most policy makers recognize that Guantánamo serves primarily a political rather than a military purpose and that the communications facilities (but not the submarine monitoring facilities) of the eastern Caribbean are obsolete. Because there will never be unanimity on any issue of inter-American relations, it is possible to identify officials who disagree with the latter assertion.[27] Their disagreement, however, is not strenuous.

Honduras and the present crisis in Central America excepted, the small number of U.S. bases in Latin America, their very restricted size and mission, and the modest budget that supports them all argue in favor of one conclusion: a rough consensus has been reached that there is little need for U.S. military bases in Latin America. This consensus is broad. It includes policy makers such as the ultraconservative James Theberge, the Nixon administration's ambassador to Somoza's Nicaragua, and the Reagan administration's ambassador to Pinochet's Chile, who has argued that "access to military facilities in the Caribbean and South America may be useful, but are [sic] by no means vital to the security of either Moscow or Washington."[28] It also includes analysts who are generally identified with the center of the Demo-

---

[27] See, for example, Margaret Daly Hayes, "The Stakes in Central America and U.S. Policy Responses," *AEI Foreign Policy and Defense Review* 4 (1982): 14; *Foreign Assistance Legislation for Fiscal Year 1982*, pt. 1, p. 154, *supra* n. 21.

[28] James D. Theberge, *The Soviet Presence in Latin America* (New York: Crane, Russak and Company, 1974), p. 93.

cratic party; they often warn that the few existing bases expose the United States to potential risks—riots and demonstrations, terrorist attacks, the taking of hostages—without providing commensurate benefits for U.S. security. Historian Richard Millett, for example, told Congress in 1985 that "there are always problems, there are always complications, and down the road we will certainly encounter this in Honduras in a way that I suspect would outweigh any short-range advantage from the presence of U.S. bases."[29]

MILITARY SUPPORT

There is much posturing and some outright prevarication on the issue of military support from Latin American armed forces. Two subjects in particular seem to encourage U.S. policy makers to be disengenuous. The first is their assessment of the fighting capabilities of Latin American militaries. "If a limited war involving the superpowers became a reality, Latin American military forces would be significant assets," is an often-heard assertion.[30] The second is their assessment of the close ties between the U.S. and Latin American militaries, ties that they believe can be exploited in a time of need. One U.S. military officer wrote, for example, that the System of Cooperation among the Air Forces of the Americas, an obscure, semiformal form of U.S. Air Force technical assistance, is "an organizational structure substantially motivated by the mystique of brotherhood and close bonds of friendship among the military aviators of the hemisphere."[31]

No one should accept either assertion as an accurate reflection of policy makers' attitudes toward the role of the Latin American military. In private interviews, U.S. officials agree almost unanimously that military assistance from Latin America has never been particularly important in the achievement of U.S. national security goals.[32]

[29] U.S. Congress, House, Committee on Foreign Affairs, Subcommittee on Western Hemisphere Affairs, *U.S. Relations with Honduras—1985*, 99th Cong., 1st Sess., 1985, p. 44.

[30] See, for example, Robert L. Scheina, "South American Navies: Who Needs Them?" *Proceedings of the U.S. Naval Institute* 104 (February 1978): 65.

[31] Richard B. Goetze, Jr., "Transgovernmental Interaction within the Inter-American System: The System of Cooperation among the Air Forces of the Americas," Ph.D. diss., American University, 1973. The quotation is from the abstract.

[32] What is meant here by the term "military assistance" is the act of fighting alongside or aiding the United States to fight a common adversary. In the past,

To begin with, the Latin American country that actually fights alongside the United States is almost a rarity. In World War II, two Latin American states sent forces abroad. In 1944/1945, Brazil sent a 25,000-man infantry division—the Brazilian Expeditionary Force—to fight in northern Italy as part of the U.S. Fifth Army. The Brazilians saw 229 days of continuous combat, much of it in the mountains during winter, and suffered 3,000 casualties, including 500 men killed. In the process the Brazilians captured 20,000 Germans, including the entire 148th Division, the first German division to surrender in Italy. Brazilian air force and navy units also participated in the fighting. The air force fighter squadron flew missions over Italy and southern Germany, while the navy escorted Allied troops and provided general shipping protection. Nearly 400 Brazilian sailors died when the crusier *Bahia* was sunk while protecting a U.S. troop convoy.

Mexico, the second Latin American country to fight in World War II, declared war on Germany, Japan, and Italy after two of its oil tankers were sunk by U-boats in May 1942. Mexican participation in the war was modest, however. In fact, it was not until early May 1945, four months before V-J Day, that the only Mexican fighting force—the U.S.-trained 201st Fighter Squadron—arrived in the Philippines. This 300-man unit participated in fifty-nine bombing raids against Japanese installations in the Philippines and Formosa. Of much greater significance to the war effort was Mexico's January 1942 agreement with the United States to permit "reciprocal inductions" of Mexican nationals living in the United States and U.S. citizens living in Mexico. Under this agreement, 250,000 Mexican citizens served in U.S. military units during the war, and more than a thousand were killed in combat.[33]

Latin American military officers have helped the United States to achieve its security goals by intervening in domestic politics, but that is a different subject.

[33] John Child, *Unequal Alliance: The Inter-American Military System, 1938-1978* (Boulder, Colo.: Westview Press, 1980), pp. 58-59. For a more detailed history of Mexican participation in World War II, see Joseph Charles Strasser, "Uncooperative Neighbors Become Close Allies: United States-Mexican Relations, 1941-1945," Ph.D. diss., Fletcher School of Law and Diplomacy, Tufts University, 1971, especially pp. 72-99; Donald F. Harrison, "U.S.-Mexican Military Collaboration during World War II," Ph.D. diss., Georgetown University, 1976; R. A. Humphreys, *Latin America and the Second World War*, 2 vols. (London: Athlone, 1981), 1: 119. The standard sources of information concerning the nonnaval aspects of World War II in Latin America are Stetson Conn, gen. ed., *The Western Hemisphere: The Framework of Hemispheric Defense*, vol. 12, pt. 1 of *The United States Army in World War II* (Washington, D.C.: Office of the Chief of Military History, Depart-

Other forms of Latin American participation in World War II were extremely modest, and some analysts suggest that it was for purely selfish purposes. Kane, for example, reports that "the success of the American hemispheric defense effort depended upon the possibility of bribing the Latin American military establishments with weapons" through the Lend-Lease program.[34] For whatever reasons—and there was undoubtedly a combination of reasons that varied from country to country—several Latin American nations agreed to provide the Allies with noncombat assistance during World War II. In November 1940, for example, four months *before* the Lend-Lease Act ended restrictions on arms shipments to Latin America, Cuba reached an agreement with the United States permitting an enlarged perimeter for the Guantánamo naval base and improvements for the airport at Camaguey, from which U.S. B-25 bombers would soon fly antisubmarine missions. After the war began the Cuban navy was placed under the U.S. commander of the Gulf Sea Frontier and given responsibility for helping to protect Allied convoys between Cuba and Port Everglades, Florida. The Cuban navy is credited with destroying one German U-boat. In addition, Mexico opened its ports to ships of any American nation at war and even approved the use of Mexican air bases for use by U.S. forces, provided that ground crews were civilians—in this case, employees of Pan American World Airways. Brazil, Colombia, Ecuador, and the Central American countries also agreed to permit the United States to use air and naval bases. Further away from home, both Chile and Peru granted base rights to the United States, and Washington hastily arranged for the construction of drydocks at Talcuahano and Callao.[35] Today this support may seem modest and, in a sense, it was. But during world War II, all of this noncombat aid from Latin America was considered useful, if not vital, to the U.S. war effort.

With the notable exception of Brazil, neither the United States nor Latin American nations wanted Latin Americans to partici-

ment of the Army, 1964), and Stetson Conn, gen. ed., *The Western Hemisphere: Guarding the United States and Its Outposts*, vol. 12, pt. 2 of *The United States Army in World War II* (Washington, D.C.: Office of the Chief of Military History, Department of the Army, 1964).

[34] William Everett Kane, *Civil Strife in Latin America: A Legal History of U.S. Involvement* (Baltimore: The Johns Hopkins University Press, 1972), p. 146.

[35] Humphreys, *Latin America and the Second World War*, 1: 92; Anthony Whitford Gray, Jr., "The Evolution of United States Naval Policy in Latin America," Ph.D. diss. American University, 1982, pp. 216-238.

pate in combat. From the Latin American perspective the low level of combat participation reflected the widely held belief that World War II was a European or North Atlantic war. Today the many policy makers who refer to Pan American military solidarity often forget the logic that kept much of Latin America neutral during World War I and a modest participant in World War II. From the U.S. perspective the limited Latin American participation in World War II also reflected a U.S. desire to keep Latin American militaries out from underfoot. One candid U.S. War Department memorandum noted that "in attaining our objective, we should concentrate on those countries of the most immediate concern to us. Our objective does *not* comprise expectations on our part of being able to use Latin American forces as effective allies in the war."[36] This attitude remains common among U.S. security officials.

Colombia was the only Latin American country to fight on the side of the United States in Korea. Although the bravery of the 1,000-man Colombian infantry battalion remains a subject of considerable pride for many Colombians, it is generally accepted that Colombian participation was not a reflection of the desire of President Laureano Gómez to assist U.S. forces but of his more pressing need to remove a group of Liberal army officers from the country during the tragic years of *la violencia*.[37] Colombia also sent two frigates to Korea to perform blockade duties and support troops with naval bombardments.

Argentina and Venezuela sent two destroyers each to participate in the 1962 quarantine of Cuba during the missile crisis, and several Latin American countries participated in the OAS peacekeeping force that was established following the U.S. invasion of the Dominican Republic in 1965.[38] Finally, Jamaica and several small eastern Caribbean ministates contributed token forces to the U.S. invasion of Grenada in 1983.

Overall, this low level of assistance seems somewhat surprising in light of the institutional structures linking the U.S. military to the Latin American armed forces. These structures began infor-

[36] Child, *Unequal Alliance*, p. 32.
[37] Robert H. Dix, *Colombia: The Political Dimensions of Change* (New Haven: Yale University Press, 1967), p. 114. On this political issue and on the Colombian battalion in general, see Ernesto Hernández, *Colombia en Corea* (Bogotá: Imprenta de las Fuerzas Armadas, 1953), especially pp. 16-17.
[38] Robert L. Scheina, "Latin American Naval Purpose," *Proceedings of the U.S. Naval Institute* 103 (September 1977): 117-118.

mally in the 1920s with the establishment of the initial U.S. military advisory groups and were given formal recognition in 1942 with the establishment of the Inter-American Defense Board (IADB) to coordinate Latin America's participation in World War II.[39] The IADB subsequently evolved into a part of the Organization of American States. It currently has the formal responsibility for preparing to defend the Western Hemisphere from attack, but no one today appears to take that responsibility any more seriously than they did a generation earlier when, in the minutes of the Standing Liaison Committee for February 24, 1943, a U.S. admiral gave his candid evaluation of IADB participation in the war effort. IADB personnel, he asserted,

> have never served a useful purpose so far as we are concerned. . . . Admiral Johnson (Senior U.S. Navy Delegate to the IADB) comes in every once in a while and asks me what we can give it to do. I have wracked my brain for something for them to do. He has suggested a number of things. . . . I am willing to give them anything we can with the understanding that they simply make suggestions and that we don't have to pay any attention to them.[40]

As Jack Child has noted, "in World War II multilateral approaches were symbolic, token, and political in nature."[41]

The situation has not changed in the intervening years. IADB members approach their formal purpose—"to study and recommend to the Governments of the American States the measures necessary for the collective self-defense of the hemisphere against aggression"—like museum curators, carefully watching over a document called the General Military Plan and its several annexes, which are supposed to govern military cooperation and coordination in the Western Hemisphere but which everyone—absolutely everyone—recognizes as meaningless to U.S. security. Assignment to the Council of Delegates of the IADB, a diplomatic

[39] For a discussion of the early years of the IADB, see Paul Tompkins Hanley, "The Inter-American Defense Board," Ph.D. diss., Stanford University, 1964.

[40] Quoted in Child, *Unequal Alliance*, p. 43.

[41] John Child, "From 'Color' to 'Rainbow': U.S. Strategic Planning for Latin America, 1919-1945," *Journal of Inter-American Studies and World Affairs* 21 (May 1979): 256. A retired army officer and now a professor at American University, Child is unquestionably the best-informed student of inter-American military relations.

post that involves virtually no work whatever, has always been considered a vacation by Latin Americans.

In practice, the present functions of the IADB are to operate the Inter-American Defense College (IADC) at Ft. Leslie McNair in Washington, D.C., and to provide an informal means of communication between Latin American military officials stationed in Washington, on the one hand, and the U.S. military, on the other. An early, somewhat-cryptic assessment by one U.S. military officer was that the IADC is "an organ of growing importance . . . for influencing those who will have the responsibility in the decades that lie ahead of selecting from the available alternatives those that will assure the survival and independence of the Peoples of the Americas."[42] In reality, assignment to the IADC is a sinecure for Latin American military officers and a career dead-end for members of the U.S. military.[43]

This low level of assistance also seems surprising in light of the Inter-American Treaty of Reciprocal Assistance, universally known as the Rio Treaty. Signed in 1947 by the United States and the twenty existing Latin American republics, the Rio Treaty was the first of the series of mutual defense agreements that the United States crafted in the early years of the Cold War. As with all such pacts, the Rio Treaty commits each signatory to come to the defense of any other signatory in the event of an attack by an outside force. Specifically, Article 3 of the Rio Treaty states that "the High Contracting Parties agree that an armed attack by any State against an American State shall be considered as an attack against all of the American States and, consequently, each one of the said Contracting Parties undertakes to assist in meeting the attack. . . ." But, Article 3 continues, "each one of the Contracting Parties may determine the immediate measures which it may individually take" until a Meeting of Consultation that shall take place "in order to agree on the measures which must be taken in case of aggression to assist the victim of the aggression or, in any case, the measures which should be taken for the common defense

[42] Douglas W. Davis, "The Inter-American Defense College: An Assessment of Its Activities," Ph.D. diss. University of Maryland, 1968. The quotation is from the abstract. Colonel Davis was saying, I believe, that it is a good idea to be friends with Latin American military officers because they determine public policy. Davis's dissertation makes clear that the term "survival and independence" is a euphemism for "anti-Communism."

[43] Founded in 1962, the IADC each year provides forty to sixty Latin American colonels and lieutenant colonels with ten-month training courses.

and for the maintenance of the peace and security of the Continent." At the Meeting of Consultation a two-thirds vote is required to take any of a number of specific measures, ranging from the recall of ambassadors to the use of armed force. All of these measures are mandatory "with the sole exception that no State shall be required to use armed force without its consent."[44]

In practice, the primary purpose of the Rio Treaty is to serve, in conjunction with the OAS's Inter-American Committee on Peaceful Settlement, to preserve or restore the peace among Latin American states.[45] This role should not be denigrated, but neither should it be confused with the original purpose of the Rio Treaty—and our concern here—which is to provide a unified front against external attack.[46]

There are two instances of an inter-American force: the 1962 missile crisis in Cuba and the 1965 invasion of the Dominican Republic. Both of these were authorized by the OAS at the insistance of the United States. Although the peacekeeping force in the Dominican Republic was commanded by a Brazilian general and contained units from Brazil, Costa Rica, El Salvador (three staff officers), Honduras, Nicaragua, and Paraguay, there was never any question about the control of this operation. Alongside the 1,743 troops from Latin America were 11,000 U.S. armed forces.[47]

On the rare occasions when a Latin American state has sought to invoke the collective defense (as opposed to the peacekeeping) provisions of the Rio Treaty, U.S. policy has proven to be the deciding factor in determining the response of the Organization of American States. Perhaps the best recent example of this U.S. dominance occurred in 1982, when Argentina sought assistance in

[44] For the full text of the Rio Treaty, see 60 Stat. 1831. The quotations are from Articles 3, 6, and 20; Article 8 contains the list of specific measures to be taken.

[45] The Inter-American Committee on Peaceful Settlement was created in 1970 to replace the Inter-American Peace Committee. OAS peacekeeping functions can be triggered by a meeting called under the auspices of either the Rio Treaty or the Inter-American Committee on Peaceful Settlement.

[46] For a detailed list of Rio Treaty and OAS peacekeeping activities between 1948 and 1973, see U.S. Congress, House, Committee on Foreign Affairs, *Inter-American Relations: A Collection of Documents, Legislation, Descriptions of Inter-American Organizations, and Other Material Pertaining to Inter-American Affairs*, 93rd Cong., 1st Sess., 1973, pp. 48-53.

[47] The Inter-American Peace Force (IAPF) for the Dominican Republic was established by the 10th OAS Meeting of Consultation of Ministers of Foreign Affairs on May 6, 1965. The IAPF replaced the 23,000 U.S. troops that had invaded the island a week earlier but it was only formally activated in late June, long after the end of the fighting. Argentina, Colombia, and Venezuela promised to contribute troops, but this was prevented due to domestic opposition.

repelling a British counterattack in the Falklands/Malvinas War. The boeotian Argentine military government delayed its request for a Meeting of Consultation for nearly three weeks after seizing the islands, waiting to see if the British would respond with force. When Argentina finally made a formal request under the Rio Treaty, the United States tried to prevent the meeting from being called. President Reagan made a much-publicized comment that it would be "advantageous" for the OAS to delay holding the meeting, and the U.S. ambassador to the OAS proposed instead that the organization voice its support of a unilateral U.S. mediation effort: "At a time when Secretary Haig is engaged in efforts to promote a peaceful solution, it seems to my government particularly inappropriate to seek a resolution within the framework of the Rio Treaty."

U.S. objections were ignored, however, and the Permanent Council of the OAS voted 18 to 0 (the U.S. abstained) to convene a Meeting of Consulation immediately. The actual meeting was held in Washington on April 26 to 28, but only after the United States had made it clear that it considered Argentina the aggressor and that any hemispheric response would have to be limited to symbolic support for a sister republic. The Meeting of Consultation passed a resolution supporting Argentine sovereignty over the islands and then adjourned.

In addition to the IADB and the consultations provided by the Rio Treaty the United States solicits cooperation from Latin American militaries through a variety of aid and training programs. Probably the most important of these is the International Military Education and Training (IMET) program, an umbrella name for several training programs operated by the Department of Defense. Tens of thousands of Latin American officers and enlisted personnel have received U.S. training under the IMET program. The United States also conducts joint training exercises with Latin American militaries; an extreme example is the ongoing series of maneuvers with the Honduras army along the Nicaraguan border in the 1980s. Many of these exercises are meant to communicate a political message, and in this sense they are the contemporary equivalent of gunboat diplomacy.

The single exception to this generalization is UNITAS, an annual training exercise involving the U.S. Navy and the South American navies.[48] Each year since 1960 the U.S. Navy has sent a

[48] No one seems to know the origin or the precise meaning of the term UNITAS.

small task force—a few destroyers and a submarine—to circumnavigate South America and engage in exercises with countries along the route. These lengthy (five months in 1984) exercises focus upon antisubmarine warfare tactics, but the relatively rudimentary electronic equipment used by most Latin American navies makes these exercises more of a demonstration of U.S. capabilities and techniques rather than a joint exercise. Prior to the Argentine confrontation with Great Britain in 1982, one observer noted that "the cultural exchanges may be of more value to the South Americans than is their ASW training with the U.S. Navy, for they face virtually no submarine threat."[49] Emphasis on the noncombat aspects of UNITAS may decrease in the future at the request of the Latin American navies, for the sinking of the Argentine cruiser *Belgrano* by a British submarine during the Falklands/Malvinas War indicates the need for something more than a cultural exchange.[50]

In addition to the U.S. interest in military cooperation from Latin America in general, many policy makers believe that the militaries of particular countries contribute disproportionately to U.S. national security. First, of course, is Brazil, but that is most frequently because of the nation's strategic geography, a subject discussed in the chapter on sea lines, rather than its military strength. Brazil is thought to provide a valuable "launching platform" to the Atlantic, although time and technological changes have made the hump of Brazil far less significant today than it was in World War II.[51] On the issue of the Brazilian military's ability to contribute to U.S. security, policy makers hold mixed views. The

---

Scheina offers the improbable suggestion that it might mean United International Antisubmarine Warfare Exercises. Scheina, "Latin American Naval Purpose," pp. 116-119.

[49] On at least two occasions UNITAS has been used to communicate a political message. In 1977, the Carter administration ordered the navy to bypass Chile as an indication of dismay over the Pinochet government's human rights record. In 1981, the Reagan administration launched its rapprochement to General Pinochet by inviting Chilean participation.

[50] In addition to these exercises with the United States, there are a number of fairly regular maneuvers involving inter-Latin American military cooperation. Brazil conducts joint naval exercises with both Argentina and Uruguay, for example, and Panama and Venezuela also exercise together. See Stockholm International Peace Research Institute, *World Armaments and Disarmament: SIPRI Yearbook, 1982* (Cambridge, Mass.: Oelgeschlager, Gunn and Haig, 1982), p. 422.

[51] Margaret Daly Hayes, "Security to the South: U.S. Interests in Latin America," *International Security* 5 (Summer 1980): 144; U.S. Department of State, *Congressional Presentation, Security Assistance Programs, FY1983*, p. 431.

Department of Defense sees Brazil as "particularly important because of its . . . present and growing military capability," while the Department of State fails to include this capability in a list of U.S. interests in the country.[52]

On the subject of the Brazilian military's *willingness* to assist the United States, there is surprising uniformity of opinion, perhaps best captured in 1982 by one of the most competent Latin Americanists in the U.S. Foreign Service, retired ambassador John Crimmins:

> Brazil has shown, certainly publicly, no readiness to enlist in a significant way with us. The current formulators and directors of Brazilian foreign policy are as bent as ever on their standing guidelines that call for the primacy of economic interests, the retention of maximum freedom of action and the avoidance of obligations that are not clearly in Brazil's interests—as that interest is defined solely by Brazil.[53]

Experts also agree on another subject: Brazil is a growing arms producer, manufacturing 60 percent of its own equipment, some of which is quite sophisticated, and a long list of exports to Third World countries. Brazil, in fact, has quietly become the tenth largest arms exporter in the world and the first among Third World exporters of major weapons. In the 1980s, Iraq, Iran, and Libya have been particularly large customers. Success in exporting moderately sophisticated hardware (light tanks, counterinsurgency aircraft) has encouraged Brazilians to raise their sights to the markets held by Northern industrial countries, including the markets for such highly sophisticated items as jet trainers and air-to-air missiles.[54] Other Latin American countries are also emerging as significant arms producers. Several countries (Argentina, Chile, Colombia, Ecuador, Mexico, Peru, and Venezuela) produce major

[52] U.S. Joint Chiefs of Staff, "United States Military Posture for FY1982," printed in U.S. Congress, House, Committee on Armed Services, *Hearings on Military Posture and H.R. 2614 and H.R. 2970*, 97th Cong., 1st Sess., 1981, pt. 1, p. 109; U.S. Department of State, *Congressional Presentation, Security Assistance Programs, FY1983*, p. 431.

[53] U.S. Congress, House, Committee on Foreign Affairs, Subcommittee on Inter-American Affairs, *United States-Brazilian Relations*, 97th Cong., 2d Sess., 1982, p. 39.

[54] For details on the types of exports and their destinations, see Stockholm International Peace Research Institute, *World Armaments and Disarmament*, pp. 405-406; Robert L. Scheina, "Latin American Navies," *Proceedings of the U.S. Naval Institute* 110 (March 1984): 34.

weapons, defined as aircraft, armored vehicles, missiles, and warships. Ten Latin American countries produce small weapons.[55]

The countries on the west coast of South America, particularly Chile, are also thought by some officials to have a significant role in U.S. defense. This role may simply be in providing ship repair and refueling facilities,[56] but it may also include an active combat role. "With its long coast and offshore islands," asserts the Department of State, "Chile is strategically located to protect sea, land and air lines of communications—especially should transit of the Panama Canal be interrupted."[57] It is with nonsense such as this that the Department of State created its reputation for shoddy analysis—there is no significant land or air route that the Chileans are strategically located to protect. As for sea routes, it may be true, as retired General (and Reagan administration official) Vernon Walters remarked in 1981, that the Chilean navy "has a long and distinguished history."[58] But the Chileans' reputation as the finest sailors in Latin America grew out of the War of the Pacific a century ago, which featured minibattles such as Iquique (two Chilean ships), and has remained untested ever since. The Chilean navy, like most Latin American navies, suffers from a lack of equipment. There are only two aircraft carriers in Latin America— one Argentine, one Brazilian—and both are World War II vintage. In an interview in late 1982, the Chilean chief of naval staff, Vice Admiral Poisson, noted that any navy worth its salt has at least one aircraft carrier—"I think every sailor has in the back of his mind the idea of having a carrier-based task group"—but recognized that financial difficulties made such a purchase impossible.[59]

Moreover, General Walters may be correct to assert that "Chile can make a very serious contribution in an area where we do not presently have the means of doing anything."[60] But the reason the United States does not have the means to do anything is because

[55] Stockholm International Peace Research Institute, *World Armaments and Disarmament*, p. 395.

[56] See, for example, Hayes, "Security to the South," p. 144.

[57] U.S. Department of State, *Congressional Presentation, Security Assistance Programs, FY1983*, p. 435.

[58] U.S. Congress, House, Committee on Foreign Affairs, Subcommittees on International Economic Policy and Trade and on Inter-American Relations, *U.S. Economic Sanctions Against Chile*, 97th Cong., 1st Sess., 1981, p. 58.

[59] Quoted in Scheina, "Latin American Navies," p. 34. Used carriers are not too expensive, but the necessary aircraft and operating costs prohibit a purchase.

[60] *U.S. Economic Sanctions Against Chile*, p. 58, *supra* n. 58.

U.S. policy makers have no reason to *want* to do anything. The west coast South America sea lines of communication are almost universally acknowledged to be the least important of those surrounding Latin America. And even if there were a need for the United States to use force off the coast of Chile, there is no evidence to suggest that Washington would ask the Chilean armed forces for assistance. Not one U.S. national security official interviewed for this study expressed any confidence in the Chilean navy's ability to maintain the safety of West Coast sea lines.

What these statements by policy makers such as General Walters and representatives of the Department of State indicate, in fact, is a search for an acceptable reason to reestablish close relations with the repressive Pinochet government. These statements, all of which came during the course of the congressional debates in 1981 and 1982 over a request to reinitiate military and economic aid programs in Chile, are prime examples of how a national security consideration can be employed to manipulate U.S. foreign policy debates. It is with statements such as these that the Reagan administration attempted to dismantle the Carter human rights policy of dissociating the United States from the Pinochet dictatorship. Rather than attempt to convince Congress that the Carter policy was incorrect on its merits, the Reagan administration tried to scare Congress with a national security argument.

A similar use of the same national security argument occurred in 1981/1982, when the Reagan administration wished to end the prohibition on military aid to the third country that is alleged to contribute disproportionately to U.S. national security, Argentina. This time the Department of Defense led the administration's fight on Capitol Hill:

> In many ways Argentine and U.S. security interests coincide in the South Atlantic and the United States could probably expect active Argentine assistance if we are able to develop the state-to-state relations such cooperation would require. If we are not able to demonstrate that the United States is willing to be a reliable security partner, then we will not be able to rely on obtaining Argentine cooperation in our pursuit of all U.S. interests, including those of security as well as human rights.[61]

[61] U.S. Congress, Senate, Committee on Foreign Relations, *Foreign Assistance Authorization for Fiscal Year 1982*, 97th Cong., 1st Sess., 1981, p. 465.

Then the administration told Congress why Argentine military support is important to U.S. security: "Argentina is capable of providing forces which could augment U.S. capability for ocean surveillance, convoy escort and underway replenishment as well as facilities for logistic and maintenance support."[62] Rebuffed in 1981, a year later the administration tried again, using the same argument for ending the military aid restrictions imposed by Congress: the need for Argentine cooperation in "the defense of vital lines of communication in the South Atlantic."[63]

It is unclear to what South Atlantic sea lines the administration was referring. The only vital sea lines in the South Atlantic are those carrying oil from the Persian Gulf around the Cape of Good Hope to Europe and the Caribbean, but the Argentine coast is not only south of the southern tip of Africa but also lies about four thousand miles to the west; in other words, the truly vital South Atlantic sea lines are far out of reach of the Argentine armed forces.[64] The line referred to as "vital" must be that which goes around the southern tip of South America and, as we will see in Chapter 6, most geopolitical analysts and policy makers acknowledge that this route is of relatively little significance to U.S. security interests.[65]

But, accepting for the moment that this route might be important, then the question becomes: *Could* the Argentine navy help defend it from attack? If Argentina's performance in the 1982 Malvinas/Falklands War can be taken as an indicator, surely the answer is no. Against a far less powerful foe than the Soviet Union, the Argentines suffered a humiliating defeat in the very waters they would be expected to defend.

The performance of Argentina's four submarines provides an excellent illustration of why the Argentines are all but helpless in a confrontation with any sophisticated adversary. Of the four, two were ancient U.S. "Guppy" class vessels that had been sold to the Argentines after being declared obsolete by the U.S. Navy. One of these, the *Santiago del Estero*, had even been declared obsolete by the Argentines. Decommissioned in September 1981, the *San-*

[62] Ibid.

[63] U.S. Department of State, *Congressional Presentation, Security Assistance Programs, FY1983*, p. 415.

[64] U.S. Congress, House, Committee on Foreign Affairs, Subcommittee on Western Hemisphere Affairs, *U.S. Policy toward Argentina*, 98th Cong., 1st Sess., 1983, pp. 38-39.

[65] See, for example, Hayes, "Security to the South," p. 144.

tiago del Estero sat in the Mar del Plata navy yard serving as a stationary training vessel. During the war, it was moved to a hideout in Puerto Belgrano in the hope that Great Britain would worry that it had gone on patrol. The second Guppy, the *Santa Fe*, was assigned the task of transporting technicians to reinforce the Argentine unit occupying the South Georgia Islands. As it neared its destination, the *Santa Fe* was spotted by the British, attacked by helicopter gunships, and nearly sunk. The crew nursed the crippled vessel into the port of Grytviken and abandoned ship. Once the British had reestablished control of the islands, they towed the *Santa Fe* into deep water and scuttled her.

The remaining two submarines were of the modern "Type 209" class, designed and built in sections in West Germany, then shipped to Argentina for assembly in the the mid-1970s. In this construction process the submarines became something of a joke. Thus just prior to the initiation of hostilities, one of these ships, the *Salta*, was undergoing major repairs, which were then rushed to completion. In a shakedown cruise in Golfo Nuevo the *Salta* made such excessive noise that she returned to port, and the war ended before she could be fixed. This left only the *San Luis*, which was fully operational but manned by an inexperienced crew. She made three attacks upon what were thought to be British ships—her crew heard rather than saw two of the three targets and simply assumed they were British, not a comforting thought for U.S. Navy personnel who might someday engage in joint defense efforts with the Argentines. Since the *San Luis'* fire control computer was out of service and no one on board knew how to fix it, the captain was forced to rely upon a manual guidance system. As luck would have it, that system did not work either—the wires guiding the torpedos broke. One can only imagine the exasperation of the hapless Argentines, as the torpedos churned around at hull-depth for a few minutes and then exploded, killing only the unfortunate fish that happened to be in the vicinity. Eventually the *San Luis* returned to port and sat out the final days of the conflict.[66]

What this dismal performance suggests is that the acquisition of sophisticated weaponry by relatively unsophisticated armed

---

[66] Robert L. Scheina's fascinating "Where Were Those Argentine Subs?" *Proceedings of the U.S. Naval Institute* 110 (March 1984): 115-120 contains a verbatim transcript of an interview with the skipper of the *San Luis*. See also Robert L. Scheina, "The Malvinas Campaign," *Proceedings of the U.S. Naval Institute* 109 (May 1983): 98-117.

forces reduces warfare to a question of luck. With their Exocet missiles (which are relatively unsophisticated missiles), the Argentines were lucky; with their submarines, fate dictated the opposite result. In either case, one would be justified in worrying if U.S. officials were to rely in any meaningful way upon the fighting capabilities of the Argentine (or, indeed, of any Latin American) armed forces. In the case at hand, an honest assessment of Argentina's military capabilities would be that the nation's military is extremely capable of maintaining internal order through the use of terrorist tactics repugnant to the core values of civilized human beings. Beyond that the bloated Argentine military has plenty of hardware—by 1983 it had replaced its 1982 losses—but none of the skills required of an effective fighting force.

Even if these sea lines were vital to U.S. security, and even if the Argentine military had the capability of defending them, *would* they? There is a breathtaking naïveté in the above assertions by officials of the Reagan administration that "in many ways, Argentina and U.S. security interests coincide" and that we "could probably expect active Argentine assistance" if Congress would only permit the U.S. military to develop friendly relations by repealing the human-rights-related legislation prohibiting military assistance. The fact is that Argentina has *never* pursued a foreign policy that accommodated the United States. From time to time Argentina has cooperated with the United States—by sending ships to participate in the quarantine of Cuba and by providing military intelligence training to Central American governments, for examples—but these are exceptions to the rule of ignoring U.S. urgings while pursuing a policy dictated by Argentine self-interest. This policy became evident at the first Pan American conference in Washington in 1889, is perhaps best characterized by the understandable reluctance to enter World War II until March 27, 1945, hardly a month before Hitler committed suicide, and was most recently exemplified by the refusal to participate in the U.S. grain embargo established in the wake of the Soviet Union's invasion of Afghanistan. Indeed, Argentina took advantage of the U.S. embargo to increase its grain exports to the Soviet Union.

This is not to suggest that Argentina should have adopted different policies but only to argue that it is folly to believe that Argentina will assist the United States in any security crisis. It might, but history indicates that the chances are better that it might not. As Falcoff has argued, "Argentina has its own cultural traditions and history, its own sense of self, its own aspirations for the future, and it is extremely reluctant to allow others—especially Americans, whom Argentines of all political stripes do not particularly like or admire—to dictate, or appear to dictate, their course of action."[67] Argentina's leaders (and those of Brazil, Chile, and almost every other country in Latin America) will do what they believe is best for their own interests, and these interests, as history so clearly demonstrates, should not be expected to coincide with those of the United States.

THE conclusions, then, are two. First and most obvious, Latin American militaries are not very proficient at fighting. As Kane has observed, "the idea that the uncoordinated and inefficient Latin American militaries could defend themselves against conventional armed attack by one of the major powers falls within the realm of the comic."[68] Understandably, therefore, in Pentagon contingency planning Latin American armies and air forces are not given any role other than national defense.[69] It is highly doubtful that the one task assigned to major Latin America states—the protection of sea lines in the Atlantic—could or would be fulfilled, for Latin American navies (with the possible exception of Brazil) have virtually no antisubmarine warfare capabilities.[70]

Second, even if Latin American militaries could fight, it is doubtful that they would fight on the side of the United States. Jordan and Taylor speak for most U.S. national security officials when they conclude that "it is highly unlikely that Brazil or its Spanish-speaking neighbors would enter into a conflict involving NATO and the Warsaw Pact."[71] To many policy makers the reluctance of Latin American militaries to aid the United States is not simply a reflection of the desire to steer clear of a "European" war but rather a reflection of antipathy toward the United States. This

[67] Falcoff's comment is made in *U.S. Policy toward Argentina*, p. 99, *supra* n. 64, where there are similar comments by the respected Argentinist, Gary Wynia, pp. 44, 48. See also Joseph S. Tulchin, "Two to Tango," *Foreign Service Journal* 59 (October 1982): 18-23. For an interesting analysis of Soviet-Argentine relations, see Aldo César Vacs, *Discreet Partners: Argentina and the USSR since 1917* (Pittsburgh: University of Pittsburgh Press, 1984).

[68] Kane, *Civil Strife in Latin America*, p. 161.

[69] Amos A. Jordan and William J. Taylor, Jr., *American National Security: Policy and Process* (Baltimore: The Johns Hopkins University Press, 1981), p. 454.

[70] Scheina, "Latin American Naval Purpose," pp. 116-119; Jorge I. Domínguez, "The United States and Its Regional Security Interests: The Caribbean, Central, and South America," *Daedalus* 109 (Fall 1980): 117.

[71] Jordan and Taylor, *American National Security*, p. 456.

belief, broadly accepted by U.S. officials, has never been captured better than in the words of Ronald Reagan:

> Joint defense . . . brings to mind a picture of friendly allies going forward, shoulder to shoulder, in friendly camaraderie, the Americans voicing, probably, their customary marching chants, such as the well-known "Sound Off. One, two." The Panamanian Guardia Nacional will be chanting the words that they use in their present training. They march to these words: "Muerte al gringo! Gringo abajo! Gringo al paradon!" Translation: "Death to the gringo! Down with the gringo! Gringo to the wall!" Now that should reassure us gringos about the kind of cooperation we might have under the new treaties.[72]

What Mr. Reagan believed in 1978 about the Panamanian National Guard is widely believed today by U.S. policy makers about the Argentines, Brazilians, Chileans, Peruvians, and nearly every other fighting force in the region. Only those armed forces whose very existence is dependent upon continuous infusions of U.S. aid—El Salvador and Honduras are the principal current examples—are believed to be ready to support the United States in a crisis. They, however, are unable to provide anything more than symbolic assistance.

[72] Mr. Reagan's statement is in William F. Buckley, Jr., and Ronald Reagan, "On Voting Yes or No on the Panama Canal Treaties," *National Review* 30 (February 17, 1978): 213-214.

· 6 ·

SEA LINES OF COMMUNICATION

FREEDOM of the seas is among the most enduring and significant principles of United States foreign policy. In times of peace the world's maritime routes—or, to use the term employed by national security officials, sea lines of communication (SLOCs)—belong to no nation; all nations are free to use the sea for military, commercial, or recreational purposes. When international tensions rise, however, the continued use of essential sea lines becomes a fundamental problem of national security. An assistant secretary of defense summarized perfectly the Latin American aspect of this problem when he warned that "radical regimes, working with Cuba or the Soviet Union, greatly complicate the defense of the region, add to instability, and threaten our lines of communication through the Caribbean and the South Atlantic."[1]

The purpose of this chapter is to analyze the significance of the various sea lines surrounding Latin America to the security of the United States. These routes cannot be analyzed as a single unit, for they vary greatly in virtually every respect, including their importance to U.S. security. It is common, therefore, to divide the subject into four parts: the west coast of South America, the South Atlantic, the Caribbean, and the Panama Canal.

Before beginning to discuss each part separately, it should be noted that a fifth part—the west coast of Mexico and Central America—is rarely considered a "Latin American" sea line. There are two reasons for this. First, the economic significance of this route is fairly low. Of the three U.S. trade routes linking the United States and the Pacific coast of Central America and Mexico, two are classified as "nonessential." The third includes all of the west coast from San Diego to Tierra del Fuego. This permits it to be classified as essential not because of the trade from Mexico and Central America but because of the shipment of nonferrous metals and fertilizer from Bolivia, Chile, and Peru.[2] Second, re-

[1] U.S. Congress, House, Committee on Foreign Affairs, *Foreign Assistance Legislation for Fiscal Year 1983*, 97th Cong., 2d Sess., 1982, pt. 6, p. 16.

[2] On this point, see U.S. Department of Transportation, Maritime Administration, *United States Oceanborne Foreign Trade Routes* (Washington, D.C.: Department of Transportation, September 1984), pp. vii-viii, 90. For record-keeping purposes, the U.S. Maritime Administration has identified and numbered sixty-four

sponsibility for the protection of shipping in the area rests with the navy's 3d Fleet, which must provide security for the U.S. West Coast and for the extraordinarily important SLOCs linking the U.S. mainland (including Alaska) to Hawaii and U.S. allies in Asia and the Pacific. In comparison, the route along the west coast of Mexico and Central America is insignificant, and as a result it is such a minor concern of U.S. security officials that the literature, which is not large, simply fails to acknowledge its existence.[3]

THE WEST COAST OF SOUTH AMERICA

Of the Latin American sea lines, those that follow the west coast of South America are the least important to the United States. From the perspective of U.S. economic security, the west coast sea lines do not constitute a major ocean trade route, although they are used to ship several important raw materials to the United States from South America: tin from Bolivia and copper from Chile and Peru are particularly significant. No west coast country is the principal supplier of any raw material to the United States, however, and, with the exception of a very small amount of petroleum from Ecuador and Peru, none of the essential strategic raw materials discussed in Chapter 4 comes from the region.

To U.S. policy makers, then, the primary concern about the security of west coast South America sea lines centers upon their availability for military transport. Here their use is highly specialized. These SLOCs would be essential in the event that the United States needed (a) to move a major component of the navy from the Atlantic to the Pacific or vice versa without (b) sailing through the Indian Ocean and either the Suez Canal or around the Cape of Good Hope. This use of west coast SLOCs is due entirely to the

---

standard U.S. ocean trade routes. The term "essential" refers to commercial rather than security concerns. Section 211 of the Merchant Marine Act of 1936 requires that the president (through the Maritime Administration) designate which routes are essential "to the promotion, development, expansion, and maintenance of the foreign commerce of the United States." The west coast lines linking the United States to Mexico and Central America are numbered 25, 71, and 72. The last two are "nonessential."

[3] The contemporary disregard of these west coast sea lines was not evident during the early years of World War II when U.S. officials feared a Japanese attack upon the Panama Canal and southern California. On this point, see Joseph Charles Strasser, "Uncooperative Neighbors Become Close Allies: United States-Mexican Relations, 1941-1945," Ph.D. diss., Fletcher School of Law and Diplomacy, Tufts University, 1971, pp. 80-81.

supercarrier orientation of the U.S. Navy, which deploys its ships in what are known as Aircraft Carrier Battle Groups (CVBGs). These consist of an aircraft carrier and up to twenty-five support and defense vessels: cruisers, guided missile frigates, tenders, etc. Since all of the navy's carriers[4] are too large for the locks of the Panama Canal, any battle group moving from ocean to ocean must either sail around the world in the opposite direction or temporarily divide: one part transits the canal and then waits about fifteen days while the carrier and, depending upon the perceived threat, perhaps ten of its escorts go around South America.

The interoceanic transfer of U.S. naval units occurs infrequently. The principal reason the United States maintains a four-fleet, multiocean navy is to reduce the amount of time needed to react to a security crisis. By design, therefore, the navy rarely moves a carrier from the 2d Fleet (Atlantic) or the 6th Fleet (Mediterranean) to stations in areas protected by the 3d Fleet (eastern Pacific) or the 7th Fleet (western Pacific). In the entire sixteen-year period from 1953 to 1969, there were a total of thrity-six interoceanic transfers of vessels of the U.S. Navy that could not use the Panama Canal.[5] Even during the Vietnam War, only four ships— the carriers *Annapolis*, *Boxer*, *Enterprise*, and *Independence*—had to avoid the Canal in passing from the Atlantic to the Pacific theaters. Each of these four carriers avoided the west coast South America SLOCs by sailing eastward to Vietnam. In a future crisis in the eastern Atlantic or the Mediterranean, the shortest route for a carrier stationed in the western Pacific would be to reverse this route, sailing westward and avoiding the west coast South America sea lines.[6]

[4] The navy's goal is to have fifteen aircraft carriers by 1990. The *Theodore Roosevelt* became the fourteenth in 1986, and the *Abraham Lincoln* will become the fifteenth in 1989. A final Nimitz-class carrier, the *George Washington*, will replace the aging *Coral Sea* in 1991.

[5] U.S. Atlantic-Pacific Interoceanic Canal Study Commission, *Interoceanic Canal Studies, 1970* (Washington, D.C.: GPO, 1971), annex II, p. 9.

[6] One navy captain has written that the navy has specific reasons to avoid transiting the tip of South America. Merchant ships almost always sail through Chile's Strait of Magellan, which is 230 miles shorter than the Cape Horn route and which avoids the notoriously foul weather and huge icebergs—some up to 20 miles in length—that plague the more southerly route. But the Chilean government requires that a Chilean pilot steer all ships through the Strait, and this author asserts that the U.S. Navy is reluctant to turn its carriers over to these pilots. See Raymond A. Komorowski, "Latin America—An Assessment of U.S. Strategic Interests," *Proceedings of the U.S. Naval Institute* 99 (May 1973): 155. The official navy position is that U.S. ships "use pilots throughout the world," and therefore "the navy has no

In the event that the Panama Canal were closed during a security crisis the west coast South America SLOCs would acquire some additional importance, but not much. Closure of the canal would add about eight thousand miles to an Atlantic-Pacific transit, with the precise distance depending upon destination and departure point. This would require at least fifteen days for ships sailing at twenty knots or better. The added distance and the long period of vulnerability to submarine attack on the high seas would discourage the use of this route for any but the lowest priority cargo.[7] The United States would be far more likely to move essential cargo by air or by land across the isthmus or across the continental United States. Depending upon the contingency, some naval units that would normally pass through the canal might sail around South America, but it is difficult to estimate the size of such a movement without first specifying the precise contingency. To take a worst-case example, if the canal were closed and the United States became involved in a major war in which it had to retreat from the Pacific in order to concentrate all its forces in the defense of the North Atlantic Treaty Organization, then a significant part of the 7th Fleet might use the west coast South America SLOCs as it moved from the western Pacific to the northern Atlantic. If, on the other hand, the United States simply wanted to show some muscle around an eastern Caribbean island, then no naval vessel would need to transit these sea lines.

In short, most analysts and policy makers agree that the strategic importance of west coast South America sea lines is quite modest.[8] This route pales to insignificance when compared to the strategic importance of the other maritime routes surrounding Latin America.[9]

---

policy which would prohibit a carrier from using the Straits." Letter, Office of Information, U.S. Navy, November 14, 1984. In fact, each year navy ships use Chilean pilots as they transit the Strait during UNITAS exercises, which do not involve U.S. aircraft carriers. Although the navy does not maintain records on this subject for long periods, Pentagon officials reported in 1984 that no carrier had transited the Strait of Magellan or the tip of Tierra del Fuego for more than a dozen years.

[7] U.S. Atlantic-Pacific Interoceanic Canal Study Commission, *Interoceanic Canal Studies, 1970*, annex II, p. 4.

[8] See, for example, Margaret Daly Hayes, "Security to the South: U.S. Interests in Latin America," *International Security* 5 (Summer 1980): 144.

[9] To suggest that this maritime route has relatively little importance to U.S. security is not to assert that it is *economically* unimportant. To several South American countries it is of the greatest importance, of course. It is also significant to the Japanese, since many of the raw materials Japan purchases from the United States

The South Atlantic

When policy makers discuss U.S. strategic interests in South Atlantic sea lines, they do not refer to routes running to or along the east coast of South America. The South Atlantic sea lines of concern are those that reach from the Persian Gulf, through the Indian Ocean, around South Africa, and up the Atlantic to the United States and Western Europe. The east coast of South America simply establishes the extreme western boundary of these shipping routes.[10] Through the South Atlantic flows nearly all of the petroleum the United States imports from the Middle East (7.6 percent of total petroleum imports in 1985) and most of the petroleum the United States imports from Africa (18.4 percent of total imports in 1985). A much larger proportion of European oil imports also passes through the same waters.[11]

Historically, there is ample reason to be concerned about the South Atlantic, for the region was a major theater of naval operations during World War II. Early in the war German U-boats inflicted considerable damage on Allied shipping in the region. In response, the United States adopted the technique of convoying shipping and established heavy patrols in the triangle between Trinidad, the Cape Verde Islands, and Cape São Roque. Soon there-

---

are shipped in large bulk carriers that cannot transit the Panama Canal. For example, about forty of these carriers regularly use the Strait of Magellan when carrying coking coal from Hampton Roads, Virginia, to the Far East. On this economic use, see U.S. Congress, House, Committee on Merchant Marine and Fisheries, Subcommittee on the Panama Canal, *Sea-Level Canal Studies*, 95th Cong., 2d Sess., 1978, p. 188; U.S. Congress, Library of Congress, Congressional Research Service, *Current Information on the Republic of Panama, The Panama Canal, and the Panama Canal Zone* (Washington, D.C.: CRS, December 1977), p. 9.

[10] On policy makers' understanding of the petroleum orientation of these SLOCs, see U.S. Congress, House, Committee on Foreign Affairs, *Foreign Assistance Legislation for Fiscal Year 1982*, 97th Cong., 1st Sess., 1981, pt. 2, p. 117; U.S. Congress, House, Committee on International Relations, Subcommittee on Inter-American Affairs, *Arms Trade in the Western Hemisphere*, 95th Cong., 2d Sess., 1978, p. 33; Clarence A. Hill, Jr., "United States Strategic Interests in the Southern Hemisphere and the Need for a Common Maritime Defense," in *Argentine-United States Relations and South Atlantic Security*, ed. Z. Michael Szaz (Washington, D.C.: American Foreign Policy Institute, 1980), p. 3; Robert S. Leiken, "Eastern Winds in Latin America," *Foreign Policy*, no. 42 (Spring 1981), p. 98; Hayes, "Security to the South," p. 140.

[11] American Petroleum Institute, *Basic Petroleum Data Book: Petroleum Industry Statistics* (Washington, D.C.: American Petroleum Institute, May 1986), Section IX, Table 4c. These figures are subject to rapid changes. In 1981, the Middle East provided 28 percent of total U.S. imports, and Africa supplied 30 percent.

after, Admiral Doenitz moved most of his U-boats to the North Atlantic, and the attacks subsided.[12] The South Atlantic maintained its importance thoughout most of the war, however. The United States established a number of military bases in Brazil, including airfields at Natal, Belem, Fortaleza, Recife, Bahia, Rio, and on British-controlled Ascension Island, twelve hundred miles east of Recife. For a period the air route from the United States to the hump of Brazil and then across the Atlantic became a principal link to British and American forces fighting in North Africa and Italy. Naval facilities were also developed in Recife and Bahia, and from 1941 to 1945 the U.S. 4th Fleet was based in Brazil to protect the South Atlantic.

Recalling this historic importance of the South Atlantic, many U.S. policy makers were alarmed when the development of the Soviet blue water navy in the 1960s and 1970s coincided with the establishment of fairly close Soviet relations first with the government of Guinea and then, after Soviet-Guinean relations had cooled, with a Soviet-leaning government in Angola.[13] In 1970, Soviet naval units spent about 200 ship days in the South Atlantic; in 1980, the Soviets spent 2,600 ship days in the region.[14] This Soviet presence in the South Atlantic and in West Africa obviously increased the potential vulnerability of United States and NATO-bound oil shipments and led to a renewed concern over the security of South Atlantic sea lines. As in World War II interest focused upon the northern part of the South Atlantic, specifically upon the Atlantic Narrows, a fourteen-hundred-mile wide stretch of ocean between Cape São Roque and the west coast of Africa. Because this "narrows" covers an immense area, it is extremely difficult to defend.

In the early 1980s, interest focused upon a cooperative defense,

[12] Komorowski, "Latin America—An Assessment of U.S. Strategic Interests," p. 157; Hill, "United States Strategic Interests in the Southern Hemisphere," p. 3.

[13] For a brief period Soviet naval reconnaissance aircraft enjoyed stopover privileges at a Guinean airbase, and Soviet naval units called occasionally at Conakry. In the 1980s, however, the Soviets were no longer sending military aircraft to Guinea, and they never had a base in the country.

[14] U.S. Department of Defense, Annual Report to the Congress, Caspar W. Weinberger, Secretary of Defense, Fiscal Year 1983 (Washington, D.C.: GPO, 1982), p. II-23; U.S. Congress, Senate, Committee on Foreign Relations, Foreign Assistance Authorization for Fiscal Year 1982, 97th Cong., 1st Sess., 1981, p. 465; Foreign Assistance Legislation for Fiscal Year 1982, pt. 7, p. 77, supra n. 10; Arms Trade in the Western Hemisphere, p. 33, supra n. 10; Foreign Assistance Legislation for Fiscal Year 1982, p. 117; Leiken, "Eastern Winds in Latin America," p. 95.

with the United States enlisting the assistance of Argentina, Brazil, and South Africa.[15] From the onset a South Atlantic Treaty Organization that would link together such disparate states as Brazil and South Africa seemed highly improbable, and the discussion of South Atlantic defenses seemed to die not long after the beginning of the Reagan administration. The immediate reasons for its demise were, first, soured Argentine-U.S. relations following the Malvinas/Falklands debacle and, second, inattention from U.S. policy makers, whose eyes were riveted on the Caribbean and Central America.

The ease with which the subject was abandoned suggests that the threat to South Atlantic shipping is not perceived as serious, although there is no uniformity of opinion on this issue. Indeed, in the mid-1980s even the Department of Defense was divided, with the Office of International Security Affairs (ISA) arguing that the threat was minimal and officials from the Office of the Joint Chiefs of Staff arguing that it was substantial.[16] In general, however, in the 1980s there has been only very limited concern among Washington policy makers over the safety of South Atlantic SLOCs; in interviews most officials never mention the subject.

This lack of concern reflects one basic fact: it is difficult to devise a scenario in which the Soviet Union would attack the United States or NATO by sinking crude oil carriers in the South Atlantic. In the first place, everyone agrees that it is unrealistic to forecast a submarine campaign against U.S. shipping anywhere in conditions short of general war. If such a war were to occur, the Soviets would be unlikely to strike in the South Atlantic; rather, they would take the far easier approach of exerting a choke hold at the Strait of Hormus, near the mouth of the Persian Gulf. Through this strait passes as much as half of Europe's and, importantly, an even larger proportion of Japan's petroleum imports.[17] With the

[15] See, for example, U.S. Joint Chiefs of Staff, "United States Military Posture for FY1982," printed in U.S. Congress, House, Committee on Armed Services, *Hearings on Military Posture and H.R. 2614 and H.R. 2970,* 97th Cong., 1st Sess., 1981, pt. 1, p. 109; U.S. Department of State, *Congressional Presentation, Security Assistance Programs, FY1983* (Washington, D.C.: Department of State, 1982), p. 415; *Foreign Assistance Authorization for Fiscal Year 1982,* p. 465, *supra* n. 14; Komorowski, "Latin America—An Assessment of U.S. Strategic Interests," p. 158.

[16] Discussion with Colonel James Gravette, Office of International Security Affairs, Department of Defense, and Colonel Ricardo Flores, Office of the Joint Chiefs of Staff, Department of Defense, Washington, D.C., September 12, 1984.

[17] American Petroleum Institute, *Basic Petroleum Data Book*, Section X, Table 5.

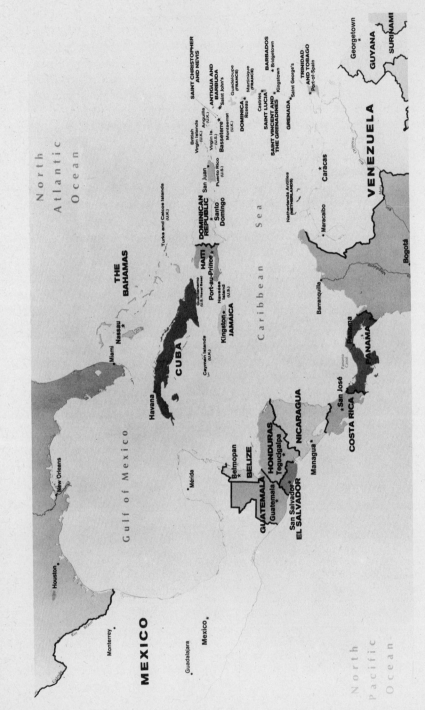

MAP 6.1. The Caribbean Region

Persian Gulf blocked, there would be no need to ferret out each tanker after it rounded the Cape of Good Hope and headed north. If, on the other hand, there were no war and the Soviets simply wished to harass the West, then there is little reason to be concerned about the South Atlantic sea lines; all the Soviets could do there is sink ships and, as a consequence, go to war.

Because no plausible scenario can be created in which the Soviet Union would launch isolated attacks upon U.S. shipping in the South Atlantic, concern about the safety of South Atlantic sea lines is minimal among U.S. national security officials—greater than the level of concern over the west coast South America SLOCs but completely insignificant when compared with the high level of concern over the Caribbean.

THE CARIBBEAN

The father of U.S. naval doctrine, Alfred Thayer Mahan, minced no words about the Caribbean: "One thing is sure. In the Caribbean Sea is the strategical key of two great oceans, the Atlantic and Pacific; our own maritime frontiers."[18] The Caribbean Sea and the Gulf of Mexico together constitute much of the southern border of the United States. Lines of communication reach out from Galveston/Houston, New Orleans, Mobile Bay, and Tampa Bay into the Gulf of Mexico and then, depending upon direction, through at least one narrow passage—the Straits of Florida to the east or the Yucatan Channel to the south—and into the Caribbean or the Atlantic. The Caribbean Sea, in addition, constitutes the eastern approach to the Panama Canal. As one navy officer asserted in 1977, "the huge oceanic oval formed by the Greater and Lesser Antilles contains the strategic choke points through which 13 major trade routes lead directly to the canal."[19] All canal traffic must pass through the Caribbean and at least one narrow passage: the Yucatan Channel, the Windward Passage between Cuba and Haiti, the Mona Passage between Puerto Rico and the Virgin Islands, or be-

[18] Quoted in Ephraim R. McLean, Jr., "The Caribbean—An American Lake," *Proceedings of the U.S. Naval Institute* 67 (July 1941): 947. I have been unable to find Mahan's original statement, but the statement quoted by McLean (now deceased) sounds very much like something Mahan would have said.

[19] Paul B. Ryan, "Canal Diplomacy and U.S. Interests," *Proceedings of the U.S. Naval Institute* 103 (January 1977): 48. The Greater Antilles consist of Cuba, Hispaniola, Jamaica, and Puerto Rico. The Lesser Antilles consist of the chain of islands sweeping southeast from Puerto Rico to the coast of Venezuela and are typically divided at the Martinique Passage into the Leeward and Windward Islands.

199

tween any of the several major passages in the Lesser Antilles that constitute the eastern edge of the Caribbean.[20]

Through these sea lines pass a considerable portion of total U.S. seaborne trade. Of the sixty-four standard U.S. ocean trade routes, five begin or end in the Caribbean, which is defined as the sea area between the northeastern border of Mexico and the southern border of French Guiana. Two of these five are of particular importance. The trade route linking the U.S. east coast and the Caribbean accounted for 15.7 percent of total U.S. oceanborne imports in 1982, while the trade route linking U.S. Gulf ports and the Caribbean accounted for 19.2 percent—together these two routes provide 35 percent of all U.S. oceanborne imports. The importance of these two routes is due primarily to the transport of a strategic raw material: petroleum. Ninety percent of the tonnage on the east coast route and 81 percent of the tonnage on the Gulf route consist of petroleum, most of which comes from Mexico, the second largest (Japan is first) U.S. oceanborne trading partner.

In addition, thirty-three of the remaining fifty-nine maritime routes pass through the Caribbean. This is due largely to the importance of U.S. Gulf ports. In 1982, four Gulf ports (Gramercy, La., Houston, New Orleans, and Baton Rouge) accounted for one-fourth of total U.S. seaborne trade, a figure that rises to over half of all U.S. trade if the lesser Gulf ports (Corpus Christi, Tampa, Texas City, Beaumont, Mobile, Galveston, etc.) are included. Unlike the west coast South America sea lines, there can be no doubt that the Caribbean is a major economic lifeline for the United States.[21]

During the early days of World War II, ships plying the Caribbean trade routes proved to be extraordinarily vulnerable to Axis attacks. From February to September 1942, German U-boats conducted Operation Neuland in the southern Caribbean, concentrating upon the areas around the Netherlands Antilles, where Venezuelan oil was refined and shipped north, and around Trinidad, a

[20] The Windward and the Mona share prominence as the principal passages between the Caribbean and the North Atlantic. The principal routes linking the South Atlantic and the Caribbean are the Galleons Passage south of Grenada and the St. Lucia and St. Vincent passages on either side of St. Lucia. Of these, the twenty-seven-mile-wide St. Vincent Passage is the most heavily used.

[21] U.S. Department of Transportation, *United States Oceanborne Foreign Trade Routes*, pp. 186-187, 209, 212, 288, 292, 302-307; U.S. Department of Transportation, Maritime Administration, *U.S. Oceanborne Foreign Trade Routes* (Washington, D.C.: Department of Transportation, April 1983), pp. 184, 321.

key shipping center.[22] In 1942, German submarines sunk 336 Allied ships in the Caribbean and Gulf of Mexico. The United States soon became proficient at protective measures, however, especially by introducing the system of convoying coastal shipping, and the amount of damage inflicted by submarines in the Caribbean decreased dramatically. By early 1943, the worst was over, and by August 1943, all but a few U-boats had been withdrawn from the Caribbean. In 1943, the number of Allied ships destroyed in the Caribbean dropped to 35, and in 1944 the number was 3.[23] Despite the success of defensive measures Hitler's navy clearly demonstrated the vulnerability of Caribbean maritime routes.

Many policy makers believe that the situation would be even worse today. In 1983, Secretary of State Shultz told an audience that the Soviet-Cuban alliance makes the Caribbean more perilous than during the heyday of Nazism, for in the 1940s, "Germany had no bases in the Caribbean, not even access to ports for fuel and supplies."[24] Similarly, President Reagan argued that "Cuba is home to a Soviet combat brigade, a submarine base capable of servicing Soviet subs, and military air bases visited regularly by Soviet military aircraft that control [patrol?] our shores. If the Nazis during World War II and the Soviets today have recognized that the Caribbean and Central America are vital to our interests, don't you think that it's about time that we recognized that, too?"[25]

So much has changed since the mid-1940s, however, that many

[22] Trinidad was one of the world's largest shipping centers during World War II, with the Gulf of Paria handling traffic between the United States and the east coast of South America. Bauxite supplies from the Guianas were of particular importance, but the U.S. Great Lakes ore carriers that were assigned to the route during the war could not navigate their shallow rivers. To solve the problem, Port of Spain became a transshipment point, with smaller ships ferrying ore from the mines. See Samuel Eliot Morison, *History of United States Naval Operations in World War II*, vol. 1, *The Battle of the Atlantic, September 1939-May 1943* (Boston: Little, Brown, 1947), pp. 144-146.

[23] Stetson Conn, gen. ed., *The Western Hemisphere: Guarding the United States and Its Outposts*, vol. 12, pt. 2 of *The United States Army in World War II* (Washington, D.C.: Office of the Chief of Military History, Department of the Army, 1964), pp. 431, 437.

The number of U-boats in the Caribbean was small, and their effectiveness was reduced by their limited fuel supply. Some German submarine tankers were stationed in the area for a brief period in 1942. See Karl Doenitz, *Memoirs*, trans. R. H. Stevens (London: Weidenfeld and Nicolson, 1959), p. 219.

[24] Speech to the Dallas World Affairs Council and the Chamber of Commerce, Dallas, Texas, April 15, 1983.

[25] *Weekly Compilation of Presidential Documents* 19 (July 25, 1983): 1,011.

policy makers no longer accept the experience of World War II as a sufficient guide to policy. It is true that some officials continue to assert the importance of the Caribbean on the basis of geographic proximity; because it is close to home, they argue, it is important. To David Jordan, for example, "it is an elementary fact of geopolitics that the sealanes of the Caribbean are indispensable to the U.S. military and economic position."[26] To others, however, geographic proximity may remain an elementary fact, but it is a fact that requires reinterpretation in our era of intercontinental missiles and nuclear submarines.

Nearly all U.S. officals agree, however, that it is extremely important for the United States to be free to utilize the Caribbean. Two specific uses are generally cited for this importance. First, nearly all policy makers recognize that the United States transports a significant proportion of its imports, especially petroleum, through the Caribbean. At the 1982 unveiling of his Caribbean Basin Initiative, President Reagan noted that "the Caribbean region is a vital strategic and commercial artery for the United States." In particular, he added, "nearly half of our trade, two-thirds of our imported oil, and over half of our imported strategic materials pass through the Panama Canal or the Gulf of Mexico."[27]

[26] David C. Jordan, "U.S. Options—and Illusions—in Central America," *Strategic Review* 12 (Spring 1984): 56. A political scientist, Jordan was the Reagan administration's ambassador to Peru.

[27] *Weekly Compilation of Presidential Documents* 18 (March 1, 1982): 219. The Reagan administration had considerable difficulty deciding exactly how much the United States depended upon Caribbean sea lines. On another occasion in 1983, President Reagan reversed his figures and said that "two-thirds of our foreign trade and nearly half of our petroleum pass through the Caribbean." *Weekly Compilation of Presidential Documents* 19 (July 25, 1983): 1,011. New numbers appeared in the President's speeches in 1984: "half" and "nearly half" were used in the same *sentence* of one speech. *Weekly Compilation of Presidential Documents* 20 (May 14, 1984): 677. In the specific case of petroleum imports, the President tended to vacillate among "nearly half," "half," "more than half," and "two-thirds." *Weekly Compilation of Presidential Documents* 19 (March 28, 1983): 445; 19 (May 2, 1983), p. 608; 19 (July 25, 1983), p. 1,011; 20 (May 14, 1984), p. 677; 22 (June 30, 1986), p. 866. Assistant Secretary of State Thomas Enders settled upon "nearly one-half." See U.S. Congress, House, Committee on Foreign Affairs, Subcommittees on International Security and Scientific Affairs and on Inter-American Affairs, *U.S. Arms Transfer Policy in Latin America*, 97th Cong., 1st Sess., 1981, pp. 5-7. Meanwhile, in one forty-eight-hour period, Secretary of State Shultz vacillated between "three-quarters" and "45 percent." Compare his statement before the Senate Committee on Finance, April 13, 1983, with his speech to the Dallas World Affairs Council and the Chamber of Commerce, Dallas, Texas, April 15, 1983. In general,

Second, concern over the safety of Caribbean sea routes has also focused upon the need to resupply Europe during a national security crisis. In 1982, the Department of Defense noted that "in wartime, half of NATO's supplies would transit by sea from Gulf ports through the Florida Straits and onward to Europe. Much of the petroleum shipments and important reinforcements destined for U.S. forces in Europe would also sail from Gulf ports. The security of our maritime operations in the Caribbean, hence, is critical to the security of the Atlantic Alliance."[28] Jeane Kirkpatrick used the term "most" rather than the DOD's "much": "The Soviet Union understood the strategic importance of the region, through which most sealanes pass with most of the oil and other strategic materials the United States would supply to Europe in case of an emergency."[29] Secretary of State Shultz was able to quantify NATO's dependence on these sea lanes: "In a European war, 65 percent of our mobilization requirements would go by sea from gulf ports through the Florida Straits and onward to Europe."[30]

Any evaluation of the strategic importance of Caribbean sea lines must begin by recognizing their unquestionable economic significance. These are major trade routes. The U.S. economy would suffer, and some of it would suffer substantially, if the Caribbean were closed even briefly to U.S. shipping. Nevertheless, the

---

the Department of Defense preferred the lower figure of 45 percent, while Mr. Reagan's most conservative supporters used the higher figure of three-quarters. Contrast, for example, U.S. Department of Defense, *Annual Report to the Congress, Caspar W. Weinberger, Secretary of Defense, Fiscal Year 1983*, p. II-23, with "Soviet Penetration of the Caribbean," mimeographed, Heritage Foundation National Security Record No. 22 June 1980, p. 1; and L. Francis Bouchey, "Reagan Policy: Global Chess or Local Crap Shooting?" *Caribbean Review* 11 (Spring 1982): 20.

[28] U.S. Department of Defense, *Annual Report to the Congress, Caspar W. Weinberger, Secretary of Defense, Fiscal Year 1983*, p. II-23.

[29] Speech to the Royal Institute for International Affairs, London, April 9, 1984.

[30] Speech to the Dallas World Affairs Council and the Chamber of Commerce, Dallas, Texas, April 15, 1983. For the views of other policy makers, see U.S. Joint Chiefs of Staff, "United States Military Posture for FY1982," p. 109; Mickey Edwards, "Soviet Expansion and Control of the Sea-Lanes," *Proceedings of the U.S. Naval Institute* 106 (September 1980): 50; Ernest Graves, "U.S. Policy toward Central America and the Caribbean," in *The Central American Crisis: Policy Perspectives*, ed. Abraham F. Lowenthal and Samuel F. Wells, Jr., Working Paper No. 119, Latin American Program, The Wilson Center, Washington, D.C., 1982, p. 17; Jack L. Roberts, "The Growing Soviet Naval Presence in the Caribbean: Its Politico-Military Impact upon the United States," *Naval War College Review* 23 (June 1971): 33.

analysis of the *national security* importance of the Caribbean sea lines is not equivalent to the much broader discussion of the convenience, efficiency, profitability, or desirability of using the Caribbean as a trade route. The analysis of vital national security concerns is necessarily narrower. In the event of a closure U.S. security officials face two principal problems: the protection of petroleum shipments and the need to resupply Europe during a crisis.

Protecting Petroleum Shipments

The discussion of protecting petroleum supplies passing through the Caribbean is much more complex than might be imagined. It is even difficult to estimate the size of the flow. The higher estimates used by policy makers in the 1980s were invariably off the mark, often by a fairly wide margin. The figure of "three-quarters of our imported oil" that was cited by Secretary Shultz and others is simply wrong, and the lower figures—44 percent was the lowest used by a member of the Reagan administration—are accurate only if part of the oil from Alaska is classified as "imported" and if the short trip for Mexican oil across the Gulf of Mexico is considered a trip through the Caribbean which, of course, it is not. Overall, however, the importance of the Caribbean to the U.S. petroleum trade cannot be denied. Slightly more than half of all U.S. petroleum "imports" (in 1983, the exact figure was 53.6 percent) are either loaded or discharged (or both) in the Gulf of Mexico or the Caribbean.[31]

The interesting point, however, is that any of these figures could

[31] In making its calculations, the U.S. Energy Information Administration considers oil shipped from Alaska to Puerto Rico and the Virgin Islands to have been exported. Any petroleum that is shipped from these two sites to the United States (and virtually all of the Virgin Islands' production is shipped to the U.S. east coast) is therefore considered to have been imported. U.S. Congress, Senate, Committee on Foreign Relations, Subcommittee on East Asian and Pacific Affairs, *Export of Alaskan Crude Oil*, 98th Cong., 1st Sess., 1983, p. 386.

This method of calculation causes some strange figures to appear in policy makers' speeches. It permits the claim, for example, that the Virgin Islands supplies 4.9 percent of total U.S. petroleum imports, whereas in reality there is not a single oil well in the Virgin Islands and the refinery there operates almost exclusively on oil from Alaska, which cannot logically export anything to the United States since it is a part of the country. See American Petroleum Institute, *Basic Petroleum Data Book*, Section IX, Table 8b.

easily be much lower. They have become as high as they are for peculiar reasons, two of which merit consideration here.

The first concerns petroleum shipped to U.S. Gulf ports from Alaska's North Slope oil fields. From 1969 to 1973, a bitter controversy occurred in Congress over the request by a consortium of oil companies for permission to build a forty-eight-inch diameter pipeline from the North Slope, whose ports are icebound much of the year, across the fragile Alaskan wilderness to the ice-free southern port of Valdez. The pipeline's proponents emerged victorious over concerned environmentalists and Midwest consumer groups, who preferred a route through Canada. Worried about a recurrence of the Arab oil embargo, Congress opted for the Alaska route because it could be completed several years before a route across Canada. To ensure that U.S. citizens benefited from the pipeline, Congress added a provision—Subsection 28(u)—to the authorizing legislation that effectively prohibited the export of Alaskan crude oil. This provision greatly disturbed the oil companies that were developing the North Slope fields. It particularly upset officials of Standard Oil of Ohio, one of the principal developers, whose distribution outlets are primarily in the Midwest.[32]

In any event, when Alaskan oil began flowing through the $7.7 billion pipeline in June 1977, its direction was toward the West Coast of the United States, where about half of it was refined and consumed.[33] A number of distribution routes to other parts of the United States were explored—reversal of the Trans-Mountain pipeline that had been bringing Alberta crude oil to the Pacific Northwest, construction of a new pipeline across the northern United States, conversion of a natural gas pipeline linking Texas and southern California—but for one reason or another (expense, delays in obtaining government approval, and, most important, potential damage to the environment) these were all rejected in fa-

[32] The basic purpose of this extraordinarily complex legislation is to permit a pipeline to be built across federal lands in Alaska. The Trans-Alaska Pipeline Authorization Act (Title II of PL93-153) is the specific legislation giving permission. But this legislation does not stand alone; rather, it is a lengthy amendment to the Mineral Leasing Act of 1920 (PL66-146). Section 28 of PL66-146 authorizes the use of federal lands for pipelines. Title I of PL93-153 added a new subsection, (u), to Section 28. Subsection 28(u) restricts and regulates in several ways (but does not formally prohibit) the export of Alaska pipeline oil. The practical effect of Subsection 28(u) is to make it extremely difficult but technically not absolutely impossible to export Alaska pipeline oil. As of 1986, no Alaska pipeline petroleum had ever been exported.

[33] *Export of Alaskan Crude Oil*, pp. 175, 383, *supra* n. 31.

vor of a longer route. Using both very large crude carriers (VLCCs—supertankers) and smaller tankers, the petroleum is sent from Valdez down the Pacific coast to Panama. Because supertankers, like aircraft carriers, do not fit into the canal's locks or into most U.S. Gulf ports, the oil is transshipped on smaller vessels to complete the final leg of the voyage to Gulf ports. From there it is refined or shipped as crude by pipeline to refineries in the East and Midwest. Since 1980, most of this oil has bypassed the canal and moved across the isthmus of Panama in a newly constructed pipeline, which lowered shipping costs by about fifty cents per barrel.[34] Whether by pipeline or canal, almost one-half of Alaska North Slope (ANS) oil passes through Panama into the Caribbean. An additional 10 percent of ANS crude moves by VLCC around Cape Horn to the Ameranda Hess refinery in the Virgin Islands, where it is converted to fuel oil for the U.S. East Coast market.[35]

If Congress had not required that North Slope oil be consumed in the United States, there are two reasons why little or no ANS oil would pass through the Caribbean. Geography is the first reason: if a free market existed, Alaskan oil now sent to U.S. Gulf ports would instead be sent to the nearest large consumer, Japan, for it is 3,400 miles from Valdez to Yokahama, and 6,500 miles from Valdez to Galveston via the Panama Canal or 6,100 via the Panama pipeline.[36] For their domestic clients the major oil companies would prefer to use the revenues from a sale to Japan to purchase oil located closer to home.[37] This type of exchange occurs daily in the highly sophisticated international petroleum market. Since 1976, for example, the Soviet Union has regularly fulfilled part of its agreement to supply Cuba with petroleum by arranging a swap with Venezuela. The Soviet Union delivers its petroleum to Venezuela's customers in Europe, and Venezuela delivers an equivalent amount of oil to Cuba. Similar arrangements were planned by the companies that developed Alaska's North Slope because it would reduce shipping costs and thereby lower con-

[34] *Sea-Level Canal Studies*, p. 275, *supra* n. 9. The eighty-one-mile Trans-Panama Pipeline is a cooperative venture between the government of Panama and two U.S. construction firms, Chicago Bridge and Iron Industries and Northville Industries. The line runs across Panama along the Costa Rican border, from Puerto Armuelles to Chiriqui Grande.

[35] *Export of Alaskan Crude Oil*, pp. 175, 377, *supra* n. 31.

[36] Ibid., pp. 246, 460.

[37] U.S. Congress, Senate, Committee on Interior and Insular Affairs, *The Trans-Alaska Pipeline and West Coast Petroleum Supply, 1977-1982*, 93rd Cong., 2d Sess., 1974, pp. 20-22.

sumers' costs and/or increase producers' profits; that is, a swap would make sense in the 1980s, when there is enough oil for everyone. Subsection 28(u) was written in 1973, however, when there was not enough oil for everyone and U.S. legislators wanted to ensure domestic consumption of ANS petroleum.

The second reason why ANS oil would not use the Caribbean if government regulation did not require domestic consumption is Section 27 of the Merchant Marine Act of 1920, the "Jones Act." This law reserves the U.S. intercoastal shipping trade for ships built and registered in the United States and owned and operated by U.S. citizens. ANS oil shipped to Gulf ports must be transported in Jones Act ships, the operators of which have used their monopoly position to maintain an overpriced and inefficient means of transportation. One 1983 study estimated that the cost of using Jones Act ships to transport ANS oil is twice that of using foreign-flag carriers; another study estimated that the cost of sending one barrel of petroluem from Valdez around Cape Horn to the Virgin Islands in a U.S.-flag VLCC is $1.98; in a foreign-flag VLCC, the cost would be $.81.[38]

The combined effect of these two features of the ANS oil trade (distance to Gulf ports and inefficiency of U.S.-flag carriers) is to make the use of this Caribbean trade route economically unthinkable. Estimates vary, but it is at least five times more costly to ship ANS oil to the U.S. Gulf as it is to ship it to Japan. One 1983 study concluded that the cost of moving one barrel of ANS oil to Japan would be fifty cents, while the cost of moving the same barrel to Galveston is $5.25.[39] This enormous shipping differential, which reflects little more than interest-group lobbying in Congress, has the effects of lowering oil company profits and government tax revenues (the latter by an estimated $8.5 billion over twenty years), while raising the costs to U.S. consumers.[40] In comparison to the estimated $4.00 to $5.25 cost of shipping ANS oil to U.S. Gulf ports, it costs twenty-five cents to move a barrel of Mexican

[38] *Export of Alaskan Crude Oil*, p. 155, *supra* n. 31; U.S. Congress, Senate, Committee on Energy and Natural Resources, *Sohio Crude Oil Pipeline*, 96th Cong., 1st Sess., 1979, p. 34. For a contrary view emphasizing the positive aspects of the U.S. Merchant Marine, see L. Edgar Prina, "The Merchant Marine and National Defense," *Seapower* 27 (August 1984): 67, 70.

[39] *Export of Alaskan Crude Oil*, p. 143, *supra* n. 31.

[40] U.S. Congress, Senate, Committee on Banking, Housing, and Urban Affairs, Subcommittee on International Finance, *U.S. Export Control Policy and Extension of the Export Administration Act, Part III*, 96th Cong., 1st Sess., 1979, p. 164.

oil, $1.04 to move a barrel of North Sea oil, and $1.60 to move a barrel of Persian Gulf oil to U.S. Gulf ports.[41] The use of Caribbean SLOCs for Alaskan oil is an affront to common sense.

Our purpose here, however, is not to become indignant but to note that the use of the Caribbean to transport ANC oil adds *unnecessarily* to the security vulnerability of Caribbean sea lines. If the congressional requirement of domestic consumption did not exist, the multinational oil firms would resolve this part of our national security problem in a month.[42] The ANS oil would have to be replaced by petroleum from another source, of course, and thus the security vulnerability would not end with the export of ANS oil to Japan. But the ANS-Gulf route is particularly insecure—a long Pacific voyage, transshipment facilities, and a pipeline through the Central American jungle—when contrasted with alternative sources, particularly Mexico.

There is yet another reason so much imported oil passes through the Caribbean. This reason, too, concerns the relationship between government regulation, on the one hand, and the economics of petroleum distribution, on the other. In this case, however, the issue involves two factors other than interest-group lobbying: the natural depth of U.S. coastal waters and environmental protection.

Most of the oil exported to the United States from the Middle East and Africa leaves those areas in supertankers, the behemoths that begin at about 250,000 deadweight tons (DWT) and are as large as 550,000 DWT. These VLCCs are at least 1,100 feet long, 170 feet wide, and—most important for us—when fully laden they sit as deep as 93 feet into the water. Supertankers tend to be unwieldy, particularly in close quarters and rough seas. Large leaks caused by accidents can be unmitigated disasters, as Europeans discovered in March 1978, when the *Amoco Cadiz* snapped in two on rocks in heavy seas off the coast of Brittany, spilling 1.2 million barrels of oil into the sea. The resulting oil slick covered six

[41] *Sohio Crude Oil Pipeline*, p. 35, *supra* n. 38; *Export of Alaskan Crude Oil*, p. 156, *supra* n. 31.

[42] There have been several unsuccessful attempts to repeal Subsection 28(u). Failure has been due to controversies over unrelated issues included in the repeal legislation. In 1984, for example, a House-Senate conference committee met fourteen times between April and October in an attempt to reach agreement over a proposed ban on exports to South Africa. When no agreement was possible, the bill, which contained a repeal of Subsection 28(u), died. See *Department of State Bulletin* 83 (September 1983): 57-59; *Congressional Quarterly Almanac* 39 (1983): 253-257; *Congressional Quarterly Almanac* 40 (1984): 169-170.

hundred square miles of the English Channel—in spots as much as a foot thick—blackened seventy miles of the Breton coast, and caused severe damage to the region's tourist, fishing, and oyster industries. Yet despite these repeated accidents, supertankers are widely used because they provide a much more economical means of transporting petroleum than the standard-sized crude carrier. Since 1966, when the first VLCC began service, supertankers have come to represent over 50 percent of the world's total tanker capacity.[43]

For reasons related primarily to the protection of the environment the United States has been slow to develop superports to accommodate VLCCs. In fact, no U.S. East Coast port can accommodate tankers exceeding 60,000 DWT, not because the tankers are too large, but because they sit too deep in the water. There is only one place on the U.S. East Coast where the continental shelf is not too shallow to permit VLCCs, and the citizens of that place—Palm Beach—have decided not to zone their ocean frontage for use as a petroleum depot. In the Gulf, where ports are also too shallow to permit access for the largest VLCCs, offshore facilities have been constructed in Louisiana and Texas. The Louisiana facility (LOOP) was the first; it became operative in 1981, fifteen years after it was needed.[44]

While delays (caused largely by concern over environmental damage) kept these ports from being constructed and other ports from being dredged to accommodate VLCCs, the oil industry took advantage of the nearby location of several Caribbean islands, where the drive for economic development and the need to reduce acute poverty are often viewed by local officials as sufficient reasons to override environmental concerns. These islands stand directly between the South Atlantic sea lines and the United States, thereby providing an ideal location for supertankers carrying crude oil from the Middle East and from parts of Africa. Refineries and transhipment facilities in the Bahamas, Puerto Rico, the Virgin Islands, the Netherlands Antilles (Aruba and Curacao), and Trinidad have been constructed or enlarged since the advent of supertankers. These refineries and transshipment sites reduce the need for U.S. superports.[45]

[43] Thomas S. Wyman, "Petroleum Imports and National Security," *Proceedings of the U.S. Naval Institute* 103 (September 1977): 34.

[44] Thomas D. Anderson, *Geopolitics of the Caribbean: Ministates in a Wider World* (New York: Praeger, 1984), p. 117.

[45] U.S. Central Intelligence Agency, *Major Petroleum Refining Centers for Ex-*

There are two types of refineries in the Caribbean. Older refineries such as those in the Netherlands Antilles, Trinidad, and Venezuela were constructed to refine local crude, while newer refineries, particularly in the Bahamas and the Virgin Islands, were constructed to convert Middle Eastern, African, and (to a much lesser extent) Venezuelan and Alaskan crude into fuel oil for the U.S. East Coast market. With the exception of Venezuela and Trinidad, all of the refineries rely completely on imported crude. Overall, about 80 percent of the petroleum refined in the Caribbean is shipped to the United States, most of it in the form of fuel oil.

In sum, in contrast to petroleum deposits in the Middle East, Africa, and the North Sea, or to mineral deposits in southern Africa, the use of Caribbean SLOCs for U.S. petroleum imports is a man-made problem for U.S. security. The concentration of U.S. refineries along the Gulf coast, the development of VLCCs, the failure to develop superports on the U.S. East Coast, the decision not to export ANS oil—all these acts have combined to determine the importance of the Caribbean to the U.S. petroleum industry. Today about 35 percent of U.S. petroleum imports *must* pass through the Caribbean region; this is the amount provided by Venezuela (10 percent), Trinidad (3 percent), and Mexico (22 percent). If the use of the Caribbean region for Mexican petroleum were eliminated by the construction of a pipeline, that would leave 13 percent of all U.S. crude petroleum imports to be shipped through the Caribbean—not an insignificant amount but far less worrisome than the figure of "three-quarters of our imported oil" used by President Reagan and Secretary of State Shultz.

It would be a mistake, however, to emphasize too strongly the "man-made" quality of the Caribbean petroleum security problem. Looking back at an earlier era when Texas wells supplied a remarkable proportion of domestic crude oil and Venezuela supplied almost all U.S. import needs, it is easy to understand why the U.S. refinery industry, including the U.S. pipeline network, came to be centered along the Gulf coast. Today this geographic fact must be accepted, for the political and economic costs of restructuring the oil industry, perhaps the most complex aspect of the U.S. economy, are so large as to be incalculable. Even a naive po-

port, Report No. ER 77-10140 (Washington, D.C.: CIA, April 1977), pp. 2-8, 34; Lakdasa Wijetilleke and Anthony J. Ody, "World Refinery Industry: Need for Restructuring," World Bank Technical Paper No. 32 (Washington, D.C.: World Bank, 1984), pp. 29, 32-33; Wyman, "Petroleum Imports and National Security," p. 34.

litical scientist can recognize, for example, that the residents of Palm Beach are not going to rezone their oceanfront to accommodate VLCCs, nor are oil company executives going to find a set of islands that are more conveniently located between the Middle Eastern and African oilfields and U.S. consumers. In short, supplying the United States with petroleum creates security vulnerabilities in the Caribbean that are real and unchangeable. To argue that these vulnerabilities are man-made is not to lessen the vulnerability but only to suggest that some marginal adjustments (for example, selling ANS oil to Japan) are possible. After all these adjustments have been made, however, a substantial vulnerability will remain.

Resupplying Europe

U.S. security officials plan to use Caribbean sea lines to resupply Europe during a crisis. During noncrisis periods, about 23 percent of U.S. oceanborne exports leave U.S. Gulf and South Atlantic ports for northern and southern Europe.[46] A European security crisis would surely involve a substantial increase in this percentage. It is therefore difficult to dispute such statements as: "in wartime, half of NATO's supplies would transit by sea from Gulf ports" or "in a European war, 65 percent of our mobilization requirements would go by sea from Gulf ports." But it is also difficult to confirm these figures. The difficulty lies in predicting needs in Europe during a war. What kind of war? A Soviet-sponsored guerrilla insurgency by Armenians against the Turks, or a large-scale Soviet offensive against West Germany? Fought for what period of time? With nuclear or conventional weapons? At sea as well as on land? Unless the terms of the conflict can be identified, then NATO's needs can only be based upon conjecture.

Let us accept the figure of 65 percent of U.S. mobilization requirements, however, because we know that the figure, whatever it might be, is large. The question then arises, why would U.S. national security officials arrange to move 65 percent of U.S. mobilization requirements in wartime through the ninety-mile-wide Straits of Florida, which is bounded on one side by Cuba, an ally of the Soviet Union? If the Caribbean is as perilous as many policy makers claim in their speeches, it seems all but incredible that the

---

[46] The figure of 23 percent is obtained by combining the total exports from Trade Route 13 and Trade Route 21. U.S. Department of Transportation, *United States Oceanborne Foreign Trade Routes*, pp. 211-212.

United States would *plan* to forego using to a greater extent its superb East Coast ports that front directly on the Atlantic Ocean.

The unspoken fact is that U.S. national security officials have planned to use the Caribbean and U.S. Gulf ports so heavily because they are safe. They are safe, first, because it is impossible to devise a credible scenario today that repeats the experience of 1942/1943. Who is going to attack U.S. Caribbean shipping? Certainly not the Cubans, for as William LeoGrande observes, "a concerted Cuban attack on U.S. shipping would be an act of war and would surely call forth massive U.S. retaliatory strikes. Such a maneuver by the Cubans would, therefore, be pointless unless it came at the behest of Moscow as part of an ongoing general war between the superpowers. Even then it would be suicidal for Cuba. Would the Cubans willingly commit suicide in order to sink a few freighters?"[47] As for the Soviet Union, what is the probability that they would select the Caribbean or the Gulf of Mexico to attack U.S. supply lines during a European war? Such an attack would unquestionably expand the boundaries of any European conflict, an escalation we must assume both Washington and Moscow would be trying desperately to avoid. And the attack would come at great risk—at suicidal risk—to the Soviet participants, presumably submarines, far from home and confined by narrow straits to the Gulf of Mexico and the Caribbean.

Second, even if a credible attack scenario can be devised, the United States is fully capable of protecting shipping in the Caribbean during a war in Europe. The United States was caught unprepared for German U-boats at the beginning of World War II, but the U.S. Navy learned quickly and soon cleared the Caribbean of hostile vessels. The navy is no longer unprepared for antisubmarine warfare. Using the enormous technological advances that have occurred in antisubmarine warfare and the practical lessons learned in more than two decades of preparing for Cuban-based hostilities, today the U.S. military can locate and destroy any hostile submarine in the region with both precision and speed. The navy is particularly well-prepared to protect petroleum shipping to Gulf ports. There is really only one basic route from the Middle East and Africa through the Caribbean—passing on either side of St. Lucia—and it can be defended far easier than the Persian Gulf, the

[47] William M. LeoGrande, "The Author Replies," *Foreign Policy*, no. 48 (Fall 1982), pp. 181-182.

Indian Ocean, or the South Atlantic.[48] No Soviet submarine would attack oil tankers in the Caribbean when it could select a safer area to attack the same ships. As Jorge Domínguez notes, "there is no threat to the [Caribbean] sea lanes, present or foreseeable, that could not be effectively turned back by the U.S. navy based in mainland United States."[49] This, in fact, is why U.S. security strategists plan to use the Caribbean so heavily in the event of war in Europe. They know these lines are as secure as possible.

Some policy makers believe, in short, that the Caribbean sea lines do not constitute a potential national security disaster area. This belief, however, is held only by a minority. In Washington today the largest body of opinion still conceives of the Caribbean as "an American lake," to use the popular pre-World War II terminology.[50] Nothing captures this majority conception better than the 1895 declaration of Secretary of State Richard Olney:

> Today the United States is practically sovereign on this continent, and its fiat is law upon the subjects to which it confines its interposition. Why? It is not because of the pure friendship or good will felt for it. It is not simply by reason of its high character as a civilized state, nor because wisdom and justice and equity are the invariable characteristics of the dealings of the United States. It is because, in addition to all other grounds, its infinite resources combined with its isolated position render it master of the situation and practically invulnerable as against any or all other powers.[51]

The minority of U.S. officials who reject this view argue that a half-century of profound change in such areas as weapons technology and world attitudes toward colonization, to name but two,

[48] The Galleons Passage is used by VLCCs heading for the Netherlands Antilles. A third route through the Old Bahama Channel is used only infrequently because of strong eastward currents. The Providence Channel is used by ships carrying petroleum to refineries in the Bahamas. Anderson, *Geopolitics of the Caribbean*, p. 119.

[49] Jorge I. Domínguez, *U.S. Interests and Policies in the Caribbean and Central America* (Washington, D.C.: American Enterprise Institute, 1982), p. 11.

[50] McLean, "The Caribbean—An American Lake," p. 952. For the contemporary exposition of this idea, see "Soviet Penetration of the Caribbean," p. 2.

[51] Olney's statement was part of a letter to Thomas F. Bayard, U.S. ambassador to Great Britain, July 20, 1895, regarding the Venezuela-British Guiana boundary dispute. It is reprinted in James W. Gantenbein, ed., *The Evolution of Our Latin-American Policy: A Documentary Record* (New York: Columbia University Press, 1950), pp. 340-354.

have altered forever our ability to control the Caribbean as a semi-sovereign territory. They say it is unrealistic to view the Caribbean as an American lake at this point in world history, for it is, rather, an international body of water that constitutes the border of no fewer than twenty-five sovereign states, of which the United States is by far the most powerful but still only one. The Caribbean, they argue, has become a richly diverse region that promises to test indefinitely the ability of U.S. policy makers to adjust first their attitudes and then their policies to changing circumstances.

All concerned policy makers agree, however, that the use of the Caribbean now exposes the United States to certain vulnerabilities that did not exist in the past. In the past, the United States did not import much petroleum. In the past, guided missiles and nuclear submarines did not exist. In the past—September 16, 1906—Senator Henry Cabot Lodge could write President Roosevelt that "disgust with the Cubans is very general. Nobody wants to annex them, but the general feeling is that they ought to be taken by the neck and shaken until they behave themselves."[52] President Roosevelt, in turn, sent the U.S. Marines to halt the instability on the island, and Charles Magoon was installed as governor of Cuba for three years. Today disgust with the Cubans remains very general, but it is difficult to find a policy maker willing to take them by the scruff of the neck. Tens of thousands of people would die if the United States tried to return to this part of its past.

It is in this sense that the minority of U.S. policy makers argues that the costs of continuing to conceive of the Caribbean as an American lake have risen to the point that the Olney Doctrine is, for practical purposes, unenforceable. Certainly the United States can crush Cuba, and certainly Washington's home-court advantage would be overwhelming in any Caribbean conflict involving the Soviet Union, but the *costs* of maintaining strict control over the region have risen to the point that only an uncommonly grave crisis would justify their expenditure.

Thus the minority of policy makers argues that the brief period of U.S. suzerainty in the Caribbean, lasting little more than half a century, is over, never to return in our lifetime. They advocate ac-

[52] Henry Cabot Lodge, ed., *Selections from the Correspondence of Theodore Roosevelt and Henry Cabot Lodge, 1884-1918*, 2 vols. (New York: Charles Scribner's Sons, 1925), 2: 233. Lodge also observed that "we do not want to intervene if we can help it, and yet I am inclined to think that the postponement of intervention means just so much more destruction of property and injury to the Island. It is one of the burdens we must carry and we must make up our minds to face the work."

cepting increased vulnerability in the Caribbean as part of an increasingly complex world. And, of course, they advocate creative diplomacy (plus military contingency planning) to minimize the risks involved. Their focus is upon adjusting U.S. policy to the new mix of U.S. policy resources and changing international conditions. So long as ships are free to sail the seas, these officials argue, we can never be absolutely certain that the Caribbean sea lines will remain open at all times—that was the lesson of World War II. But because of the importance of Caribbean sea lanes, we must be certain—and we are—that any attack upon U.S. shipping in the Caribbean will be met and quickly overcome.

THE PANAMA CANAL

After more than seven decades of uninterrupted heavy use the Panama Canal stands as a magnificent testimony to U.S. engineering skills, rivaled in our time perhaps only by the landing of human beings on the moon. Unlike the latter achievement, the Canal has yielded obvious benefits from the day in 1914 when it first opened to traffic. Ships sailing from New York to San Francisco save nearly ten thousand miles by using the canal rather than going around the tip of South America.

"This canal," wrote Senator Claude Swanson in 1919, "is the most valuable possession we have."[53] Today, of course, the canal has ceased to be a possession, for formal ownership passed to the government of Panama in 1979. But in the minds of many policy makers the canal retains its importance to national security. Indeed, the significance of other parts of the region is often said to be related to their proximity to the canal. To the Department of State, for example, Colombia's importance for the United States stems in large measure from "its strategic position overlooking both ocean approaches to the Panama Canal."[54] Other national security

[53] Quoted in John Major, "Wasting Asset: The U.S. Re-Assessment of the Panama Canal, 1945-1949," *Journal of Strategic Studies* 3 (May 1980): 123.

[54] U.S. Department of State, *Congressional Presentation, Security Assistance Programs, FY1983*, p. 439.

Immediately prior to World War II, the United States feared that sabotage of the canal would come from Colombia, where German pilots flew for the commercial airline, SCADTA—*Sociedad Colombo-Alemana de Transportes Aéreos*. With the cooperation of Pan American World Airways (which had purchased 84 percent of SCADTA in 1931 but never exercised managerial control because of a Colombian law prohibiting foreigners from owning a voting majority of the airline's stock), the United States convinced Colombia to nationalize the airline and form Avianca. On

analysts have even suggested that the strategic significance of all South America "is to a large degree related to U.S. interests in the Panama Canal."[55] Closer to home, U.S. involvement in the turmoil in Central America in the 1980s has been regularly justified by the importance of the canal. "The canal is critical and must be kept open and defended," Secretary of State Shultz told an audience in 1983. "Yet the security of the Panama Canal is directly affected by the stability and security of Central America."[56]

Despite these assertions most U.S. policy makers are uncertain about the military significance of the canal. Many officials believe the canal serves an important function for U.S. security, but most of them carefully stress the word "important" rather than "vital" or "essential."[57] U.S. Navy officials will admit that they do not use the canal very much. In a typical noncrisis year about twenty-five navy vessels transit the canal, and nearly all of these are smaller ships—destroyers, frigates, submarines.[58] This low level of usage, some of it simply for the practice, clearly reflects the fact that a multiocean navy is supposed to minimize interoceanic ship transfers. This fact is underscored by the navy's heavy reliance upon aircraft carriers. Since navy surface strategy centers upon the projection of airpower through the deployment of carrier task groups,

---

June 12, 1940, Colombian troops (and Pan Am pilots) seized SCADTA's equipment and dismissed all German pilots, thereby completing what the Department of State referred to as "delousing" the airline. Matthew Josephson, *Empire of the Air: Juan Trippe and the Struggle for World Airways* (New York: Harcourt, Brace, 1944), pp. 64-69, 158-159; William A. M. Burden, *The Struggle for Airways in Latin America* (New York: Council on Foreign Relations, 1943), pp. 27, 72-73.

[55] Trevor N. Dupuy, Grace P. Hayes, and John A. C. Andrews, *The Almanac of World Military Power*, 3d ed. (New York: R.R. Bowker, 1974), p. 41 and, for the West Indies in particular, p. 21.

[56] Speech to the Dallas World Affairs Council and the Chamber of Commerce, Dallas, Texas, April 15, 1983.

Panama itself has never been considered a part of Central America. During the colonial era, the isthmus played a crucial role in the shipment of goods between Peru and Spain; Panama was therefore kept in the jurisdiction of the Viceroyalty of Peru and, later, New Granada. The area of the present Central American nations, conversely, was controlled by the Viceroyalty of New Spain in Mexico through the Audiencia of Guatemala. After independence in the early nineteenth century, Panama remained a part of present-day Colombia, while the Central American states made sporadic efforts at confederation before deciding to become five independent nations in 1838. Panama's independence from Colombia dates from 1903.

[57] See, for example, Margaret Daly Hayes, "The Stakes in Central America and U.S. Policy Responses," *AEI Foreign Policy and Defense Review* 4 (1982): 14.

[58] For a typical noncrisis year, 1977, see *Current Information on the Republic of Panama, The Panama Canal, and the Panama Canal Zone*, p. 172, supra n. 9.

and since carriers cannot transit the canal, there is little reason for other navy ships to use it, and they do not.[59] It should be noted, however, that not every U.S. security official agrees with this statement. In 1980, a navy admiral wrote that in any NATO crisis the United States would need to move ninety-six ships from the Pacific to the Atlantic as quickly as possible, many of which would pass through the canal.[60]

Nonetheless, most policy makers who believe the canal is important cite as their reason the canal's potential use not to move warships but to resupply U.S. forces fighting overseas. In both the Korean and Vietnamese conflicts, for example, the canal served as a major logistical artery. In 1953, 22 percent of army supplies were shipped through the canal from East Coast or Gulf ports.[61] In 1973, the commanding general of the Canal Zone called the canal "a vital factor in our ability to deploy operational forces and logistical support rapidly between the Atlantic and Pacific Ocean areas."[62] The canal's logistical capability has led the Department of State to argue that Panama is "of paramount importance to the United States."[63]

[59] The canal locks, which look enormous, are 1,100 feet long and 110 feet wide. These dimensions limit the size of ships to about 950 feet by 106 feet, with a maximum draft of 39 feet, 6 inches. The combination of length, beam, and draft restrictions limit ship size to about 65,000 deadweight tons, although specially designed ships of up to 90,000 DWT—the PANAMAX class—have used the canal. *Missouri*-class battleships with beams of 108 feet have also been squeezed through the canal—a recent example was the moderninzed *New Jersey*, which in 1983 completed its shakedown cruise in the Pacific and then went through the canal and joined the 6th Fleet off the coast of Lebanon. A *Nimitz*-class carrier is an entirely different matter, however. It has a length of 1,092 feet, is 252 feet wide, sits 37 feet in the water, and displaces 94,000 tons. The *Enterprise*-class carrier is slightly longer (1,125 feet) but displaces only 75,700 tons.

About 10 percent of the world's commercial vessels cannot use the canal, although in recent years some shipbuilders have been decreasing the size of their ships to permit passage through the canal. There are about 1,400 oil tankers of 70,000 DWT and larger; none can use the canal. In an earlier time the S.S. *United States* was specially designed to fit within the locks but, ironically, it never made a transit. *Sea-Level Canal Studies*, pp. 180, 189, *supra* n. 9; Anderson, *Geopolitics of the Caribbean*, pp. 18-19.

[60] Hill, "United States Strategic Interests in the Southern Hemisphere," p. 6.

[61] U.S. Atlantic-Pacific Interoceanic Canal Study Commission, *Interoceanic Canal Studies, 1970*, annex II, p. 4.

[62] W. B. Rosson, "U.S. Southern Command in Latin America," *Commanders Digest* 16 (October 18, 1973): 3.

[63] U.S. Department of State, *Congressional Presentation, Security Assistance Programs, FY1983*, p. 485.

Most policy makers believe that these assessments overstate the importance of the canal. Looking at the advances made in offensive capabilities during the war and at the further innovations that were appearing on the horizon, by as early as 1949 many U.S. officials had come to view the canal as "a minor asset."[64] Today no one doubts that the canal would be reduced to rubble in a matter of minutes in the event of a full-scale war with the Soviet Union. In a conflict of lesser magnitude, an army officer wrote in 1980, "the consensus among military analysts today appears to be that the United States could survive without the Panama Canal, but its loss would require substantially greater numbers of ships and amounts of fuel, manpower, and other resources in order to be able to defend U.S. national interests abroad."[65]

On balance, then, U.S. military officials believe that the Panama Canal is a convenience, but they also know better than anyone that it is extremely vulnerable to attack now that the Soviet Union has a highly credible strategic missile capability. In adjusting to this new reality it has been the task of these security officials to arrange U.S. defenses in such a way that the canal's continued operation is no longer vital to U.S. national security.[66] Today, Thomas Anderson notes, "neither the most valuable cargo nor the most powerful naval craft currently make common use of the

[64] Major, "Wasting Asset," p. 141.

[65] John Child, "Military Aspects of the Panama Canal Issue," *Proceedings of the U.S. Naval Institute* 106 (January 1980): 47. Estimates of how much additional resources would be required depend upon the nature of the threat to U.S. security. If there were a need for a 52,000-marine amphibious force to engage Warsaw Pact forces in northern Europe, for example, the Department of Defense would require forty-five days to assemble and deliver the force if the canal were open, sixty days if it were closed. If there were a need to provide munitions to an ally in the Caribbean, the canal would not be used, since most U.S. munitions are manufactured in the East and since there is only one munitions port on the West Coast. If the needy ally were in East Asia, however, use of the canal could be much more important, depending upon the duration of the conflict and the quantity of supplies that were required.

[66] For a discussion of the canal's vulnerability by SOUTHCOM's commanding general, Wallace Nutting, see U.S. Congress, House, Committee on Appropriations, Subcommittee on Foreign Operations and Related Agencies, *Supplemental Appropriations for 1982,* 97th Cong., 2d Sess., 1982, pt. 2, pp. 323-339. See also Charles I. McGinnis, "A New Look at the Panama Canal," *Military Engineer* 66 (July-August 1974): 219-222; Amos A. Jordan and William J. Taylor, Jr., *American National Security: Policy and Process* (Baltimore: The Johns Hopkins University Press, 1981), pp. 436-439; Jorge I. Domínguez, "The United States and Its Regional Security Interests: The Caribbean, Central, and South America," *Daedalus* 109 (Fall 1980): 118.

canal. In this important sense, its strategic significance has declined sharply since about 1950."[67] That is not to argue that the canal is so unimportant as to be left defenseless but only to suggest that no facility that is so vulnerable to Soviet nuclear attack can be allowed to remain vital to U.S. security.

Since there is no defense against a nuclear attack and since a conventional attack seems improbable, U.S. forces are oriented toward the defense of the canal against guerrilla attack and riots by the Panamanian population.[68] The U.S. Senate's consent to ratification of the treaties in 1978 was in large measure a reflection of the nature of this threat, for as the Congressional Research Service noted, "no reasonable number of U.S. troops in the Panama Canal Zone could guarantee the security of the Canal in the face of open Panamanian hostilities."[69]

Any such hostilities would require a large number of participants, for despite the common belief that the canal is vulnerable to attack by a handful of guerrillas, it would take a major effort to disrupt canal traffic for any significant period. While it is true that the destruction of any pair of locks would render the canal inoperable, it is no easy task to harm these locks. Their vulnerable parts—the gates—are up to seven feet thick and weigh between four hundred and seven hundred tons each. Similarly, destruction of Gatun Dam, which contains the fresh water used to operate the canal, would permit two lakes to drain and thereby close the canal.[70] Once closed for this reason, after repairs the canal would remain closed for up to two rainy seasons while the lake refilled. It would take a tremendous effort to achieve this destruction, however, for the Gatun spillway is the largest of its type in the world—a reinforced earthen dam one-half mile thick at its base.

[67] Anderson, *Geopolitics of the Caribbean*, p. 19. For discussions of policy makers' perceptions of the military value of the canal, see James Michael Hogan, "The 'Great Debate' over Panama: An Analysis of Controversy over the Carter-Torrijos Treaties of 1977," Ph.D. diss., University of Wisconsin—Madison, 1983, especially pp. 245-301; George D. Moffett, III, *The Limits of Victory: The Ratification of the Panama Canal Treaties* (Ithaca, New York: Cornell University Press, 1985).

[68] Child, "Military Aspects of the Panama Canal Issue," p. 49.

[69] U.S. Congress, Senate, Committee on Foreign Relations, *United States Foreign Policy Objectives and Overseas Military Installations*, 96th Cong., 1st Sess., 1979, p. 205.

[70] The integrity of Gatun Dam determines the utility of Madden Dam and Madden Lake, which are "upstream" from Gatun Dam. If Gatun Dam were permanently open, Madden Dam would not be opened because then its waters, too, would simply drain into the Caribbean.

According to one expert, the canal is most vulnerable at the Gaillard Cut, a narrow eight-mile-long section that can be blocked by landslides or by a sunken ship.[71] Given the complaisance of Panamanians since the treaties were signed, however, SOUTHCOM forces assigned to defend the canal are thought to be adequate for this defense.

Although it is most common to discuss the national security aspects of the Panama Canal in terms of military (particularly naval) usage or in terms of canal defenses, anything that contributes significantly to the economic vitality of the United States also has national security ramifications. So it is that trade through the Panama Canal is commonly vested with national security significance.[72]

This trade is far from modest: about 5 percent of total annual *world* seaborne trade passes through the Panama Canal. Of this amount, about a third is U.S.-oriented.[73] On the other hand, the proportion of U.S. seaborne trade using the canal has now stabilized at a fairly modest level—about 12 percent—after declining slowly throughout the 1960s and 1970s.[74] All the countries of Central America and the west coast of South America use the canal for a much greater proportion of their trade. Peru, for example, sends about 40 percent of its trade through the canal.[75] In terms of volume the heaviest traveled canal route is between the U.S. East Coast and Gulf ports and Asia. Intercoastal U.S. trade through the canal is not large.

Thus the Panama Canal is important to world and U.S. commerce, but traffic is not so heavy that the canal retains its vital importance to the U.S. economy. Summarizing the findings of dozens of studies over the years, Leon Cole of the Congressional

[71] Child, "Military Aspects of the Panama Canal Issue," p. 49.

[72] For statements to this effect by President Reagan, see *Weekly Compilation of Presidential Documents* 18 (March 1, 1982): 219, and 19 (May 2, 1983): 608. See also John Bartlow Martin, *U.S. Policy in the Caribbean* (Boulder, Colo.: Westview Press, 1978), p. 278; Ryan, "Canal Diplomacy and U.S. Interests," p. 48; Roberts, "The Growing Soviet Naval Presence in the Caribbean," p. 37.

[73] U.S. Congress, House, Committee on Appropriations, Subcommittee on the Department of Transportation and Related Agencies Appropriations, *Department of Transportation and Related Agencies Appropriations for 1976*, 94th Cong., 1st Sess., 1975, pp. 67-68.

[74] *Current Information on the Republic of Panama, The Panama Canal, and the Panama Canal Zone*, p. 113, *supra* n. 9.

[75] *Department of Transportation and Related Agencies Appropriations for 1976*, p. 68, *supra* n. 73.

Research Service concluded that "market conditions in origin and destination countries exert much more influence on aggregate commodity and product prices than would increased levels of Panama Canal tolls or complete closure, after an interim period of adjustment in trade routes and markets." This is because "many alternative trade routes now exist for the most important products and commodities, and more would become economically competitive if the canal were closed."[76]

No generalization is possible on the subject of Latin American sea lines of communication, for these maritime routes differ greatly in their importance to U.S. security. By common consent the west coast South America routes are of minimal significance to U.S. security. The South Atlantic SLOCs are used to transport many essential imports, including petroleum from the Persian Gulf and minerals from southern Africa, but it is difficult to construct a set of circumstances that would lead the Soviet Union to use the South Atlantic to launch an attack against U.S. shipping. Such an attack is possible, but it is not probable. If the U.S. military possessed many more resources, it would probably direct some of them to the protection of South Atlantic sea routes from Africa (but not from Latin America). Since an attack along the east coast of South America is highly improbable, these sea lines are a low-priority concern to U.S. officials.

While the heavily used Caribbean sea lines should be considered essential to U.S. security, two additional comments are appropriate. First, as the discussion of petroleum shipments and NATO supply lines indicates, there is no law of nature that makes the Caribbean as important as it has become. Deliberate public policy could reduce substantially the current U.S. dependence upon the Caribbean. Second, and much more important, the reason no such policy exists is that the United States is extremely well-prepared to defend Caribbean maritime routes; a Soviet or Cuban attack upon them would be suicidal. The possibility of suicide attacks cannot be overlooked, but they are only conceivable in the context of an ongoing general war. In that event U.S. officials will be concentrating on subjects of far greater importance than Caribbean sea lines.

[76] Ibid., pp. 4, 64. For a comprehensive study of the commercial utilization of the canal, see Norman J. Padelford and Stephen R. Gibbs, *Maritime Commerce and the Future of the Panama Canal* (Cambridge, Md.: Cornell Maritime Press, 1975).

Finally, there is the Panama Canal. The canal is important to world commerce, but its economic importance is clearly no longer of major concern to U.S. security. Militarily the canal would be useful as a resupply route, particularly from the U.S. Gulf ports to a crisis center in the Pacific, but there are a variety of alternative routes. Moreover, the U.S. Navy cannot use the canal for its principal fighting units. Since World War II the Panama Canal's importance to U.S. security has decreased dramatically.

PART III

The Consequences of Instability: Strategic Denial

· 7 ·

SOVIET MILITARY BASES

SINCE the end of World War II a fundamental problem facing U.S. policy makers has been how to keep the Soviet Union out of Latin America. The problem is strategic denial, and it is part of the very core concern of postwar U.S. foreign policy: the containment of Communism. But it is more than that. Since 1823, a quarter century before publication of the *Communist Manifesto* and almost a century before the Russian Revolution, the United States has pursued a general policy of keeping other nations, whatever their ideologies or intentions, from acquiring territory and, later, influence in the Western Hemisphere. It is perhaps best, then, to think of the policy of excluding the Soviet Union from Latin America as the Cold War corollary to the Monroe Doctrine.

This is the first of two chapters that analyze why U.S. policy makers believe it is important to pursue this corollary. Policy makers were asked what they thought the Soviets would do if they were able to obtain control of all or a part of Latin America. Once again their responses varied greatly. Some officials were extremely fearful of the consequences of further Soviet access to Latin America; others seemed much less concerned. Some officials were able to describe a number of specific negative consequences of this access; others knew only that it would be bad for U.S. security. Many policy makers voiced fears that the Soviet Union would damage U.S. security by ending U.S. access to the region—the topic of the three preceding chapters—but most described a different type of potential damage as well. In addition to limiting U.S. access to Latin America, four specific consequences of Soviet access were regularly mentioned. In a sense each of these deserves a separate chapter. But two of these four consequences, while frequently cited in policy makers' public statements, turn out to be far less important to U.S. officials when they are speaking privately and off the record. They can therefore be discussed here fairly briefly.

One of the two focuses upon the damage to U.S. security that would occur if additional Soviet intelligence facilities were established in Latin America. This is a subject with which the United States has considerable experience, since the Soviet Union has long maintained an intelligence center at Lourdes, Cuba. This

225

large (twenty-eight-square-mile) facility is operated by fifteen hundred Soviet technicians, who monitor telecommunications in the eastern United States. The Soviets also station a large Bal'zam-class intelligence collection ship in the Caribbean to monitor U.S. naval activities, and since 1970 Moscow has used Tu95/Bear aircraft to conduct infrequent aviation reconnaissance flights, skirting the U.S. East Coast while flying between the Soviet Union and Cuba. Since early 1983, these reconnaissance aircraft have been accompanied on occasion by Tu142 antisubmarine aircraft.

Although foreign policy officials periodically comment on these Soviet intelligence capabilities, none of them has focused upon the issue as a major reason to deny the Soviets additional access to Latin America. In an address to the nation in 1983, for example, President Reagan used the Lourdes facility as a prime example of the negative consequences of Soviet access to Latin America. Although the President used strong language,[1] he cited this issue only in passing as part of a longer list of reasons to resist Soviet incursions. Other officials have observed that the Lourdes base could be used not only to monitor but to disrupt U.S. communications during a crisis; they have also noted, however, that U.S. countermeasures would leave essential communications unaffected and, if this failed, Lourdes could be wiped from the face of the earth in about ten minutes.[2] In this age of satellite intelligence

[1] Pointing to an aerial photograph, Mr. Reagan said: "This Soviet intelligence collection facility, less than a hundred miles from our coast, is the largest of its kind in the world. The acres and acres of antennae fields and intelligence monitors are targeted on key U.S. military installations and sensitive activities." *Weekly Compilation of Presidential Documents* 19 (March 28, 1983): 444.

[2] For examples of these observations from analysts and policy makers who specialize in hemispheric security issues, see U.S. Joint Chiefs of Staff, "United States Military Posture for FY1982," printed in U.S. Congress, House, Committee on Armed Services, *Hearings on Military Posture and H.R. 2614 and H.R. 2970*, 97th Cong., 1st Sess., 1981, pt. 1, p. 109; U.S. Department of Defense, *Annual Report to the Congress, Caspar W. Weinberger, Secretary of Defense, Fiscal Year 1983* (Washington, D.C.: GPO, 1982), p. II-23; Constantine C. Menges, "The United States and Latin America in the 1980s," in *The National Interests of the United States in Foreign Policy*, ed. Prosser Gifford (Washington, D.C.: University Press of America, 1982), pp. 53-72; Jeane J. Kirkpatrick, "U.S. Security and Latin America," *Commentary* 71 (January 1981): 29; U.S. Congress, House, Committee on International Relations, Subcommittee on Inter-American Affairs, *Arms Trade in the Western Hemisphere*, 95th Cong., 2d Sess., 1978, p. 308; Margaret Daly Hayes, "The Stakes in Central America and U.S. Policy Responses," *AEI Foreign Policy and Defense Review* 4 (1982): 14; Margaret Daly Hayes, "Security to the South: U.S. Interests in Latin America," *International Security* 5 (Summer 1980): 136.

gathering, advanced electronic countermeasures, redundant communications systems, and (when push comes to shove) very big bombs, the Soviet Union's intelligence facilities in Cuba are more of an annoyance than a genuine threat to U.S. security. The existing facility in Cuba is undoubtedly a convenience to Moscow, but it performs no vital function that cannot be performed by satellites or by Soviet naval units anchored on the high seas. Just as U.S. security officials cannot permit the highly vulnerable Panama Canal to be vital to U.S. security, so Soviet policy makers cannot permit vital installations in extraordinarily vulnerable Cuba.

The development of additional facilities elsewhere in Latin America would have significance as a symbol of Soviet expansion—the topic of Chapter 8—but not as a threat *per se* to U.S. security. In their public comments some officials express concern that Soviet landing rights anywhere in Central America would permit reconnaissance flights up the U.S. West Coast. In 1985, for example, Secretary of Defense Caspar Weinberger commented on a new Nicaraguan airfield northeast of Managua: "The considerable benefits to the Soviet Union of such a base in Nicaragua for flights along our Pacific coast are obvious."[3] In interviews, however, many officials, including some who worked for Mr. Weinberger at the Department of Defense, argue that any such flights, should they occur, would be more of a nuisance to Washington than an asset to Moscow. To these officials, flights along the U.S. Pacific coast would only provide the Soviets with information that they already had obtained by other means.

The second minor reason for resisting Soviet expansion in the Western Hemisphere involves refugees. To many U.S. policy makers, the Soviets, being oppressive, would create a wave of humanity to wash up upon the shores of the United States. In 1983, for example, President Reagan told a Mississippi audience that "a string of anti-American Marxist dictatorships" in Central America would lead to "a tidal wave of refugees. And this time they'll be 'feet people' and not 'boat people' swarming into our country, seeking a safe haven from Communist repression to our South. We cannot permit the Soviet-Cuban-Nicaraguan axis to take over Central America."[4]

Mr. Reagan's remark was repeated frequently by other policy

[3] Comments to a White House Outreach Working Group on Central America, Washington, D.C., January 30, 1985.

[4] *Weekly Compilation of Presidential Documents* 19 (June 27, 1983): 901.

makers. Secretary of State Haig, for example, sounded a similar warning: "Just think what the level of undocumented immigration might be if the radicalization of this hemisphere continues. . . . Why, it would make the Cuban influx look like child's play."[5] The speech gained Mr. Haig a headline in the next day's edition of the *New York Times*: "Haig Fears Exiles from Latin Areas May Flood the U.S." His successor warned a Texas audience of "unprecedented flows of refugees northward to this country";[6] the DSAA's General Graves foresaw "a flood of refugees and illegal immigrants larger than any we have experienced to date";[7] and Assistant Secretary of State Enders spoke of "hundreds of thousands or even millions of people who would flee" Soviet inroads in Latin America.[8] No one used this line of reasoning more frequently than Jesse Helms, chairman of the Senate Foreign Relations Subcommittee on Western Hemisphere Affairs from 1981 to 1987: "The American people had better come alive and realize that what happens in Central America, what happens in El Salvador is not long ago and far away, because if Central America falls, we are going to be flooded with refugees, 10, 15, 20 million of them coming across our borders, people who cannot speak English, who have no jobs, who have nothing."[9]

What is most interesting about these and similar statements concerning refugees is the context in which they were made. To my knowledge, in the 1980s no senior policy maker ever gave a speech that focused upon the issue of Central American refugees. Instead, the issue of Latin American refugees was raised almost in passing either (a) before an audience that could be expected to oppose further Hispanic immigration in general (for example, Texas Anglos), or (b) as part of a laundry list of reasons to support admin-

[5] Speech to the National Governors' Association, February 22, 1982.

[6] Speech to the Dallas World Affairs Council and the Chamber of Commerce, Dallas, Texas, April 15, 1983.

[7] Ernest Graves, "U.S. Policy toward Central America and the Caribbean," in *The Central American Crisis: Policy Perspectives*, ed. Abraham F. Lowenthal and Samuel F. Wells, Jr., Working Paper No. 119, Latin American Program, The Wilson Center, Washington, D.C., 1982, p. 19.

[8] Testimony before the House Foreign Affairs Subcommittee on Inter-American Affairs, March 1, 1983.

[9] U.S. Congress, Senate, Committee on Foreign Relations, *U.S. Policy toward Nicaragua and Central America*, 98th Cong., 1st Sess., 1983, p. 30. For a similar comment a year earlier, see U.S. Congress, Senate, Committee on Foreign Relations, Subcommittee on Western Hemisphere Affairs, *Human Rights in Nicaragua*, 97th Cong., 2d Sess., 1982, p. 84.

istration policy toward Latin America. Moreover, while these foreboding statements about "floods of refugees" and "millions of fleeing people" highlighted policy makers' public pronouncements, the issue rarely arose in private interviews. Although they probably did not intend to be cynical, many officials were clearly vulnerable to the charge of raising the refugee issue as a tactic to muster support for their policies.

Whatever the motivation behind these statements, Soviet intelligence gathering and Latin American refugees are clearly minor concerns of U.S. policy makers, few of whom seemed to be losing any sleep over either problem in the 1980s. That, perhaps, is because two other concerns—the establishment of Soviet military bases and the maintenance of the balance of world power (the subject of Chapter 8)—completely dominate U.S. strategic thinking about Latin America. These issues constitute the core U.S. national security concerns in the region.

POLICY MAKERS' BELIEFS ABOUT SOVIET BASES IN LATIN AMERICA

Over the course of the past quarter of a century the Soviet Union has dramatically increased its ability to project power. By the early 1980s, this fact had come to dominate the thinking of virtually all U.S. national security officials, beginning with the president. In his speech to the nation in March 1983, Mr. Reagan began by warning that "for 20 years, the Soviet Union has been accumulating enormous military might"; he then gave a detailed list of the new weapons Moscow had deployed.[10] A year later Mr. Reagan once more went before the television cameras to tell the public that "the growth of Soviet military has meant a radical change in the nature of the world we live in."[11]

To many U.S. policy makers one of the most radical of these changes has been in the nature of superpower competition in Latin America. Here the Soviet challenge has many facets, but what many officials fear most is the establishment of a Soviet military base anywhere in the hemisphere. Shortly after Mr. Reagan's election Jeane Kirkpatrick summarized this fear when she criticized the Carter human rights policy for creating a situation that "threatens now to confront this country with the unprecedented need to defend itself against a ring of Soviet bases on and around

[10] *Weekly Compilation of Presidential Documents* 19 (March 28, 1983): 443-448.

[11] *Weekly Compilation of Presidential Documents* 20 (May 14, 1984): 676.

our southern and eastern borders. In the past four years, the Soviet Union has become a major military power within the Western hemisphere."[12] At almost the same time the Joint Chiefs of Staff made a similar assessment of the Soviet threat: "Cuban military ties with the Soviet Union and the growth of Soviet air and naval presence in Cuba pose the most significant military threats to US security interests in this hemisphere."[13]

The fear of Soviet bases became a central theme in discussions of Central America. Lynn Bouchey of the ultraconservative Council on Inter-American Security warned that "Soviet naval bases are naturals for Nicaragua and as revolutionary ferment spreads, Soviet naval installations would be easy exchanges for economically bereft countries in the region."[14] In 1984, Ambassador Kirkpatrick argued that the very core goal of Washington's Central America policy was to prevent the establishment of these bases: "It is true that the United States has strategic goals in Central America. We believe it would be bad . . . for there to be installed one-party, Marxist-Leninist states integrated into the Soviet bloc and willing to have their territory serve as bases for the projection of Soviet military power in the hemisphere. That is the strategic basis of our policy."[15]

As we have seen, other policy makers regularly criticized—sometimes extremely strongly—the Reagan policy toward Latin America. But there was little public criticism on the issue of Soviet bases; indeed, in the 1980s few issues seemed to unite so many U.S. policy makers so quickly and so predictably as the threat of a Soviet military base in Latin America. Walter Mondale spoke for the mainstream of Democratic party opposition when he asserted in mid-1983 that "if the Soviet Union or the Cubans . . . were to establish a major base or military position in Central America, I think that would be a matter of very, very severe concern. I think it would go to the vital interests of our country and it would raise grave challenges."[16]

Similar statements are found throughout the writings and speeches of the majority of analysts and policy makers. Ambassa-

12 Kirkpatrick, "U.S. Security and Latin America," p. 29.
13 U.S. Joint Chiefs of Staff, "United States Military Posture for FY 1982," pt. 1, p. 109.
14 L. Francis Bouchey, "Reagan Policy: Global Chess or Local Crap Shooting?" Caribbean Review 11 (Spring 1982): 20.
15 Speech to the Royal Institute for International Affairs, London, April 9, 1984.
16 The New Republic, August 8, 1983, p. 14.

dor William Luers spoke for virtually all career foreign service officers when he observed that

> it is in the U.S. interest to make clear to the U.S.S.R. repeatedly and forcefully that certain types of military presences of the Soviet Union and its allies in the region would be unacceptable to the United States. It is also in the U.S. interest to make the necessary military and political contingency plans to assure that we can and will accomplish the withdrawal or elimination of such a Soviet military presence.[17]

Similarly, Lowenthal and Wells spoke for a majority of Washington-based analysts when, in summarizing the proceedings of a conference, they noted that "participants agreed that the United States has an 'irreducible minimum security interest' in preventing the establishment of hostile bases in countries dependent on the Soviet Union which might threaten the security of the United States." For Latin America this means in particular that "the establishment of naval facilities—whether under Soviet sovereignty or not—should be resisted by diplomatic and, if necessary, by other means."[18]

On the specific issue of Central America the position taken by Robert Osgood characterized the position of most policy makers in the 1980s: "The most essential U.S. interest and the only one worth trying to protect with direct force can best be expressed in negative terms: preventing the establishment in Central America of a base of Soviet, Cuban, or indigenous power that would threaten through military or paramilitary operations or through subversion or revolution the physical security of the Panama Canal, militarily critical sea lanes, Mexico, or the United States itself."[19] Viron Vaky, an FSO and Carter administration Assistant

[17] William H. Luers, "The Soviets and Latin America: A Three Decade U.S. Policy Tangle," *Washington Quarterly* 7 (Winter 1984): 29.

[18] Abraham F. Lowenthal and Samuel F. Wells, Jr., "Introduction," in *The Central American Crisis: Policy Perspectives*, p. 2.

[19] Robert E. Osgood, "Central America in U.S. Containment Policy," in *The Central American Crisis: Policy Perspectives*, p. 91. For a variety of similar statements, see John Bartlow Martin, *U.S. Policy in the Caribbean* (Boulder: Colo.: Westview Press, 1978), p. 278; Robert A. Pastor, "Sinking in the Caribbean Basin," *Foreign Affairs* 60 (Summer 1982): 1,042; Hayes, "The Stakes in Central America," p. 14; Hayes, "Security to the South," p. 136; Worth H. Bagley, *Sea Power and Western Security: The Next Decade*, Adelphi Papers No. 139 (London: International Institute for Strategic Studies, 1977), pp. 29-31; Richard L. Millett, "The Best of Times,

Secretary of State, noted that "there is virtually unanimous agreement—there certainly is bipartisan agreement—that Soviet-Cuban military bases or combat military presence in the isthmus would be an intolerable threat."[20]

This agreement appears to include many analysts and policy makers who fought the Reagan administration tooth and nail over its policy toward Latin America. The highly respected staff director of the House Foreign Affairs Subcommittee on Western Hemisphere Affairs, Victor Johnson, remarked publicly that he "would be quite comfortable with a region of countries with different political and economic systems, even if they were not particularly friendly, so long as they were not forces for instability and bases that Cuba and the Soviet Union could use against us."[21] Similarly, Walter LaFeber, whose writings excoriated the Reagan administration, wrote that "the United States would be wise to step aside and allow the revolutions to work themselves out, with the single provision that no Soviet-controlled bases or military personnel be allowed in Central America."[22]

While it would seem that opposition policy makers and analysts have joined the Reagan administration to form a fairly solid consensus on the need to deny the Soviet Union military bases in the Western Hemisphere, there are two reasons to believe that there may exist more dissension than these public statements would suggest. First, public pronouncements may not resemble private beliefs. Specifically, during the years of the Reagan administration, many analysts and policy makers of liberal inclinations—especially those seeking elective office—seemed to feel obliged to establish their credentials as anti-Communist "realists" in order to be taken seriously during policy debates. They would begin their public criticism of the Reagan policy with statements such as, "I am no friend of the Sandinistas, but . . ." or "I, too, am inalterably opposed to a Soviet base anywhere in this hemisphere, but . . . ." In speaking privately with these individuals the Soviet base

the Worst of Times: Central American Scenarios—1984," in *The Central American Crisis: Policy Perspectives*, p. 87; Bouchey, "Reagan Policy," p. 20.

[20] Viron P. Vaky, "A Central American Tragedy," in *Central America and the Western Alliance*, ed. Joseph Cirincione (New York: Holmes and Meier, 1985), p. 58.

[21] Victor C. Johnson, "Discussion Paper on Central America," in *The Central American Crisis: Policy Perspectives*, p. 46.

[22] Walter LaFeber, "Inevitable Revolutions," *Atlantic Monthly* 249 (June 1982): 83.

issue almost never seemed to be very important. This is not to suggest that these officials were being deceitful; rather, like skilled tacticians in any field, they seemed simply to be playing the appropriate mood music in order to achieve their goal—in this case to achieve credibility in policy-making circles dominated by the Reagan administration.

Second, even the public consensus is not perfect. Some officials and analysts whose other beliefs seem extremely similar to those of Mondale, Johnson, and LaFeber argue openly against a policy whose "irreducible minimum" is no Soviet bases. To Princeton political scientist Richard Ullman, for example, "in an era of intercontinental missiles, firing nuclear weapons from nearby bases conveys no real benefit. And there are no plausible ways in which the Soviet Union could profit from attacking North America with conventional weapons. . . . the overwhelming military advantage that the United States would enjoy in its own backyard is what makes it so untenable to argue that a Soviet or Cuban presence in Central America would seriously threaten U.S. national security. . . . If Soviet or Cuban bases were ever established in Central America, they would be used only to defend the country in which they were located—Nicaragua, for instance."[23] Similarly, Yale Ferguson, a respected academic specialist on inter-American relations, argues that "were we to face the worst case, the Soviet Union's eventually setting up a base in, say, Nicaragua, this would still give them no important military advantage in this age of intercontinental and submarine-launched missiles."[24] And, like Ferguson, former defense policy maker and Washington analyst Morton Halperin believes that "the possibility of these [Soviet] bases being of any importance is so small as not to even remotely approximate the degree of importance that would justify the current level of U.S. effort in El Salvador."[25]

In short, despite an apparent consensus it is possible to identify considerable disagreement among policy makers on the subject of

[23] New York Times, July 10, 1983, p. E21. See also Richard H. Ullman, "Paths to Reconciliation: The United States and the International System of the Late 1980s," in Estrangement: America and the World, ed. Sanford J. Ungar (New York: Oxford University Press, 1985), p. 295.

[24] Yale H. Ferguson, "The Reagan Administration's Latin American Policies: An Assessment," mimeographed, paper presented at the International Congress of the Latin American Studies Association, Mexico City, September 30, 1983, p. 5.

[25] Morton H. Halperin, "U.S. Interests in Central America: Designing a Minimax Strategy," in The Central American Crisis: Policy Perspectives, p. 24.

Soviet bases in the Western Hemisphere. No one in Washington would be pleased by the creation of such a base, but some would be much less worried than others. As we shall see in the following pages, that is because policy makers differ considerably in their beliefs about the nature of the threat the Soviets would pose from a base in the region. To some a base would be a threat to vital U.S. interests; to others it might well be a nuisance, but not much more than that.

EVOLVING U.S. POLICY

Until the early 1960s, the Soviet Union had no place to build a base in Latin America. Since the Cuban revolution, however, it has not been easy for U.S. policy makers to keep a Soviet base from being constructed somewhere in the region. In July 1960, President Eisenhower issued a carefully worded statement affirming "in the most emphatic terms that the United States will not be deterred from its responsibilities by the threats Mr. Khrushchev is making. Nor will the United States, in conformity with its treaty obligations, permit the establishment of a regime dominated by international communism in the Western Hemisphere."[26] Within a year Fidel Castro had aligned his government with international Communism, Dwight Eisenhower had left office, and John Kennedy had failed in his attempt to overthrow the Cuban government. The Soviets, for their part, then attempted to build an offensive missile base in Cuba, and the Kennedy administration had to push the world to the brink of war in order to negotiate with a much weaker Soviet Union for the satisfactory end to the Cuban missile crisis. The substance of the Kennedy-Khrushchev agreement was interpreted as a major victory for the United States: the Soviets pledged to remove their missiles and not to establish an offensive military capability in Cuba. In return the United States agreed not to invade Cuba.

For the purposes of this discussion the most significant aspect of the Cuban missile crisis is not who won but rather what Washington policy makers perceived as the nature of the Soviet threat. The Soviet missiles clearly were perceived as constituting a special type of threat: once installed they would have provided the Soviet Union with the ability to attack the United States with nuclear weapons at a time when Moscow did not yet have a reliable means

[26] *Public Papers of the Presidents of the United States. Dwight D. Eisenhower, 1960-61* (Washington, D.C.: GPO, 1961), p. 568.

of delivering nuclear weapons to the U.S. mainland. As subsequent events have suggested, had the base been for a different purpose (to gather intelligence, to refuel naval vessels, to train Cuban troops), the U.S. reaction probably would have been different because U.S. officials would have perceived a different (and lesser) threat. Certainly this is suggested by the outcome of the two subsequent tests of the Kennedy-Khrushchev agreement.[27]

The first test occurred during the Nixon administration. In July 1969, the Soviet Union sent naval units into the Caribbean for the first time. They left the region after spending a week in Havana but were followed in May 1970 by a second deployment and in September 1970 by a third. The September visit caused considerable consternation among U.S. intelligence officials, for it included a submarine tender and a large landing craft, which docked at Cienfuegos on Cuba's southern coast and promptly unloaded two nuclear submarine support barges used for the disposal of radioactive waste. A modest amount of construction at Cienfuegos served to convince skeptics in the Nixon administration that the Soviets were developing a naval base for use by Soviet submarines.[28]

High-level negotiations with the Soviet Union were undertaken at once. After the United States announced that it viewed construction of any Soviet base in Cuba with "utmost seriousness," all but one of the ships left the Caribbean.[29] Speaking of this confrontation, President Nixon revealed that an understanding had been reached on October 13, 1970, expanding the Kennedy-Khrushchev agreement by defining the term "offensive military capability" to include a naval base. Mr. Nixon has been quoted as saying that "in the event that nuclear submarines were serviced in

[27] The Kennedy-Khrushchev agreement has been interpreted in a remarkable variety of ways. After looking at the relevant documents, one member of the Kissinger Commission, William P. Clements, Jr., concluded that "there was no understanding or agreement." U.S. National Bipartisan Commission on Central America (Kissinger Commission), "Report of the National Bipartisan Commission on Central America," mimeographed [Washington, D.C.: Department of State(?), January 1984], p. 129. While Clements is correct in the sense that the United States and the Soviet Union did not reach an agreement on a detailed list of "acceptable" and "unacceptable" Soviet actions, the agreement clearly included an understanding about the installation of offensive weapons that would threaten the U.S. mainland.

[28] Barry M. Blechman and Stephanie E. Levinson, "Soviet Submarine Visits to Cuba," *Proceedings of the U.S. Naval Institute* 101 (September 1975): 31-32; Christopher A. Abel, "A Breach in the Ramparts," *Proceedings of the U.S. Naval Institute* 106 (July 1980): 48.

[29] A tugboat remained at Cienfuegos and has been there ever since.

Cuba or from Cuba, that would be a violation of the understanding."[30] Nuclear attack submarines have visited Cuba, however, as have diesel-powered submarines with nuclear-tipped missiles; what has *not* occurred is a visit by a nuclear-powered submarine carrying strategic ballistic missiles. That is apparently the Soviet understanding of the boundary agreed upon in 1970.[31]

The second test of the Kennedy-Khrushchev agreement occurred in September 1979. The fact that it severely embarrassed the Carter administration has been permitted to obscure the more important fact that the U.S.-Soviet military balance was changing rapidly. In late August 1979, Democratic senator Frank Church of Idaho was beginning what promised to be a difficult reelection campaign. In an apparent attempt to bolster his image among conservative constituents, Church called a news conference in Boise to announce that he had obtained secret documents revealing the existence of a Soviet combat brigade in Cuba. He demanded the brigade's removal, since it ostensibly violated the 1962 agreement prohibiting an offensive Soviet military capability on the island.

After some days of intense media coverage two facts became evident. First, Soviet troops had been stationed in Cuba since the early 1960s and, second, the Carter administration was having difficulty convincing the Soviets, who had grown much stronger militarily since their 1962 humiliation, to withdraw the brigade. In a televised speech on September 7, Mr. Carter said that "we consider the presence of a Soviet combat brigade in Cuba to be a very serious matter and that this status quo is not acceptable." Three days earlier, however, the President had told Senate Majority Leader Robert Byrd that "there was no way to mandate that the Soviets withdraw those troops."[32] Sensing Mr. Carter's weakness,

[30] Raymond A. Komorowski, "Latin America—An Assessment of U.S. Strategic Interests," *Proceedings of the U.S. Naval Institute* 99 (May 1973): 170.

[31] In the wake of the Cienfuegos dispute, the U.S. Navy created a surface surveillance force, Destroyer Squadron 18, to escort and observe Soviet naval activities in the Caribbean. It was decommissioned in February 1973 after fewer than two years of activity. For an interesting story of the squadron's activities, see the article by its commander, Leslie K. Fenlon, Jr., "The Umpteenth Cuban Confrontation," *Proceedings of the U.S. Naval Institute* 106 (July 1980): 40-45. For an informative discussion of the U.S.-Soviet negotiations over the Cienfugos base, see Raymond L. Garthoff, *Détente and Confrontation: American-Soviet Relations from Nixon to Reagan* (Washington, D.C.: The Brookings Institution, 1985), pp. 76-83.

[32] Jimmy Carter, *Keeping Faith: Memoirs of a President* (New York: Bantam Books, 1982), p. 263.

administration critics sought to exploit the consensus on the desirability of excluding Soviet military facilities from Latin America. With an assertive self-confidence that ignored the lack of reliable information, the Heritage Foundation announced that the combat brigade "came from East Germany and Czechoslovakia, where they guarded nuclear weapons depots and mobile missile launchers." Referring to unnamed "intelligence reports," the foundation added that the brigade's troops were "being used to service nuclear weapons from Soviet submarines."[33]

In fact, it was (and still is) difficult to determine the brigade's mission. The Soviet position is that the brigade constitutes a training unit, has no offensive capability, and therefore is not in violation of the 1962 agreement regarding offensive weapons. U.S. intelligence sources have reported that the unit consists of 2,600 soldiers divided into three infantry battalions and one tank battalion—not a normal training configuration. The brigade is equipped with forty tanks and sixty armored personnel carriers, which is not standard equipment for training units.[34] Whatever the case, on October 1, 1979, President Carter accepted the Soviet's assurances that the unit was a training brigade and that it would not be given an offensive capability. The troops remained in Cuba; Senator Church lost his bid for reelection.[35] As of 1982, the Department of State estimated that the Soviet military presence in Cuba still consisted of "a ground forces brigade of about 2,600 men, a mili-

[33] Heritage Foundation, "The Soviet Military Buildup in Cuba," *Backgrounder*, no. 189, June 11, 1982, p. 7.

[34] Stockholm International Peace Research Institute, *World Armaments and Disarmament: SIPRI Yearbook, 1982* (Cambridge, Mass.: Oelgeschlager, Gunn and Haig, 1982), p. 399.

[35] In his memoirs President Carter accuses Church of using his access to classified information, gained as chairman of the Senate Committee on Foreign Relations, for personal political purposes: "He saw an opportunity to meet some of the conservative political attacks on his liberal voting record. He called a news conference and did everything possible to escalate the report into an earth-shaking event." In Carter's view Church's behavior was "absolutely irresponsible." Carter, *Keeping Faith*, pp. 262-263. For a much more detached (and comprehensive) analysis of the combat brigade episode and its implications for international relations, see Garthoff, *Détente and Confrontation*, pp. 828-848.

Perhaps the most revealing public accounts of the combat brigade incident are by former CIA director Stansfield Turner. See his "The Stupidity of Intelligence," *Washington Monthly*, February 1986, pp. 29-33, and *Secrecy and Democracy: The CIA in Transition* (Boston: Houghton Mifflin, 1985), pp. 229-235. Like President Carter, Turner attributes the incident to domestic politics.

tary advisory group of 2,000, and an intelligence-collection facility. There also are 6,000-8,000 Soviet civilian advisers in Cuba."[36]

Although there is an extremely broad consensus among national security officials that the 1962 Kennedy-Khrushchev agreement should continue in force, the training brigade incident indicates that some types of Soviet troop deployments will have to be overlooked or defined as "nonoffensive" and therefore outside the confines of the 1962 accord. The reason for this is that the the Soviets enjoy considerable new strength. There is no longer any assurance that, should we once again stand eyeball-to-eyeball, the Soviets will blink first, as they did in 1962. Consequently, there now is considerable confusion among policy makers over whether the agreement ending the Cuban missile crisis can be generalized to include all of Latin America. On the one hand, no one argues that the Soviets should be denied military landing rights at Cuban airports, despite the fact that the aircraft flying this route are capable of transporting offensive nuclear weapons. On the other hand, there is the Caribbean ministate of Grenada, whose government decided to build a new airport.

Unfortunately, they chose the wrong contractor. President Reagan told a nationwide TV audience in early 1983 that "the Cubans with Soviet financing and backing are in the process of building an airfield with a 10,000-foot runway." This, he asserted was a "rapid buildup of Grenada's military potential," and he warned that "the Soviet-Cuban militarization of Grenada, in short, can only be seen as a power projection into the region."[37] The following October, U.S. armed forces invaded the island and eliminated the threat to U.S. security. Grenada, President Reagan told another nationwide TV audience, "was a Soviet-Cuban colony, being readied as a major military bastion to export terror and undermine democracy. We got there just in time."[38]

The Lesson of Grenada

There are many lessons to be drawn from the U.S. invasion of Grenada, but perhaps the most important is that many U.S. policy makers are willing to take extraordinary measures in order to en-

---

[36] U.S. Department of State, Bureau of Public Affairs, *Cuban Armed Forces and the Soviet Military Presence*, Special Report No. 103 (Washington, D.C.: Department of State, August 1982), p. 5.

[37] *Weekly Compilation of Presidential Documents* 19 (March 28, 1983): 445.

[38] *Weekly Compilation of Presidential Documents* 19 (October 31, 1983): 1,501.

sure that no Soviet (or Soviet-Cuban) military facility exists in Latin America. To understand these policy makers' belief that an invasion was essential to U.S. security—and there can be little doubt that this belief was sincere—we need to discuss briefly the evolution of U.S.-Grenadian relations.

The United States had no interest whatever in Grenada until 1974, the year the island gained its full independence from Great Britain. Within a few years the elected government of Sir Eric Gairy had developed a reputation for repression, intolerance, and eccentricity, which the United States observed without alarm. A former U.S. ambassador, Sally Shelton, noted in 1981 that the Gairy administration was "probably responsible for deaths of certain opposition political figures (among them the father of the current Prime Minister) as well as for vicious beatings of other opposition leaders—among whom figure two members of the current Government, including Prime Minister Maurice Bishop."[39] No one in Washington believed that democracy was destroyed in Grenada in March 1979 when Maurice Bishop and his supporters in the New Jewel Movement overthrew the despotic Gairy and his group of personal thugs, the Mongoose Gang.[40]

The new People's Revolutionary Government promptly announced that its priorities would differ substantially from those of the deposed government, whose principal goals were first, in domestic policy, the concentration of political power in the hands of Eric Gairy through methods that often lacked a due regard for legality and, second, in foreign policy, the establishment of a U.N. commission to investigate unidentified flying objects. The social reforms of the Bishop government included free milk and hot school lunches for primary school children, the reduction and then elimination of secondary school fees, an adult literacy campaign, the creation of the nation's first eye clinic, the expansion of dental clinics to all seven parishes (there had been only one on the island), an increase in medical doctors from eight to thirty-nine,

[39] U.S. Congress, House, Committee on Foreign Affairs, Subcommittee on Inter-American Affairs, *United States Policy toward Grenada*, 97th Cong., 2d Sess., 1982, p. 65. In 1977, the Carter administration named Shelton a Deputy Assistant Secretary of State for Inter-American Affairs, but she was soon asked to step down and accept the position of U.S. ambassador to the eastern Caribbean.

[40] The New Jewel Movement was formed in 1973 by the merger of Bishop's Movement for the Assembly of Peoples and the Joint Effort for Welfare, Education and Liberation (JEWEL), a splinter of the Grenada National party, the middle-class opposition to Gairy's Grenada United Labor party (GULP).

and a boost in university scholarships from three in 1978 to 209 in 1981.[41] Alongside these progressive reforms, unfortunately, was a continuation of repressive politics, including press censorship. The targets of the repression changed, however, to focus upon groups and individuals who supported the previous regime or who openly attacked the policies of the new government. Thus, Grenada under the Bishop government was characterized by the strange mixture that U.S. policy makers have such difficulty understanding in Latin America: a combination of progressive social programs and repressive politics. Grenada looked like the Cuban model.

The construction of an international airport at Point Salines reinforced the perception. In November 1979, Prime Minister Bishop announced that Cuba had agreed to provide workers and equipment to construct the airport runway and some ancillary facilities. The announced goal of the project was to increase Grenada's tourist industry, since the island's only commercial airport, Pearls, was poorly located and, with a very short runway that ends with a drop into the sea, unable to accommodate commercial jets. Anyone who ever landed at Pearls knows that Grenada needed a new airport. To reach Grenada tourists were obliged first to fly to neighboring Barbados or Trinidad then to board connecting flights on small propeller-driven aircraft. Assuming a safe arrival at Pearls on the northeast coast, a visitor then faced an hour-long ride on a bone-jarring road over mountainous terrain to reach St. George's and the Grand Anse resort area on the southwest coast. Grenadian authorities argued that revenues were lost from many potential visitors who preferred not to bother with such inconveniences. The planned airport at Point Salines was certainly not unusual for the eastern Caribbean area—at nine thousand feet the runway was similar in length to airports on St. Vincent and Trinidad and two thousand feet shorter than Grantley Adams Airport at Barbados. The positioning of various airport facilities, particularly the exposed placement of fuel depots, indicated the absence of military considerations in the airport design.

[41] *United States Policy toward Grenada*, pp. 82-83, *supra* n. 39. Grenada has no university, but it is a member of the association that operates the University of the West Indies. The university belongs to all the island commonwealth members and Belice, serving the people of these nations from campuses at Cave Hill, Barbados; Mona, Jamaica; and St. Augustine, Trinidad. The Gairy administration had halted government contributions to the university, thereby forcing individual Grenadians to bear higher costs.

While the Cubans were not the only foreign contributors to the airport, Cuba's role was the largest: one hundred pieces of heavy construction equipment and 250 technicians valued at $33.6 million, but no cash. Financial support also came from Algeria, Iraq, Libya, Syria, Venezuela, OPEC, and the European Economic Community (Grenada is an ACP state of EEC). The prime contractor was not Cuba but rather the Plessy Company, a British engineering firm. Other Cuban aid to Grenada included thirteen medical doctors and dentists, seven fishing trawlers and technical assistance in developing a fishing industry, and numerous university scholarships.[42] An easy friendship appeared to exist between Prime Minister Bishop and President Castro, and on a variety of occasions Bishop expressed his admiration of the accomplishments of the Cuban revolution. Grenada's foreign policy included formal support for the Southwest Africa People's Organization, the African National Congress, the Polisario, and the Palestine Liberation Organization. Relations were strained with the conservative neighboring governments of Barbados, Dominica, and Trinidad and Tobago. In the United Nations, Grenada regularly voted with the nonaligned nations, and—worst of all in the view of U.S. policy makers—the country refused to support a resolution condemning the Soviet invasion of Afghanistan. To U.S. officials Grenada was on the fast track of becoming another Cuba.

U.S. hostility began during the Carter administration. Despite Ambassador Shelton's later criticism of the hostile Reagan policy she participated in the formulation and implementation of several unfriendly initiatives. At the insistence of the United States, for example, in 1980 the Windward Islands Banana Growers' Association (which includes Dominica, St. Lucia, St. Vincent, and Grenada) excluded Grenada from a U.S. grant for the rehabilitation of banana trees damaged by Hurricane Allen. That same year the United States attempted to block aid for flood relief from the OAS Emergency Fund.[43] As these actions were occurring, analysts were sounding the alarm in Washington: "The Cubans surely appreciate that Grenada is strategically located by the route over which about one-half of U.S. imported oil passes," noted Robert Leiken. "Moreover, the airfield on Grenada could serve as a refueling stop for flights carrying Cuban troops to Africa or as a staging area for

[42] Ibid., pp. 7, 90.
[43] Ibid., p. 91.

241

clandestine operations by air or sea against Trinidad, Guyana, Surinam, or Venezuela."[44]

The Reagan administration intensified U.S. hostility. Early in 1981, U.S. opposition was successful in reducing or withholding loans to Grenada from the IMF and the World Bank. In June 1981, the United States offered $4 million to the Caribbean Development Bank on the condition that none of the money would go to Grenada; the Bank refused the grant. On at least three occasions in 1981 to 1982, the United States pressured EEC members not to assist in cofinancing the Point Salines airport. During his April 1982 visit to Barbados, President Reagan specifically excluded Grenada from participation in his Caribbean Basin Initiative.

In a formal sense the United States maintained diplomatic relations with Grenada, but when the Reagan administration appointed a new ambassador for the eastern Caribbean—at the time one U.S. ambassador stationed at Barbados served all the eastern Caribbean ministates—Grenada was excluded from his charge. Nor would the administration accredit a Grenadian ambassador to the United States. The Grenadian ambassador-designate, Dessima Williams, came to Washington anyway to serve concurrently as ambassador to the Organization of American States, but she was a pariah. In 1982, she noted that "in March and again in August, 1981, our Prime Minister wrote President Reagan, expressing a desire for better relations between our two countries. To this date, no reply from President Reagan has been received to these letters."[45] "If the United States has a policy toward Grenada," remarked Representative Michael Barnes in 1982, "it appears to consist of not answering Grenadian mail and avoiding being seen in the same room with officials from Grenada."[46]

The only contact between the United States and Grenada came in June 1983 when Prime Minister Bishop made an unofficial visit to the United States. He was received by National Security Advisor William Clark and Deputy Secretary of State Kenneth Dam, who is the only participant to comment on the substance of the discussion: "What he was proposing is that we move toward better relations, and we said that we would be pleased to do so. We thought that the first step would be that he should stop what was quite a campaign of attacks on the United States, and that would

[44] Robert S. Leiken, "Eastern Winds in Latin America," *Foreign Policy*, no. 42 (Spring 1981), p. 101.

[45] *United States Policy toward Grenada*, p. 13, *supra* n. 39.

[46] Ibid., p. 1.

indicate that he had a desire for better relations, and that is where the matter was left."[47] Apparently it never occurred to these officials that Bishop's visit might have placed him in an extremely difficult position at home, where he would be accused of right-wing deviationism. Nor did anyone apparently feel that the appropriate response of a great power to this initiative by a weak neighbor would be to offer a modest gesture in return. This would have permitted Bishop to return home with some evidence of success to assuage his critics. Instead, while reserving the option of continued hostility, the United States insisted upon unilateral conciliatory gestures from Grenada.

At the same time Washington was asking the Bishop government to moderate its hostile behavior the United States was increasing its pressure upon Grenada. In addition to continuing programs of economic hostility and diplomatic isolation the administration also launched what Grenadians perceived as a campaign of military intimidation. An administration spokesman dismissed these perceptions as unfounded: "Grenada has charged on numerous occasions and without a shred of evidence that the U.S. is preparing an invasion of Grenada and that various U.S. military and naval exercises in the region are part of those preparations."[48] The maneuvers in question looked pretty suspicious to most observers. The first of these, named "Ocean Venture 81," was staged from August 1 to October 15, 1981, on the Puerto Rican island of Vieques and directed against a mythical island, "Amber and the Amberdines," identified by the Department of Defense as "our enemy in the Eastern Caribbean." The *Washington Post* called it "an island suspiciously similar to Grenada" (the Grenadine islands of Carriacou and Petit Martinique are dependencies of Grenada). The second set of Ocean Venture exercises was held in 1982, and in May 1983 a third set, code-named "Universal Trek," demonstrated "how U.S. forces could land in a small Caribbean nation where a civil war is taking place," according to a U.S. Navy official.[49]

[47] U.S. Congress, Senate, Committee on Foreign Relations, *The Situation in Grenada*, 98th Cong., 1st Sess., 1983, p. 12. I have been unsuccessful in attempts to obtain more detailed accounts of the meeting.

[48] *United States Policy toward Grenada*, p. 39, *supra* n. 39.

[49] With all this training the actual invasion seemed almost anticlimatic to at least one participant. "We did a real nice job on that place," reported Commander Michael O'Brien, a Navy squadron commander. "We went down there prepared and did the job right. . . . That's the kind of real training we need." Timothy J. Christ-

Since the end of World War II covert action by the Central Intelligence Agency has been among the most important means of implementing U.S. policy. In every case where the record is available for public scrutiny U.S. diplomatic and economic hostility toward a Latin American government has always been accompanied by covert action. Latin Americans who have observed postwar U.S. policy toward the region are conditioned by history to expect that Washington's hostility will be expressed in part by covert action. Thus Grenadians undoubtedly believed that the Central Intelligence Agency was up to its old trick in Grenada: destabilization. Or, if Grenadians were ignorant of this history, they could read the newspapers. In 1983, the *Washington Post* reported that in July 1981 the CIA proposed a covert action plan against Grenada. According to the *Post*, a Republican member of the Senate Intelligence Committee described the plan as "economic destabilization affecting the political viability of the government."[50]

And so the Bishop government began to seek protection. It obtained arms to defend itself, arms that were later used by the U.S. government as proof that without the U.S. invasion "Grenada would have become a fortified Cuban/Soviet military outpost."[51] It tightened relations with Cuba and signed secret aid agreements with the Soviet Union and North Korea, agreements that were later used by the U.S. government to demonstrate how wrong critics had been when they accused the administration "of exaggerating the dangers of Cuban/Soviet activities in countries like Grenada."[52]

---

mann, "TacAir in Grenada," *Naval Aviation*, November-December 1985, pp. 8-9. For a contrasting appraisal that emphasizes the military shortcomings of the U.S. invasion forces, see Richard Gabriel, "Scenes from an Invasion," *Washington Monthly*, February 1986, pp. 34-41.

[50] *Washington Post*, February 27, 1983, p. A11. It should be noted that there is no public information on whether the CIA destabilization proposal was ever implemented. The *Post* story states that the Senate Intelligence Committee prohibited implementation, but one member of the committee said that "if they [the CIA] were going to do something, I'm not sure they would tell us. I think they would wait until it was all over and then [CIA Director William] Casey would stop somewhere at a phone booth and call the committee."

[51] The statement is by Deputy Secretary of State Kenneth W. Dam in a speech before the Associated Press Managing Editors' Conference, Louisville, Kentucky, November 4, 1983. In assessing the accuracy of Dam's perceptions, it should be noted that the principal Soviet agreement to provide Grenada with military supplies was signed in Moscow on July 27, 1982, long after the initiation of U.S. hostility.

[52] Ibid. In 1984, the Department of State published a large volume that contains

In short, what seems to have occurred in Grenada in the early 1980s is the classic self-fulfilling prophecy that has been seen so often in postwar U.S. policy toward Latin America. As in the cases of Arbenz's Guatemala, Castro's Cuba, Goulart's Brazil, Allende's Chile, and the Sandinistas' Nicaragua, in Grenada U.S. officials perceived a Latin American government exhibiting an independent foreign policy that included linkages to one of our adversaries, Cuba. The Cuban agreement to help construct an airport was perceived not as a Cuban agreement to help construct an airport but as a Soviet challenge to U.S. security. Once that perception had been established in Washington, U.S. policy (under both the Carter and Reagan administrations) was designed to protect U.S. security interests by destroying the government of Grenada. At this point the United States sought to undermine the economy, to destabilize the political system, to isolate the government internationally, and (probably) to create chaos through covert action.

As these actions by the U.S. government became apparent, the Bishop administration was forced to choose either to negotiate a *modus vivendi* or to fight back with weapons provided by adversaries of the United States. After attempts at negotiation were rebuffed—the unanswered Bishop letters to Washington in 1981—Grenada, pushed into a corner, had no choice but to turn for assistance, probably in desperation, to U.S. rivals. Then the Reagan administration used the existence of Eastern bloc assistance as proof that it was correct in pushing the Grenadians into a corner and, ultimately, invading Grenada.

"Nonsense" was the response of Reagan administration officials. In 1982, a year after the CIA destabilization plan had allegedly been presented to the Senate Intelligence Committee for approval, Deputy Assistant Secretary of State Stephen Bosworth explained that

> we are not asking for perfection from Grenada. We do, however, ask that the level and tone of Grenadian rhetoric and criticism of the United States diminish. We would think it would be useful if Grenada were to give some indication of its plans to return to a constitutional system. We are concerned

---

photocopies of these agreements along with other evidence to support the administration's claims. See U.S. Department of State, *Grenada Documents: An Overview and Selection* (Washington, D.C.: Department of State, September 1984). Further documentation is available in Paul Seabury and Walter A. McDougall, *The Grenada Papers* (San Francisco: ICS Press, 1984).

about its performance in the area of human rights. We are concerned about the extent to which it is spending its very scarce resources on an airport for which we are unable to find an economic justification. We are also concerned about its very close, and what appears to be growing ties with Cuba, and its failure to adhere to any type of legitimate non-alignment.[53]

Not everything in this statement rings true. At no time in recent history has the United States refused to accredit the ambassador of a Latin American government because it has refused to return to a constitutional system, because of its performance in the area of human rights, or because of its unwise expenditure of scarce resources on airports or any other project.[54] The *real* problem with Grenada was explained after the invasion by Deputy Secretary of State Kenneth Dam:

We now know that we had underestimated Soviet use of Cuba as a surrogate for the projection of military power in the Caribbean. Examine again what we found—well-armed Cubans called construction workers; fortifications, stockpiled weapons, secret military treaties; personnel from Eastern Europe, Africa, and East Asia, all innocently enjoying a tourist paradise no doubt. Think again about the facilities all this would have secured—the Point Salines Airport, which would have enabled a MIG-23 carrying four 1,000-pound bombs to strike and return from Puerto Rico in the north to Venezuela in the south; the Calivigny military training area; a 75,000-watt radio transmitter capable of blanketing the entire Caribbean; the potential for a deepwater harbor.[55]

[53] *United States Policy toward Grenada*, p. 47, *supra* n. 39. For the interpretation of a Carter administration official who argues that both sides were responsible for the confrontation, see Robert A. Pastor, "The United States and the Grenada Revolution: Who Pushed First and Why?" Documento de Trabajo No. 26, Centro de Investigaciones del Caribe y América Latina, Universidad Interamericana de Puerto Rico, 1986, especially p. 29.

[54] At the time of the invasion the Department of State stood by its assessment that the airport construction could not be justified on economic grounds. Three months later Secretary of State Shultz landed in an Air Force transport on the unfinished Point Salines field and announced that "it certainly is a facility that is needed here and in one way or another I'm sure it will be completed." After an infusion of $19 million from AID, the airport opened in October 1984. For the original State Department assessment, see *The Situation in Grenada*, p. 46, *supra* n. 47; for the statement by Secretary Shultz, see *New York Times*, February 8, 1984, p. 3.

[55] Speech to the Associated Press Managing Editors' Conference, Louisville, Kentucky, November 4, 1983.

In interviews no U.S. official expressed the belief that the airport in Grenada would ever have been used as a base for bombing runs on Puerto Rico or Venezuela. The fear was often expressed, however, that Point Salines would become a convenient stopover for Cubans going to places the United States did not want them to go (for example, Africa) or for incoming arms shipments to Latin America. In his speech to a joint session of Congress in April 1983, President Reagan observed:

> We're all aware of the Libyan cargo planes refueling in Brazil a few days ago on their way to deliver "medical supplies" to Nicaragua. Brazilian authorities discovered the so-called medical supplies were actually munitions and prevented their delivery. You may remember that last month, speaking on national television, I showed an airfield being built on the island of Grenada. Well, if that airfield had been completed, those planes could have refueled there and completed their journey.[56]

The available evidence does not permit us to confirm or reject Mr. Reagan's and Mr. Dam's assertions. Too many questions remain unanswered. One group of questions concerns the evolution of U.S.-Grenadian relations. Could wiser U.S. leadership have prevented the schism with Grenada by, for example, offering to help build the airport rather than treating it as a "Soviet-Cuban militarization"? Or were Maurice Bishop and his followers in the New Jewel Movement so committed to alignment with the Soviet Union and Cuba that no amount of U.S. patience or U.S. assistance would have kept the Grenadians from permitting their new airport to be used to threaten U.S. security interests? We will never know the answer to these questions.

A second group of unanswered questions concerns U.S. direct involvement in creating or helping to create the chaotic conditions that justified an invasion. We know that the U.S. government was hostile toward the Bishop government, and there is some evidence that covert action was part of this hostility, but what kind of hostility? Was the CIA active in Grenada? If so, was the CIA involved only in economic destabilization, or did it initiate yet another of its assassination plots? Or did the CIA simply work to create schisms among members of Bishop's New Jewel Movement? Whom did the CIA support in Grenada? The people

---

[56] *Weekly Compilation of Presidential Documents* 19 (May 2, 1983): 609.

who killed Prime Minister Bishop? Who arranged for Grenada's governor general to appeal to the Organization of Eastern Caribbean States (OECS) for assistance? Who arranged for the OECS to request a U.S. invasion?[57] How did faraway Jamaica become involved? These are all parts of a single broader question: To what extent did the United States create the circumstances that led to the rationale for the U.S. invasion? Someday we will know much more about these questions than we do at present.[58]

All that seems certain today is that in invading the island of Grenada the United States demonstrated its firm commitment not to permit either a Soviet military presence in the Western Hemisphere or a Cuban military presence beyond the island of Cuba.

[57] The U.S. position is that "in October 1983, Grenada's eastern Caribbean neighbors proved their democratic mettle when they acted—without hesitation and with the support of other democratic nations, including the United States—to restore order in Grenada after the country had fallen prey to a bloody power struggle among its Marxist-Leninist leaders." U.S. Department of State, Bureau of Public Affairs, *Democracy in Latin America and the Caribbean*, Current Policy No. 605 (Washington, D.C.: Department of State, August 1984), p. 12. Within the U.S. government it was the Department of State and not the Department of Defense that championed the invasion. The director of policy planning for State's Bureau of Inter-American Affairs remarked with some pride that "the action in Grenada—and I don't accept the word intervention—was not originated in the Pentagon. . . . It originated here [in the Department of State] at the request of Grenada's neighbors who had no military force of their own." Discussion with Luigi Einaudi, Washington, D.C., September 12, 1984.

[58] Clearly, it will be some time for these answers to emerge, for in the face of overwhelming public support for the invasion, there was no possibility of an investigation by Congress. Sensing the public mood, the members of Congress who might have sponsored an investigation rushed instead to endorse the administration's action: "There was no longer any alternative to force," wrote Representative Michael Barnes, chairman of the House Foreign Affairs Subcommittee on Western Hemisphere Affairs. *Washington Post*, November 9, 1983, p. A19. Without an initiative by Congress we must now await the declassification of documents by twenty-first-century historians. In this sense the congressional reaction to Grenada closely parallels its reaction to the U.S.-sponsored coup in Guatemala in 1954, when members of Congress vied with one another to demonstrate their anti-Communist credentials to a public conditioned by McCarthyism. Congressional timidity in both the Guatemala and Grenada cases contrasts dramatically with the assertive behavior of Congress in the mid-1970s, when several committees produced remarkably revealing analyses of U.S. policy toward Chile. While we await the research by historians, political scientists have begun to produce descriptive studies whose conclusions rest heavily upon personal policy preferences in lieu of facts. See, for example, Alexander Haywood McIntire, Jr., "Revolution and Intervention in Grenada: Strategic and Geopolitical Implications," Ph.D. diss., University of Miami, 1984.

THE NATURE OF THE THREAT

The beliefs of most policy makers are strongly influenced by geography: because Latin America is located close to the United States, a Soviet base in Latin America would increase Moscow's ability to attack the United States. From the time of the Monroe Doctrine this geographic fact has been a prominent feature of U.S. strategic thinking. In 1890, Alfred Thayer Mahan wrote that "it should be an inviolable resolution of our national policy that no foreign state should henceforth acquire a coaling position within three thousand miles of San Francisco—a distance which includes the Hawaiian and Galapagos Islands and the coast of Central America. For fuel is the life of modern naval war; it is the food of the ship; without it the modern monsters of the deep die of inanition."[59] Two decades later the Lodge Corollary to the Monroe Doctrine expressed a similar fear of foreign naval bases near the United States. By 1940, the airplane had replaced the naval vessel as the vehicle carrying the threat to the United States. But the airplane, too, was subject to inanition. That is why President Roosevelt called a joint session of Congress to warn that Mexico might soon come under the influence of Axis powers, and "Tampico is only two and a quarter hours to St. Louis, Kansas City, and Omaha."[60] Today no one uses the word "inanition," and the intercontinental range of many weapons has made geography less significant. But to most analysts and policy makers geography remains important. To Robert Tucker, "what we cannot view with equanimity is the states of Central America entering into a relationship with either the Soviet Union or its proxy, Cuba, that resembles the relationship between the Soviet Union and Cuba. Geographical proximity has not lost its significance."[61]

The irony is that although it has no military base in Latin America, for nearly two decades the Soviet Union has been able to use its increasingly capable naval and air forces to maintain a modest but regular presence in the Caribbean.[62] The Soviet Union now

[59] Alfred Thayer Mahan, *The Interest of America in Sea Power: Present and Future*, 12th ed. (Boston: Little, Brown, 1918), p. 26.

[60] *Congressional Record* 86 (May 16, 1940): 6,243.

[61] Robert W. Tucker, *The Purposes of American Power: An Essay on National Security* (New York: Praeger, 1981), p. 182.

[62] There is some question whether the Soviet Union currently "has" a base in Cuba or only "uses" Cuban bases. In a 1986 speech urging congressional support for the Nicaraguan Contras, President Reagan asserted that "the Soviet Union already uses Cuba as an air and submarine base in the Caribbean." When debating

has a modern blue-water navy, and it uses it to patrol in waters that had until recently never seen a Soviet naval vessel. In the 1970s, the USSR permanently stationed three ships in the Caribbean: one tugboat, one ship to support Soviet space communications, and one intelligence collector (or trawler). In 1978, the U.S. Navy stated that these ships did not pose a threat to U.S. security or to the security of U.S. allies in the area.[63] Since 1969, the Soviet Union has also regularly sent small naval task groups into the Caribbean; typically they consist of two or three combatants accompanied by a supply ship. Occasionally one of the ships is a submarine but never a nuclear-powered strategic missile submarine. The frequency of these deployments has been quite low—twenty-two occurred in the fourteen-year period between 1969 and early 1983.[64] A typical visit by a four-ship task group (a guided-missile cruiser, an escort frigate, a diesel-powered submarine, and a tanker) occurred from November 1982 through February 1983 and, like most visits, included something new but not dramatically new. This time it was the refueling of an intelligence collector off the coast of Florida and a move to within fifty miles of the Mississippi Delta, the closest ever to the United States mainland. There were also the expected port visits to Havana and Cienfuegos, along with the now-routine antisubmarine exercises with Cuban units—the Soviet version of UNITAS.[65]

Policy makers differ in their interpretation of the messages

whether to support the overthrow of the Sandinista government in Nicaragua, lawmakers therefore would be well-advised, he said, to "think how Cuba became a Soviet air and naval base." *Weekly Compilation of Presidential Documents* 22 (June 30, 1986): 866.

[63] *Arms Trade in the Western Hemisphere*, p. 308, *supra* n. 2. For many years the Soviet Union has also stationed its trawlers off major U.S. ports and military facilities, and they bristle with activity during U.S. space shots or missile tests. See Harriet Fast Scott and William F. Scott, *The Armed Forces of the USSR*, 2d ed. (Boulder, Colo.: Westview Press, 1981), p. 166.

[64] U.S. Congress, House, Committee on Foreign Affairs, Subcommittee on Inter-American Affairs, *Impact of Cuban-Soviet Ties in the Western Hemisphere, Spring 1980*, 96th Cong., 2d Sess., 1980, p. 7; U.S. Department of Defense, *Annual Report to the Congress, Caspar W. Weinberger, Secretary of Defense, Fiscal Year 1983*, p. II-23; *New York Times*, February 14, 1983, p. 10. For a list of the first fourteen deployments, see U.S. Congress, House, Committee on International Relations, Subcommittee on International Political and Military Affairs, *Soviet Activities in Cuba—Parts VI and VII: Communist Influence in the Western Hemisphere*, 94th Cong., 2d Sess., 1976, p. 4.

[65] Donald C. Daniel and Howard M. Hansel, "Navy," *Soviet Armed Forces Review Annual* 7 (1982-1983): 221, 235; *New York Times*, February 14, 1983, p. 10.

being communicated by these Caribbean deployments. Few would even guess at the message of the fourteenth deployment (May to June 1975) because the ships sailed to Cuba not from the Soviet Union but from Boston, where they had just completed a courtesy call as part of the U.S. bicentennial celebration. The other visits were not so overtly benign, however. Some policy makers agree with Barry Blechman, who a decade ago interpreted the deployments as probes, designed to test the U.S. reaction and, much more important, eventually to reduce U.S. credibility:

> The Soviet submarine visits to Cuba, by gradually encroaching upon previous "understandings," more significantly pose a political challenge to U.S. security. Potentially, and if successful over the long term, this sort of activity could help to bring into question, in the eyes of Soviet decision-makers and leaders of other nations, the credibility and impact of U.S. statements, warnings, and other forms of verbal behavior. And without such credibility, the fabric of the U.S. posture in world affairs could be undermined seriously.[66]

Most policy makers accept the first part of this interpretation— that the deployments are probes to test U.S. reactions—but they tend to reject the second part—that the probes are designed to reduce U.S. credibility in world affairs. Since the United States has never asserted the right to restrict maritime access to the Caribbean, and indeed defends vigorously the principle of freedom of the seas, it is difficult to understand why the United States would lose credibility when an adversary does something that Washington admits the adversary has a right to do.

Most policy makers tend to think of these deployments as serving a variety of purposes for the Soviet Union, with the particular emphasis on one purpose or another dependent upon the circumstances at the time of the deployment. Policy makers frequently suggest, first, that the Soviets are making a statement about international freedom of the seas. The Caribbean and the Gulf of Mexico do not belong to the United States, of course. Beyond the limits of coastal states' territorial seas, these are international waters for the Soviet Union and anyone else to use. This is precisely the position taken by the United States on the use of the Black Sea and the Baltic Sea, which are perhaps of greater geopolitical and eco-

---

[66] Blechman and Levinson, "Soviet Submarine Visits to Cuba," p. 38.

nomic importance to the Soviet Union than the Caribbean is to the United States.

Second, some officials interpret the Soviet deployments as a carefully calculated *quid pro quo*: the United States cruises into "their" waters, and they cruise into "ours." Using this interpretation, Washington's decision to deploy two U.S. destroyers in the Black Sea in June 1969 is thought by at least one analyst to have provoked the first Soviet naval deployment in the Caribbean a month later.[67] Similarly, the Soviet deployment that came so close to the Mississippi may have been in response to deployments of a U.S. frigate, the *Caron*, which cruised to within twelve miles of Gdansk, Poland, in August 1980, during the birth of Solidarity and then, during a subsequent cruise, came to within about fifteen miles of the Soviet naval base at Murmansk.[68]

While many U.S. officials worry about these Soviet deployments, and while a few debate among themselves the meaning of each visit, there is nothing they can do about them. Barring the outbreak of war, the Soviet Union will undoubtedly continue to exercise its right to send warships into the Caribbean. Accepting this reality, U.S. officials have sought not to stop the deployments but rather to minimize their potential effectiveness. This is the reason why a Soviet naval *base* in Latin America is so feared: it would increase the effectiveness of Soviet units. Morton Kaplan argues (with some hyperbole) that

it is virtually impossible to mount a global naval policy without extended bases, for the absence of bases makes costing

[67] Abel, "A Breach in the Ramparts," p. 47. The comparison of the Black Sea and the Caribbean Sea is not accurate, for access to the Caribbean is totally free, while the Montreux Convention governs passage through the Turkish Straits and, therefore, access to the Black Sea. Article 13 requires that vessels of war give Turkey eight days' notice prior to passage, and "it is desirable that in the case of non-Black Sea Powers this period should be increased to fifteen days." More important, Article 14 limits the maximum aggregate tonnage that may use the Straits at any one time to 15,000 tons, except those ships making courtesy visits to Turkish ports in the Straits. No similar restrictions apply to the use of the Caribbean.

To assert its right to send warships into the Black Sea, the United States typically sends two small vessels (destroyers or frigates) through the Straits in the spring and fall of each year. For a discussion of this use of the Black Sea and the reaction of the Soviet Union, see Harry N. Howard, *Turkey, the Straits, and U.S. Policy* (Baltimore: The Johns Hopkins University Press, 1974), pp. 275-276.

[68] Jeffrey Richelson, *The U.S. Intelligence Community* (Cambridge, Mass.: Ballinger Publishing Company, 1985), p. 129. On the general issue of reciprocity in the Caribbean, see Jorge I. Domínguez, *U.S. Interests and Policies in the Caribbean and Central America* (Washington, D.C.: American Enterprise Institute, 1982), p. 9.

[*sic*] astronomical. The maintenance of an American base posture in the Caribbean and the exclusion of further Soviet basing involves in a vital way the entire cost structure of the American military establishment. . . . Extended basing on the part of the Soviet Union sharply reduces the cost of Soviet naval threat and by correlation sharply increases the costs of American defensive operations.[69]

U.S. security officials have been particularly anxious about the problems that would occur if the Soviet Union were to station part of its submarine fleet at a base in Latin America.[70] Because of the central role assigned to strategic nuclear submarines (SSBNs) in deterrence theory (and, of course, practice), neither the United States nor the Soviet Union is comfortable when its strategic submarines enter foreign ports. U.S. officials are properly horrified by the thought of an SSBN being attacked or—worst of all—captured, and foreign ports inevitably present a relatively high risk of security breaches. Thus U.S. SSBNs make foreign port visits on an infrequent basis, often to communicate a message to the Soviet Union. In April 1963, for example, a Polaris strategic submarine, the *Sam Houston*, made a port call at Izmir, Turkey, to demonstrate as clearly as possible that the United States was committed to the defense of its NATO ally, even though it was withdrawing its land-based Jupiter missiles.[71] In 1983, a typical year, U.S. SSBNs made three visits, all to safe foreign ports: Bremerhaven, West Germany; Portland, England; and Agadir, Morocco.

Unlike the Soviet Union, however, the United States maintains three nuclear submarine bases overseas: on the island of Guam, at Rota in southwest Spain, and at Holy Loch in Scotland. The U.S. Navy controls these bases completely. If Moscow were to establish a similar strategic submarine base in Latin America, including facilities for replacement crews, the Soviet Union would gain two significant strategic advantages now enjoyed exclusively by the United States.

[69] Morton A. Kaplan, "American Policy toward the Caribbean Region: One Aspect of American Global Policy," in *Issues in Caribbean International Relations*, ed. Basil A. Ince et al. (Lanham, Md.: University Press of America, 1983), p. 58.

[70] For representative statements of these fears, see Jack L. Roberts, "The Growing Soviet Naval Presence in the Caribbean: Its Politico-Military Impact upon the United States," *Naval War College Review* 23 (June 1971): 37; Michael K. Mcc-Gwire, "Soviet Maritime Strategy, Capabilities, and Intentions in the Caribbean," in *Soviet Seapower in the Caribbean: Political and Strategic Implications*, ed. James D. Theberge (New York: Praeger, 1972), pp. 39-58.

[71] Abel, "A Breach in the Ramparts," p. 34.

First, Soviet submarines would increase their on-station time. This is particularly true for the older SSBNs and attack submarines attached to the Northern Fleet, the most important element of the Soviet navy, that must presently use facilities along the isolated Murman coast, east of Norway and far above the Arctic Circle. No one knows precisely how much a base in Latin America would increase Soviet on-station time, but there are many estimates. In 1972, Herbert Scoville argued that from their existing bases "it would take Russian [missile] submarines a minimum of six days in the Atlantic and eight days in the Pacific to reach the nearest launch stations, so that the transit time to and from home ports, in many cases a quarter to a third of the duration of the patrol, seriously degrades the operational readiness of the Russian fleet."[72] In 1970, the Department of Defense estimated that a base in Latin America would permit a 33 percent to 40 percent increase in on-station time; in 1971, the range was broadened to from 20 percent to 50 percent.[73] Other estimates include "nearly twice as long," "a factor of two or more," and 25 percent to 40 percent.[74]

The Soviet submarine fleet and its missile capability have improved greatly since these estimates were made in the 1970s, but no one suggests that the Soviets have been able to overcome completely the disadvantages of a lack of forward bases. Whatever the correct figure might be in the mid-1980s, a Soviet submarine base in Latin America would clearly increase fleet efficiency. It would, for example, reduce the reaction time for Soviet attack submarines assigned to trail U.S. aircraft carriers based at Norfolk, Virginia, and Mayport (near Jacksonville), Florida, and the U.S. SSBNs based at Charleston, South Carolina, and Kings Bay, Georgia.[75] With a base in Latin America the same number of Soviet submarines could provide more military capability, or the number of Soviet submarines could be reduced at no loss of capability. This latter

[72] Herbert Scoville, Jr., "Missile Submarines and National Security," *Scientific American* 226 (June 1972): 24.

[73] *Washington Post*, December 6, 1970, pp. K1, K17; Roberts, "The Growing Soviet Naval Presence in the Caribbean," p. 36.

[74] James D. Theberge, "Soviet Policy in the Caribbean," in *Soviet Seapower in the Caribbean*, p. 10; Donald W. Mitchell, "Strategic Significance of Soviet Naval Power in Cuban Waters," in *Soviet Seapower in the Caribbean*, p. 35; MccGwire, "Soviet Maritime Strategy, Capabilities, and Intentions in the Caribbean," p. 50.

[75] Blechman and Levinson, "Soviet Submarine Visits to Cuba," p. 34. The round trip from Soviet Northern Fleet headquarters to the waters off Virginia takes between two and three weeks. A round trip from Cuba to Virginia would take about three days.

possibility could become quite significant in the event of a U.S.-Soviet arms control agreement that called for a reduction in the number of submarine-launched ballistic missiles.

All of this assumes, however, that Soviet SSBNs are stationed fairly close to U.S. shores, far from their Northern and Pacific fleet ports. This has not been the case with the newest generation of Soviet submarines, which have more powerful missiles. When Scoville was writing the article from which the above quotation was taken, the Soviet's *Yankee*-class SSBNs, their best, carried missiles with a thirteen-hundred-mile range. Advances in Soviet missile technology have permitted their new generation of *Delta*-class SSBNs to destroy any target in the United States by firing its missiles from forty-two hundred miles away in the Barents Sea (north of the Kola Peninsula) and the Sea of Okhotsk (off the Pacific coast of the USSR). The Soviets still deploy their *Yankee*-class submarines, however, and the Soviet fleet of attack submarines will always have to patrol far from home, but the single most significant threat the Soviets possess is their *Delta*-class SSBNs, and to these vessels Latin America is absolutely irrelevant. There is no better example of how changes in technology have required the reevaluation of Latin America's geopolitical significance. Anything ever written or said in the pre-*Delta* era about a Soviet SSBN base (but not an attack submarine base) in Latin America is simply out of date.[76]

The second important advantage to the Soviets of a base in Latin America would be the increased ability to deploy submarines secretly. All Soviet strategic submarines are based at Northern Fleet facilities on the Murman coast or at Pacific Fleet bases on the western coast of Siberia. Facilities are also available at Soviet naval bases in the Baltic Sea and the Black Sea, but their use by submarines is limited. Three NATO nations—Denmark, West Germany, and Norway—can blockade the Baltic, and the exit from the Black Sea to the Mediterranean is through the Dardanelles, the Sea of Mamora, and the Bosporus, an international waterway under the sovereignty of another NATO member, Turkey. Passage through this waterway is governed by the 1936 Montreux Convention, which requires (Articles 12 and 13) not only that prior noti-

---

[76] Compare, in particular, the maps in Scoville, "Missile Submarines and National Security," pp. 20-21, which was written in 1972, with the maps in Joel S. Wit, "Advances in Antisubmarine Warfare," *Scientific American* 244 (February 1981): 39-40. It is a comparison that underscores extraordinarily well the speed of technological innovation in modern military hardware.

fication be given with respect to the passage of warships but that "submarines must travel by day and on the surface."[77] This requirement deprives submarines of their principal tactical advantage, stealth.

As the Soviet's Northern Fleet (non-*Delta*-class) submarines leave their bases to go on patrol, they must pass through the so-called Greenland-Iceland-United Kingdom Gap. This is hardly a narrows—the widest opening in the gap, the distance between Iceland and the Danish Faroe Islands, is about 400 miles—but it does permit NATO units to identify all submarine traffic. Thereafter, the U.S. Ocean Surveillance Information System employs a variety of electronic marvels to track Soviet submarines on patrol.[78] As a matter of policy the Department of Defense refuses to comment on its antisubmarine tracking capability, but it is widely believed that the U.S. Navy could continuously monitor the position of every Soviet submarine if it wished to allocate the enormous resources that would be required.[79]

In the event of a major national security crisis it would be difficult to overestimate the value of this tracking capability, and it is for this reason that a Soviet submarine base in Latin America is viewed with alarm by many policy makers. U.S. defenses could be adjusted to accommodate such a base and maintain existing capabilities, but the adjustment would be expensive and the results could never quite equal the security of having no Latin American base about which to worry. An additional Soviet submarine base would inevitably increase the probability of error on the part of U.S. trackers, an error with an extremely high potential cost.

On the other hand, there are even greater potential costs to *knowing* the precise location of every Soviet SSBN: in a crisis situation this knowledge in the hands of the United States would en-

[77] 173 L.N.T.S. 215.

[78] Richelson, *The U.S. Intelligence Community*, pp. 139-150; Blechman and Levinson, "Soviet Submarine Visits to Cuba," p. 34; Komorowski, "Latin America—An Assessment of U.S. Strategic Interests," p. 158.

[79] In 1986, the Soviet Union had 62 SSBNs. Although only a small proportion of these submarines is on patrol at any time, the task of continuously monitoring their positions (plus the positions of the hundreds of Soviet attack submarines) would clearly place an extraordinary burden on the U.S. Navy. Whether the United States actually expends these resources is a closely guarded secret. Letter, Office of Information, Department of the Navy, Washington, D.C., July 23, 1984. Wit, "Advances in Antisubmarine Warfare," and Richelson, *The U.S. Intelligence Community*, are the most informative published discussions of the U.S. antisubmarine capability.

courage the Soviet Union to use its submarine-launched missiles in a preemptory first strike. SSBNs have been designed to serve as second-strike weapons; the logic of SSBN deterrence is based upon the assumption of their ability to survive a first strike. If the Soviets believed that U.S. antisubmarine capabilities made their SSBNs' survival impossible (or even improbable), in a crisis they would have no alternative but to launch their missiles preemptively. Thus there is considerable reason to believe that the United States does not want to know (or, perhaps, does not want the Soviet Union to know if, in fact, we know) the precise location of patrolling Soviet SSBNs.[80] In the context of this facet of strategic theory a Soviet submarine base in Latin America cannot be considered a threat to U.S. security.

Much of what has been said about the national security implications of Moscow's use of Latin American naval facilities is also asserted (but with less intensity) about the activities of Soviet military aviation in the region.[81] Since April 18, 1970, Soviet Tu-95/Bear naval reconnaissance aircraft, the largest Soviet naval aircraft, have been flying about once every other month from the Soviet Union's Murman coast, around the North Cape, down the Norwegian Sea and across the Atlantic to Cuba's airbase at San Antonio de los Baños, southwest of Havana. The Tu-95s generally fly in pairs and remain in Cuba for several weeks before returning to the Soviet Union.[82] The Tu-95 has an eight-thousand-mile range and a four-and-a-half-ton cargo capacity; with relatively minor modifications it can be converted from reconnaissance and surveillance service to duty as a strategic bomber carrying air-to-surface missiles. Although the Tu-95 is an older manned aircraft, and therefore highly vulnerable to attack by U.S. forces, it could, and probably would, launch its cruise missiles outside of U.S. airspace, giving the United States a much shorter reaction time than it would have in the event of an attack by land-based ICBMs. Nonetheless, it is extremely difficult to imagine the Soviets using this weapon as the basis of a first strike on the United States.

[80] I am indebted to Richard Ullman for calling my attention to this aspect of deterrence theory.

[81] Kirkpatrick, "U.S. Security and Latin America," p. 29; Arms Trade in the Western Hemisphere, p. 308, supra n. 2.

[82] Stockholm International Peace Research Institute, World Armaments and Disarmament: SIPRI Yearbook, 1982 (Cambridge, Mass.: Oelgeschlager, Gunn and Haig, 1982), p. 399; Abel, "A Breach in the Ramparts," p. 48; Roberts, "The Growing Soviet Naval Presence in the Caribbean," p. 31.

There simply is no way to surprise the United States in any significant way with a pair of manned turboprop bombers.

Most U.S. security officials therefore assume that the Soviet flights have two roles. The first is to collect information.[83] This is generally thought to be a minor function, however, as the flights are too infrequent and the alternative means of collection (primarily satellites) probably provide superior data. The second is symbolic—to demonstrate Soviet-Cuban solidarity and to assert the freedom of international airspace in exactly the same way that Soviet naval units assert the freedom of the seas by using the Caribbean.

Were the Soviet Union to establish a permanent airbase somewhere in the Caribbean region, however, it would substantially complicate U.S. national security planning. The United States would have to reinforce its existing air defense and radar warning systems directed toward the south, but this is something that the air force has already been doing for several years. A Soviet airbase would also grant the Soviets a number of options in the region that they now lack, including (to use a worst-case analysis) the ability to deploy ground forces on short notice and to provide air support of ground combat activity by allies of the Soviet Union.

This worst case, however, is also highly improbable. Far from home and surrounded by the very heart of U.S. military power, the Soviet ground and air forces would be extremely vulnerable. Given the Soviet experience in neighboring Afghanistan, it is almost inconceivable that Moscow would commit combat forces to any conflict in Latin America. Thus, as in the case of a Soviet submarine base, a Soviet airbase in Latin America could be interpreted as more than a minor irritant to U.S. security planners, but such an interpretation would be based upon assumptions that the Soviet Union would take steps that are, at most, highly unlikely if not inconceivable.

In summary, most discussions of the nature of the threat from Soviet bases in Latin America assume that the threat would be traditional: missiles, bombers, and other hardware placed at these bases would be used against the United States or U.S. allies. My conclusion is that while this type of traditional threat can never be dismissed as impossible, it would involve such a high level of geo-

---

[83] See, for example, the Reagan administration's position in U.S. Department of State, *Cuban Armed Forces and the Soviet Military Presence*, p. 5.

258

political adventurism on the part of the ever-cautious Soviets that it must be regarded as highly improbable.

There is, however, an additional, quite different, use for these bases: as decoys. Indeed, some officials argue that the principal goal of Soviet strategic policy toward Latin America is to tie up U.S. resources that Washington would prefer to deploy elsewhere. This is often referred to as the "economy of force" argument and serves as the basis for many discussions of Soviet policy. Typical is the belief of a group of analysts from the RAND Corporation, who argued before the Kissinger Commission that "it remains a *strategic imperative* that the United States prevent threats from arising in Central America that would require the diversion of military and other resources, to the detriment of U.S. strength and flexibility elsewhere. . . . A secure Central America enhances the U.S. ability to attend to the global power struggle."[84] One of the authors of the RAND study argues that "the ability of the United States to act as a world power in a global balance-of-power system is greatly enhanced by the exclusion of that system and its related threats from the [Caribbean] Basin. Otherwise, instability and insecurity in the Basin may divert the United States to an extent that constrains its ability to play its global role from a position of strength."[85] One Reagan administration official suggested in 1982 that increased Soviet arms shipments to Cuba were for the purpose of strengthening "Cuba's ability to project power [which] will ultimately oblige the United States to divert military resources, until now better employed elsewhere."[86] Other policy makers argued in the early 1980s that "the major Soviet objective in Central America appears to be the achievement of 'freedom of action' in the area of their central strategic thrust. They seek to disrupt the American 'strategic rear' and to bog down the United States politically and militarily. This . . . seems to be the chief Soviet objective in the region."[87]

Many policy makers believe that this decoy activity may be the

[84] Edward González et al., *U.S. Policy for Central America: A Briefing*, Report No. R-3150-RC (Santa Monica, Ca.: RAND Corporation, March 1984), pp. v, 4.

[85] David Ronfeldt, *Geopolitics, Security, and U.S. Strategy in the Caribbean Basin* (Santa Monica, Ca.: RAND Corporation, November 1983), p. 8.

[86] Myles R. R. Frechette, "Letter to the Editor," *Foreign Policy*, no. 48 (Fall 1982), p. 177.

[87] U.S. Congress, House, Committee on Foreign Affairs, Subcommittee on Inter-American Affairs, *U.S. Policy Options in El Salvador*, 97th Cong., 1st Sess., 1981, pp. 121-122; Osgood, "Central America in U.S. Containment Policy," p. 92.

most deleterious effect of any Soviet military base in Latin America. U.S. resources are far from infinite, and the cost of guarding a decoy is going up constantly. These costs and other resource limitations were clearly illustrated in July 1983, when President Reagan ordered U.S. military maneuvers in Honduras and along both coasts of Nicaragua in an effort to stem the alleged flow of arms to guerrillas in El Salvador. The Pentagon responded that there would have to be a delay until November. U.S. forces were needed elsewhere. Between July and October, U.S. ground, sea, and air units were scheduled to engage in maneuvers in Eygpt, the Sudan, Somalia, and the NATO countries, where an annual exercise simulates the reinforcement of Europe from the United States. The air force lacked cargo planes, the navy lacked ships, and the Joint Chiefs of Staff lacked money until the beginning of the 1984 fiscal year. In the meantime U.S. resolve would have to be demonstrated by the presence of aircraft carriers. Even then the financial drain on U.S. military resources was significant if not staggering: in 1983, the navy estimated that it cost $963,558 a day to deploy a small carrier battle group in Central American waters.

The obvious point is that the United States cannot do everything all the time. To give the Pentagon one additional task—in this case the task of neutralizing any potential Soviet aggression from bases in Latin America—may someday require the sacrifice of another foreign policy goal. To treat any major Soviet base as if it were an inert decoy, however, would be so risky as to be unthinkable. It is far better, these policy makers argue, not to let a Soviet base be built in the first place.

Cuba As a Surrogate Military Base

Complicating this analysis of Soviet bases in Latin America is a widely held belief that, as Assistant Secretary of State Enders remarked in 1982, "we must be clear about Cuba. It is a Soviet surrogate."[88] To many in Washington, Cuban bases are the equivalent of Soviet bases. The attitude of the Department of Defense is typical: "An estimated 6,000 to 8,000 Soviet civilian advisers are in Cuba and allow the Soviet masters to monitor closely their Caribbean island."[89] Officials in the Department of State seem more

[88] Testimony before the Senate Judiciary Subcommittee on Security and Terrorism, March 12, 1982.

[89] U.S. Department of Defense, *Annual Report to the Congress, Caspar W. Weinberger, Secretary of Defense, Fiscal Year 1983*, pp. II-23.

ambivalent about the Soviet-Cuban relationship. Some use the "Soviet masters" language that dominates the thinking of the Pentagon;[90] others seem less likely to strip Cuba of its autonomy: "While Cuba is increasingly dependent on the U.S.S.R. and subject to Moscow's manipulation, it would be erroneous to regard it as merely a coerced Soviet satellite."[91] On balance, most officials in the Department of State appear to perceive Cuba as having considerable autonomy but never independence. Some members of Congress also conceive of Cuba in this role; others adhere to the view of Cuba as a Soviet lackey.[92]

Under either the guidance or the domination of Moscow, the Cuban military is perceived as a formidable fighting force: large, well-equipped, and increasingly professional.[93] This perception is probably accurate. The overall strength of the Cuban armed forces is second to Brazil in Latin America. In total military spending Cuba is third to Brazil and Argentina; thus on a per capita basis Cuba has the largest and most expensive armed forces in Latin America.[94] The Cuban army consists of about 225,000 personnel equipped with modern small and medium arms and about 650 tanks. The Cuban air force of 16,000 personnel operates over 200 aircraft that range from the very modern (MiG-23) to the quaint (An-2M biplanes used for crop-dusting).

The air force is composed primarily of fighter aircraft. In speak-

[90] U.S. Department of State, *Cuban Armed Forces and the Soviet Military Presence*, p. 2.

[91] Statement by Kenneth N. Skoug, Jr., at the Carnegie Endowment for International Peace, Washington, D.C., December 17, 1984 (mimeographed).

[92] For an example of this latter position, see the statement by Senator Howard Baker in U.S. Congress, Senate, Committee on Foreign Relations, *Foreign Assistance Authorization for Fiscal Year 1982*, 97th Cong., 1st Sess., 1981, p. 15. For the most strident expression of this view of Cuba as a Soviet puppet, see the statement by a representative of the American Legion in U.S. Congress, House, Committee on Foreign Affairs, Subcommittee on Inter-American Affairs, *U.S. Policy toward El Salvador*, 97th Cong., 1st Sess., 1981, p. 223. For a more balanced and informative analysis of the same issue, see Robert A. Pastor, "Cuba and the Soviet Union: Does Cuba Act Alone?" in *The New Cuban Presence in the Caribbean*, ed. Barry B. Levine (Boulder, Colo.: Westview Press, 1983), pp. 191-209.

[93] U.S. Joint Chiefs of Staff, "United States Military Posture for FY1982," pt. 1, p. 109.

[94] Stockholm International Peace Research Institute, *World Armaments and Disarmament*, p. 398. Per capita expenditure is not the only appropriate comparison, of course. If the comparison is between the strength of superpower threat and the size of a nation's armed forces, for example, the Cubans for long had the smallest and least expensive armed forces in Latin America. In this category Cuba slipped to third smallest behind Grenada and Nicaragua in the early 1980s.

ing of these aircraft and comparing them to other Latin American air forces, an official of the Defense Intelligence Agency reported that "most of them are defensive; some with a limited offensive capability. There are other air forces which have a better or larger long-transport element than the Cubans do. But as far as defending the homeland, I would say the Cubans would be the best." The official then was asked if Cuban airpower threatened the United States. His response: "I would say it is a very modest threat. The only real threat are the . . . Mig-23's. The other aircraft are not a threat. The Mig-23's themselves in any conventional type of role would be a very modest threat. In any nuclear role, you would not expect them to be used because the other capabilities are so much greater that the Soviets have."[95]

The Cuban navy is the smallest of the three services with about 11,000 personnel, but it was substantially modernized in the late 1970s and early 1980s. Cuba received its first two submarines in early 1979, and four more were delivered in 1983. All six are diesel-powered, torpedo-armed attack submarines formerly used by the Soviet navy. In the early 1980s, the Cuban navy also received a Soviet frigate, but the core of its surface fleet remains small missile and torpedo patrol boats—something akin to the PT boats of World War II—armed with modern weaponry, including surface-to-surface missiles.[96]

In addition to these forces, the Cubans also have deployed surface-to-air (antiaircraft) missiles at coastal defense sites. The Cuban ready reserve of perhaps 750,000 civilians is believed to be capable of rapid mobilization and would be of considerable significance in the event of an invasion of Cuba. As an offensive force, however, the reserves are insignificant.[97]

[95] *Impact of Cuban-Soviet Ties*, p. 14, *supra* n. 64.

[96] U.S. Department of Defense, *Annual Report to the Congress, Caspar W. Weinberger, Secretary of Defense, Fiscal Year 1983*, p. II-22; *New York Times*, March 28, 1983; Abel, "A Breach in the Ramparts," p. 50.

A frigate is the smallest of the major surface combatants. In the U.S. Navy, frigates are designed primarily as escort vessels. In terms of displacement, a modern frigate is about half the size of a destroyer and a third the size of a cruiser.

[97] The figure of 750,000 is obtained by adding the members of the Cuban Youth Labor Army (100,000) and the Civil Defense Force (100,000), by estimating the total strength of various smaller groups (border guards, revolutionary police, etc.) at 50,000, and finally by including a 500,000-member People's Militia created in 1981 to help defend the island in the event of a U.S. invasion. These figures are extremely rough estimates. Wayne S. Smith, "Dateline Havana: Myopic Diplomacy," *Foreign Policy*, no. 48 (Fall 1982), pp. 163-164.

Given this level of military strength, what do policy makers and analysts believe are the capabilities of the Cuban armed forces? First, there are the worst-case scenarios. In 1982, for example, the Heritage Foundation noted "the increasing Soviet offensive capability in Cuba [that] surfaced in 1979 when batteries of modified SA-2 anti-aircraft missiles were identified by air reconnaissance in Cuba. These large missiles, often equipped with nuclear weapons, can be employed quickly in a surface-to-surface mode by the simple addition of a booster. They have an operational range in excess of 150 miles and could be used against ground targets."[98] Similarly, a 1984 edition of the White House *Digest* reported that "Cuba has become a significant military force with the potential for delaying the reinforcement of NATO in time of general war. Given the conventional imbalance that exists between NATO and the Warsaw Pact, such a delay could be decisive."[99]

Few officials believe that Cuba would ever attack the United States directly, however.[100] Most only smile when an interviewer asks if they believe that Cuba would launch an attack on U.S. territory with missiles having a 150-mile range or attempt to block U.S. ships resupplying NATO. For Cuba, they say, an attack on U.S. shipping would be nothing short of suicidal. Barring an invasion by the United States, the nearly unanimous view of U.S. security officials is that Cuba will sit out any war in Europe. Cuba is not believed to be capable of inflicting damage on the United States.

The real issue is not whether the Cubans will attack the United States but whether they will use their military strength elsewhere in Latin America as they have in Africa. Here there are substantial

---

[98] Heritage Foundation, "The Soviet Military Buildup in Cuba," p. 7.

[99] "The Strategic and Economic Importance of the Caribbean Sea Lanes," mimeographed, White House *Digest*, April 4, 1984, p. 4. The *Digest* was the product of the Assistant to the President for Public Liaison, Faith Ryan Whittlesey, whose principal task was to conduct "Outreach Working Groups" on Central America, at which the *Digest* was distributed as background material. The fifty-third such group, which met on May 4, 1984, was fairly typical. The topic was religious persecution in Nicaragua, the audience was composed primarily of interns from congressional offices, and Ms. Whittlesey began with the presentation of a comparison that set the intellectual tone for the entire session: "In the Soviet Union all but a tiny percentage of churches have been closed and religious affiliation routinely brings the loss of precious privileges and sometimes brings more serious persecution. In Nicaragua, the self-admitted Marxist-Leninist leaders of the government are following the same path."

[100] The single printed exception to this view among policy makers is Frechette, "Letter to the Editor," p. 177.

differences of opinion among U.S. policy makers. To Myles Frechette, a career official who served the Reagan administration both as coordinator of Cuban affairs and later as chief of the U.S. Interests Section in Havana, "the Soviet-aided build-up of the Cuban armed forces since 1975 has transformed an essentially home defense force into a force capable of projecting power in the Caribbean."[101] To the Joint Chiefs of Staff, however, the amount of power Cuba can project is not particularly large: "Under optimum conditions and using all aircraft in the commercial and air force inventories, Havana could move several thousand lightly armed troops in a single airlift to most countries in the Caribbean basin in one day. An operation of this size could not include logistical support and would have to be reduced if service and service support elements are included."[102] The Department of State agrees with the Pentagon's assessment: "Cuba does not have the ability to conduct an outright invasion of another country in the region except for the Caribbean microstates. Nor does Havana possess sufficient amphibious assault landing craft or aircraft capable of transporting heavy equipment."[103]

In interview after interview U.S. officials acknowledged the strength of Cuban armed forces but generally minimized the probability that these forces might become involved in active fighting elsewhere in the Caribbean. They reasoned that any Cuban troop commitment would be more than matched by a U.S. troop commitment.[104] "To imagine the U.S. reaction to a Cuban military deployment," said one CIA analyst, "just look at Grenada."

To most U.S. officials the "Cuban problem" is not a question of armed invasions from Soviet-supplied bases in Cuba but rather a problem related to the training and support Cubans provide to radical governments and to armed insurgents in Africa and Latin America. The discussion of this problem would lead us far from the topic of this chapter and into the subject of Cuban internation-

[101] Ibid. For a similar view a few years earlier, see Caesar D. Sereseres, "Inter-American Security Relations: The Future of U.S. Military Diplomacy in the Hemisphere," *Parameters* 7 (1977): 55.

[102] U.S. Joint Chiefs of Staff, "United States Military Posture for FY1982," pt. 1, p. 199.

[103] U.S. Department of State, *Cuban Armed Forces and the Soviet Military Presence*, p. 1.

[104] This conclusion was also reached by Domínguez, *U.S. Interests and Policies in the Caribbean and Central America*, p. 9.

alism. That is a subject for others.[105] What is important to note here are U.S. policy makers' perceptions of Cuba's behavior. In general, the fundamental perception is of annoyance, at times bordering on outrage, that Cuba so persistently supports the groups that Washington opposes. When Washington supports an existing government such as that of El Salvador, Cuba supports the insurgents; when Washington supports insurgents such as the Nicaraguan Contras, Cuba supports the existing government.

In general, policy makers seem to believe that Cuba, a Class C country, should not have a Class A foreign policy. The United States is Class A, and its military aid program is therefore seen as "natural," while Cuba's is not. The United States should be expected to send Peace Corps volunteers to Latin America; Cuba, a poor country, cannot afford charity, and therefore its health and education workers are generally perceived as agents of Soviet subversion. Nowhere in inter-American relations is the presumption of hegemony more in evidence, and nowhere is there more evidence of policy makers' frustration. Like a persistent mosquito, Cuba is a constant nuisance that Washington is unable to swat. In this sense Cuba is an ideal asset for the Soviet Union—a mental military base located in the heart of Washington, D.C. No one in Moscow could create a better decoy.

THE history of human aggression can be written in terms of the declining importance of geographic proximity to the effectiveness of human destruction. Not very long ago one had to stand next to one's adversary in order to inflict physical harm. But as humans learned the value of throwing rocks and other ever-more-sophisticated projectiles, the need to be near an adversary has steadily decreased. The speed of this evolution has become absolutely breathtaking in the years since World War II; today one person in the Oval Office or the Kremlin can probably eliminate the entire human species.

All this has reduced dramatically the military significance of Soviet bases in Latin America. Closeness simply does not matter as much as it once did. Most startling has been the reduction in the value of a Soviet submarine base in the Caribbean region. Just a few years ago—in the mid-1970s—the establishment of such a base would have been extremely worrisome. Now giant leaps in

[105] An insightful collection of essays on this subject is Barry B. Levine, ed., *The New Cuban Presence in the Caribbean* (Boulder, Colo.: Westview Press, 1983).

265

Soviet missile technology have rendered Latin America much less important, both to the Soviets and the Pentagon. But that is not to say that geographic proximity has lost all its significance as technology has changed the character of modern warfare. Forward bases remain a convenient military asset, the significance of which will always be debatable. One thing, however, is certain: forward bases are not as significant as they were a few years ago. Moreover, the *nature* of their significance is also changing rapidly. Pentagon officials no longer worry about a Soviet first strike from Cuba; now their concerns focus upon strategically placed airfields for refueling and related logistical support, as in the case of Grenada. Technological changes on the horizon will no doubt modify this focus, too, in the years immediately ahead.

There is much more to the issue of Soviet bases than an assessment of the continuously evolving relationship between military technology and geography. Of far greater significance are policy makers' assessments of Soviet intentions. Why, exactly, does Moscow want a military base in Latin America? To bomb the United States? To support guerrillas in Central America? To demonstrate Soviet capabilities? To upset the balance of power? To destroy U.S. credibility? Or, conversely, would the purpose of such a base be to serve as a balance for the U.S. bases that have ringed the Soviet Union since the end of World War II? Or would a base simply reassure an isolated ally, Cuba, of the Soviet commitment in much the same way that the United States reassures South Korea? It seems clear that policy makers will never reach a consensus on these questions. Let us note, however, that assessing the intentions of an adversary has always been a—perhaps *the*—fundamental problem of international relations.[106] Only occasionally are the intentions of an adversary crystal clear. In Latin America in particular we should probably not expect to see something like Pearl Harbor; rather we will see airports like the one in Grenada.

U.S. officials reach differing conclusions about the intended use of a Soviet base in Latin America after conducting a personal mental calculus: they ask themselves what the Soviets would do with such a base. Their answers are diverse not only because of differing perceptions about changing weapons technology or Soviet motivations but because policy makers differ in their beliefs about

[106] Glenn Snyder and Paul Diesing, *Conflict Among Nations: Bargaining, Decision Making, and System Structure in International Crises* (Princeton, N.J.: Princeton University Press, 1977), p. 254; Hans J. Morgenthau, *Politics Among Nations: The Struggle for Power and Peace* (New York: Knopf, 1978), pp. 67-68.

such abstract concepts as power, ideology, and aggression. In the most fundamental sense everything depends on each policy maker's view of human nature.[107] On this we must always expect diversity.

The importance policy makers attach to Soviet military bases in Latin America depends, then, upon factors that are largely beyond the realm of geopolitical analysis. The investigation of these factors involves an analysis of policy makers' beliefs about Latin America's role in the balance of world power, the topic of the following chapter.

[107] Louis J. Halle, *American Foreign Policy: Theory and Reality* (London: George Allen and Unwin, 1960), p. 320.

· 8 ·

LATIN AMERICA AND THE GLOBAL
BALANCE OF POWER

IMAGINE for a moment that the worst has happened: it is the year 2000, and all of Latin America is in the hands of governments directly allied with Cuba and closely linked to the Soviet Union. Soviet naval vessels patrol at will in the Caribbean, and the 1962 Kennedy-Khrushchev agreement barring Soviet offensive weapons from Cuba has become irrelevant, since several other Latin American governments have offered Moscow bases throughout the region.

What are the implications for U.S. security? The Communist countries of Latin America, like their Eastern European comrades, remain anxious to trade with the United States. But it's not the same. The color of Pepto-Bismol, for example, is now a stomach-churning red instead of soothing free-market pink, and Estee Lauder's luminescent eye shadow, now manufactured in the People's Republic of Guatemala, has lost its cachet—a Communist beauty aid is, of course, a contradiction in terms. To be on the safe side new sources have been found for a few essential raw materials, but they are located further away, and their supply is both irregular and more costly. The Panama Canal and Caribbean sea lanes are much more insecure, despite the fact that the defense budget has been increased at the expense of domestic programs in order to cover the costs of new surveillance equipment and force deployments that defend U.S. shores from a possible attack originating in the Caribbean. Overall, the outlook is bleak.

But history is not without examples of other nations that have managed to survive and even prosper under similarly threatening geostrategic conditions; ironically, one such example is the Soviet Union. If the threat from Latin America were limited to the tangible results of "losing" the hemisphere—increased insecurity of sea lines and the establishment of Soviet bases, for example—then it is possible that we could cope.

Anxious administration officials announce, however, that there is one other problem that seals our doom: the loss of Latin America has been interpreted around the world as a major destablizing shift in the global balance of power. As country after country has been peeled off the U.S. sphere of influence, Washington officials have been unable to generate the will to act. While they argued over whether the human rights activists or the gunboat diplomats lost El Salvador, they lost Guatemala. While they argued over who lost Guatemala, they lost Mexico. And so it went, country by country, until nothing was left, including, most crucially, America's credibility as the leader of the West. Sensing Washington's weakness in its own sphere of influence, it is now only a matter of time until the Soviets strike the Middle East, NATO, and Japan. Because we have lost our credibility, no one will believe our solemn warnings that the holocaust of nuclear war will be the inevitable result of these pending attacks upon truly vital U.S. interests.

Then, without firing a shot, U.S. allies in the Middle East capitulate to domestic opposition groups aligned with the Soviet Union. Recalling the destruction of World War II, Japan and the European NATO countries soon agree to their own Finlandization. Shorn of its allies, the United States now stands alone in the world. In Washington U.S. officials finally stop quibbling long enough to recognize that there is no longer any reason for a nuclear war. We have already lost.

A few years ago—in the 1980s—the survival of America hinged on the ability of the United States to hold Central America, and rather than hold we let it go so easily. If only we had not failed to heed President Reagan's warning in his address to a joint session of Congress in April 1983, for then there was still time to avoid this unparalleled calamity:

> If Central America were to fall, what would be the consequences for our position in Asia, Europe, and for alliances such as NATO? If the United States cannot respond to a threat near our own borders, why should Europeans or Asians believe that we're seriously concerned about threats to them? If the Soviets can assume that nothing short of an actual attack on the United States will provoke an American response, which ally, which friend will trust us then? . . . The national security of the Americas is at stake in Central America. If we cannot defend ourselves there, we cannot expect to prevail elsewhere. Our credibility would collapse, our alliances would crumble, and the safety of our homeland would be put in jeopardy.[1]

[1] *Weekly Compilation of Presidential Documents* 19 (May 2, 1983): 613, 614.

Although many in Washington did not realize it at the time, Central America was Washington's last stand.

THE ULTIMATE SIMPLIFICATION

Readers who have patiently plowed through the preceding chapters in search of the reason why U.S. policy makers worry so much about instability in Latin America need read no further. There were many answers offered, and at least one—Soviet military bases—was clearly overstated but partially accurate. There was one additional answer, however, and it was the only one that mattered passionately to large numbers of officials. With all sincerity they held that the global balance of power was at stake in Central America. The U.S. loss of Central America would cause a dangerous, perhaps uncorrectable tilt in favor of the Soviet Union. To most policy makers, upon this fragile balance rests the world's hopes for peace.

It was only a matter of time until Central America was saddled with this significance. To understand why it occurred in the 1980s we need first to recognize that U.S. policy makers have always assigned more importance to other regions of the world. As Morton Kaplan observes, since World War II "the close attention of the United States is and must be focused on Western Europe, the southern rim of Europe, and the Far Eastern rim. These areas by virtue of geography, population, human skills, and economic and military power are the crucial areas in the contest between the United States and the Soviet Union."[2] That explains why, in a typical year, several individual chapters in the Brookings Institution's annual *Setting National Priorities* are awarded to critical nations or areas—the Soviet Union, China, Japan, NATO, and the Middle East all generally receive a chapter of their own—and why the fate of Latin America, conversely, is to be tucked away in an obscure corner of the chapter on North-South relations.

U.S. officials take no chances with Europe because NATO's high priority as a vital alliance means that a miscalculation by Washington or by an adversary can lead fairly easily to nuclear war. But policy makers are often willing to accept extra risks in relatively unimportant areas like Central America, much of Africa, and Southeast Asia, for a miscalculation there may involve

[2] Morton A. Kaplan, "American Policy toward the Caribbean Region: One Aspect of American Global Policy," in *Issues in Caribbean International Relations*, ed. Basil A. Ince et al. (Lanham, Md.: University Press of America, 1983), p. 53.

an important but not necessarily a vital ally. In this way low priority places like Latin America often come to serve as a laboratory, where outside forces experiment with new ideas in international relations. "The obvious risks of confrontation in hot areas like Europe move violent conflicts toward the grey areas," writes Stanley Hoffmann, "where the threat of resorting to nuclear weapons is not very credible and where, consequently, the imperatives of prudence are less compelling."[3]

Most of Latin America in general and Central America in particular have traditionally fit almost perfectly into the "grey area" category. Not surprisingly, then, when President Carter decided on a potentially risky new foreign policy emphasis upon the protection of human rights, U.S. policy makers selected Latin America as a principal focus of the experiment. And, again not surprisingly, many Washington officials noted, as Howard Wiarda did, that the Carter administration was conforming to the long-established pattern of using "Latin America as a laboratory for policies that, for fear of retribution or other dangerous consequences, we would not dare try out in other, higher priority areas."[4]

As we have seen, the incoming Reagan administration had its own foreign policy agenda to implement, one that focused upon the reassertion of U.S. power. El Salvador was a particularly tempting site for such a demonstratation. "Because the war in El Salvador looks like an easy victory," wrote critic William LeoGrande, "it provides a perfect opportunity for the new administration to demonstrate its willingness to use force in foreign affairs, its intent to deemphasize human rights, and its resolve to contain the Soviet Union."[5] Scott Thompson, who would become a minor Reagan appointee, wrote that "El Salvador is the place to take a stand against a further spreading of Soviet-Cuban influence within the U.S. security sphere. . . . The United States can choose to win. Nations sitting on the fence will cheer once the United States achieves a victory. They will see merit in American policy. They will be heartened to see that the United States has stopped aban-

[3] Stanley Hoffmann, *Dead Ends: American Foreign Policy in the New Cold War* (Cambridge, Mass.: Ballinger Publishing Company, 1983), p. 5.

[4] Howard J. Wiarda, "Conceptual and Political Dimensions of the Crisis in U.S.-Latin American Relations: Toward a New Policy Formulation," in *The Crisis in Latin America: Strategic, Economic, and Political Dimensions*, ed. Howard J. Wiarda (Washington, D.C.: American Enterprise Institute, 1984), p. 23.

[5] William M. LeoGrande, "A Splendid Little War: Drawing the Line in El Salvador," *International Security* 6 (Summer 1981): 27.

doning even its pockmarked friends."[6] Like the Carter administration, the Reagan administration quickly settled upon the gray area of Central America as an appropriate place to implement its policy agenda.

In addition to its position as a relatively low-risk site, Latin America also provides U.S. policy makers with the benefit of being relatively powerless to deflect Washington's initiatives. As nowhere else in the world, the United States can simplify complex reality in Latin America. Given the minuscule intrinsic importance of a place like El Salvador to the United States, any administration is free to assign it whatever importance it thinks is appropriate, or (as is most common) to assign it no importance whatsoever. During the initial years of the Carter administration, Central America's reality was defined in terms of human rights. U.S. policy makers said the region consisted of one country with a good human rights record (Costa Rica), one with an indifferent record that could stand some improvement (Honduras), and three with very poor records (El Salvador, Guatemala, and Nicaragua). Good, bad, or indifferent, human rights was the subject. In Washington debates over Central America, human rights became reality.

During the Reagan administration, Central America's reality was defined in terms of the East-West struggle. One country was on "their" side (Nicaragua), and the remaining four were on "ours"; two of the four (El Salvador and Guatemala) were under pressure from internationally linked Communist insurgents. "The administration—as all administrations have the power to do—has in effect defined the terms of the policy debate," wrote career diplomat Viron Vaky. "If one acts as if Central America constitutes a full-blown East-West confrontation, it soon will be."[7] The Reagan administration could have drawn the line against Communism in many places around the world; indeed, the administration tried for a while in Lebanon, but it clearly lacked the power to define the complex Lebanese reality in terms of the simpler East-West struggle, realized its impotence, and soon withdrew U.S. forces. Central America was the best site; there Wash-

[6] W. Scott Thompson, "Choosing to Win," *Foreign Policy*, no. 43 (Summer 1981), p. 83.

[7] Viron P. Vaky, "A Central American Tragedy," in *Central America and the Western Alliance*, ed. Joseph Cirincione (New York: Holmes and Meier, 1985), pp. 57-58, 60; Jorge I. Domínguez, *U.S. Interests and Policies in the Caribbean and Central America* (Washington, D.C.: American Enterprise Institute, 1982), p. 18.

ington could minimize the risk of escalation and maximize the power of the United States.

It is one thing for an administration to change the focus of the debate; it is another to generate support for the administration's position in the debate. As with human rights during the Carter administration, the Reagan administration was forced to defend its decision to use Central America as a site to confront Communism. There is a body of opinion in Washington, much of it clearly influenced by the Vietnam experience, that argues it is unwise to grant extraordinary foreign policy significance to places like El Salvador that seem, at least to the uninitiated, to be intrinsically unimportant. For all its safety Central America posed a special type of risk: both U.S. adversaries (who have to be convinced that Washington is serious) and the U.S. public (whose members have to pay for U.S. foreign policy) might not believe an assertion that an apparently unimportant place is important—the assertion might lack credibility.

And so the first task that the administration faced in the early 1980s was to establish the credibility of U.S. policy by identifying something of significance in the Caribbean region. This task was made urgent because many critics were arguing that Latin America was unimportant to the United States. There were statements similar to that of Jerome Slater, who argued earlier "that the Caribbean is in fact of no great significance to the United States. The only importance of the Caribbean today is psychological—the Caribbean is important because we think it is."[8] Similarly, many critics concluded, as did Jorge Domínguez, that "the objective interests of the United States in Central America outside of Panama are very modest—so modest, indeed, that U.S. interests in Central America should be defined almost exclusively as subjective interests."[9]

Representatives of the Reagan administration—in this case Ambassador Jeane Kirkpatrick—responded to these assertions with statements such as "I believe that the area is colossally important to the U.S. national interest. I think we are dealing here not with an ideological conception and certainly not with some sort of remote problem in some far-flung part of the world. We are dealing

[8] Jerome N. Slater, "The Dominican Republic, 1961-66," in *Force Without War: U.S. Armed Forces as a Political Instrument*, ed. Barry M. Blechman and Stephen S. Kaplan (Washington, D.C.: Brookings Institution, 1978), p. 337.

[9] Domínguez, *U.S. Interests and Policies in the Caribbean and Central America*, p. 13.

with our own border when we talk about the Caribbean and Central America and we are dealing with our own vital national interests."[10] These remarks were buttressed by the early Cold War position, still quite popular in Washington, that everything, everywhere is important all the time: "There are two world powers today: The United States and the Soviet Union. . . . there is no matter of political significance in the world that does not affect their security or their ability to protect themselves, their allies, their friends, or those whose interests become tied to theirs for *ad hoc* reasons."[11]

This debate over the significance of Central America continued for some time, with lengthy discussions about raw materials, sea lines, foreign military bases, and refugees. Eventually, however, it became clear that the administration had won this crucial part of the debate: the overwhelming majority of U.S. policy makers came to agree that Central America was more or less important, if not colossally so. This was accomplished in part by tapping the twin forces of geography and U.S. history, forces that are too strong to permit many officials to entertain the notion that Latin America might not matter very much to U.S. security.

This, however, was only half the reason for the administration's success. The other half occurred when the administration convinced many very reluctant officials—particularly key members of Congress—that Central America was a *symbol* of U.S. power and resolve. Day after day, week after week, the Reagan administration said that Central America is the place where the United States had to stop Communism from swallowing up U.S. allies. Then—and here is the crucial transition—having said so many times that Central America was important to U.S. security, Mr. Reagan went on television to tell the American people that since he, our spokesman, had drawn the line, we were stuck with it. If the rest of the world were to believe that we are unable to draw a line in our own backyard—if we cannot keep Central America on our side of the balance of power—"our credibility would collapse, our alliances would crumble, and the safety of our homeland would be put in jeopardy." It was the ultimate simplification.

---

[10] U.S. Congress, Senate, Committee on Foreign Relations, Subcommittee on Western Hemisphere Affairs, *Human Rights in Nicaragua*, 98th Cong., 1st Sess., 1982, p. 77.

[11] Kaplan, "American Policy toward the Caribbean Region," p. 51.

It was both brilliant and predictable. It was brilliant because it managed to provide an acceptable answer to the very difficult question, "Why should we consider a place like El Salvador to be important?" It was predictable because Mr. Reagan verbalized, and thereby converted into public policy, what most Washington officials have always believed about Latin America: because of geographic proximity and the history of inter-American relations since the Monroe Doctrine, the region is in our sphere of influence and, regardless of its intrinsic value to U.S. security, we cannot afford the perception of weakness that would exist if, after Cuba, another country is permitted to shift the world balance of power.

It is of the utmost importance to recognize that this perception of Latin America's role in the global balance of power is held by a clear majority of policy makers. This is partially the inevitable result of a policy-making system that assigns Latin America to the category of "grey area." Senior officials concentrate primarily upon other, more important aspects of U.S. foreign policy, and since 1945 that has meant East-West relations. Thus when instability flares in Latin America and attracts the attention of these senior officials, they quite naturally tend to perceive the instability in terms that are familiar: the Cold War balance of power. This, for example, was almost certainly the reason why Henry Kissinger became interested in (and alarmed over) the election of Salvador Allende in Chile.

This perception by senior policy makers is encouraged, in turn, by lower-level officials who specialize in Latin America. They know that the region will pass unnoticed by their superiors unless Latin America can be integrated into the mainstream concerns of U.S. foreign policy. For whatever reasons (and there surely are many), there has always been a group of Latin Americanists in the policy-making process, officials like Nestor Sánchez and Constantine Menges, who spend their careers writing reports and cables that interpret every campesino's belch as a threatening shift in the global balance of power. Most Latin Americanists outside the policy-making process have become inured to these continuous alarms, but there always seems to be a new group of senior officials whose interest is aroused by this interpretation of instability because, given their East-West focus, it seems accurate.

The ease with which this transition from complex political instability to simplified security threat is accomplished indicates much more than the structural peculiarities of the policy-making process, however. It indicates how strongly U.S. policy makers are committed to Latin America's simplified role as neither more nor less than a part of the U.S. sphere of influence. On no other subject of inter-American relations is there such intensity of feeling. As the following quotations sweep across a broad (but not all-emcompassing) segment of the political spectrum, they provide the best response to why many policy makers perceive a threat when they perceive instability in Latin America:

> Beyond the issue of U.S. security interests in the Central American-Caribbean region, our credibility world wide is engaged. The triumph of hostile forces in what the Soviets call the 'strategic rear' of the United States would be read as a sign of U.S. impotence.
>
> National Bipartisan Commission on Central America[12]

> A lot will depend on how Central America comes out. If we cannot manage Central America, it will be impossible to convince threatened nations in the Persian Gulf and in other places that we know how to manage the global equilibrium.
>
> Henry Kissinger[13]

> [The effect of failing to support allies such as Somoza is that] everywhere our friends will have noted that the U.S. cannot be counted on in times of difficulty and our enemies will have observed that American support provides no security against the forward march of history.
>
> Jeane Kirkpatrick[14]

> When engaged in a conflict for global stakes, what may appear as a marginal interest will be invested with a significance it would not otherwise have, for almost any challenge is likely to be seen by the challenger and by third parties as a test of one's will. . . . In Central America there are no vital raw materials or minerals whose loss might provide the basis for legitimate security concerns. Yet Central America bears geographic proximity to the United States, and historically it has long been regarded as falling within our sphere of influence.

[12] U.S. National Bipartisan Commission on Central America (Kissinger Commission), "Report of the National Bipartisan Commission on Central America," mimeographed [Washington, D.C.: Department of State(?), January 1984], p. 93.

[13] *Public Opinion* 6 (April-May 1983), p. 54.

[14] Jeane J. Kirkpatrick, "Dictators and Double Standards," *Commentary* 68 (November 1979): 36.

... If ... the Soviet Union observes our passivity to events in our own backyard that signal the loss of American control, what conclusions might it draw about our probable passivity in other, far more difficult areas?

Robert Tucker[15]

What clearly makes the Central American problem much more difficult for us is its relationship, whether we like it or not, to the American-Soviet rivalry. ... The United States cannot afford to lose, because of its rivalry with the Soviet Union, and whether we like it or not that rivalry is a fact. A loss would have widespread ramifications for ourselves, for others, for perceptions of international affairs that intangibly merge (and inevitably so) with the realities of international politics.

Zbigniew Brzezinski[16]

The inability of the United States to prevent a political war on its doorstep would reduce the confidence of allies and others, such as Persian Gulf regimes, looking to the United States for protection.

Constantine Menges[17]

Latin America has long figured as a cornerstone of U.S. polit-ical projection in the world. While the inter-American system is not as closely knit today as it once was, ... the hemispheric alliance still figures importantly in the East-West balance. ... [Latin America's] continued participation in the U.S. alliance structure is therefore important to the United States. Failure to achieve their collaboration would represent a net loss in U.S. weight in the international balance of power. ... The de-cline of U.S. pre-eminence in the region—an area traditionally in the U.S. sphere of influence—and of its ability to deny in-

---

[15] Robert W. Tucker, *The Purposes of American Power: An Essay on National Se-curity* (New York: Praeger, 1981), pp. 144-145, 176-177, 180.

[16] Zbigniew Brzezinski, "Strategic Implications of the Central American Crisis," in *Central America and the Western Alliance*, p. 109.

[17] Constantine C. Menges, "Central America and the United States," *SAIS Re-view* (Summer 1981): 14. See also Constantine C. Menges, "The United States and Latin America in the 1980s," in *The National Interests of the United States in For-eign Policy*, ed. Prosser Gifford (Washington, D.C.: University Press of America, 1982), p. 67. During part of the Reagan administration, Menges was a staff Latin Americanist on the National Security Council.

terference in the region by other powers, threatens to be interpreted as an indication of U.S. weakness in absolute terms.

Margaret Daly Hayes[18]

The United States cannot afford to wear blinders ignoring Cuban and Soviet efforts in the region. We must consider the serious consequences of any perception of weakness in an area
acknowledged to be basic to U.S. security and how our European allies in NATO might question our resolve in Europe if
we appear indifferent to the spread of communism in our own
backyard.

Samuel Dickens[19]

The emergence of Latin America as a target for Soviet penetration testifies not only to Soviet opportunism in exploiting
favorable developments in that as well as other Third World
regions and to growing Soviet capability to involve itself in
distant areas, but also to the importance which Moscow at-
taches to Latin America in the East-West balance of power.

Goure and Rothenberg[20]

Our failure to respond adequately to externally supported attempts to overthrow governments committed to reforms and
the electoral solutions would cause other friendly countries
to doubt our ability to help them resist assaults on their
sovereignty.

Walter J. Stoessel, Jr.[21]

---

[18] Margaret Daly Hayes, "Security to the South: U.S. Interests in Latin America,"
International Security 5 (Summer 1980): 134-135. Almost exactly the same words
may be found, along with a similar analysis, in Margaret Daly Hayes, "The Stakes
in Central America and U.S. Policy Responses," *AEI Foreign Policy and Defense
Review* 4 (1982): 15. Hayes was a member of the staff of the Senate Committee on
Foreign Affairs and later a Washington representative of the Council of the Americas, a business lobby.

[19] U.S. Congress, House, Committee on Foreign Affairs, Subcommittee on Inter-
American Affairs, *U.S. Policy toward El Salvador*, 97th Cong., 1st Sess., 1981, p.
210. Colonel Dickens, U.S. Army (retired), was a Washington representative of the
American Legion.

[20] Leon Goure and Morris Rothenberg, *Soviet Penetration of Latin America*
(Miami: Center for Advanced International Studies, University of Miami, 1975),
p. 5.

[21] U.S. Congress, Senate, Committee on Foreign Relations, *The Situation in El
Salvador*, 97th Cong., 1st Sess., 1981, p. 5. Ambassador Stoessel was the Reagan
administration's first Undersecretary of State for Political Affairs, the highest
ranking State Department position reserved for a career foreign service officer.

A Marxist-Leninist victory in El Salvador and perhaps subsequently elsewhere in that fragile, violent region would erode U.S. prestige, encourage the Soviet-Cuban superpower further to promote revolution in the Western Hemisphere, convey to the Western Europeans and to the Soviets that indeed the correlation of forces had shifted away from U.S. power.

William Luers[22]

An image of weakness or incompetence, of an ability to effectively influence the course of events in an area so close, so traditionally dominated by Washington, and so weak in its own right, should be avoided.

Richard Millett[23]

With statements such as these, Central America was accepted as a symbol. Because it was a place of almost no intrinsic importance, the objective was not to "save" Central America (although that was a necessary byproduct); rather, the objective was to demonstrate to the world that the United States was once again serious about containment. Stopping instability in Central America was the *means*, not the goal, of U.S. policy. After such monumental debacles as Vietnam and the Iranian hostage crisis U.S. policy makers (and, the polls tell us, U.S. citizens) felt the need to reassert their resolve to contain Soviet expansionism, and Central America was a logical, even convenient, place to force a test of wills. The United States enjoyed a long-established primacy in the region, geographically close to home and with the Soviet Union half the globe away. It was a place where the United States could win. And—again the crucial transition—once Washington had publicly awarded Central America its central symbolic role in U.S. security policy, then there was no turning back. Rhetorical commitment served *to create* a vital interest, an interest in not having to back down and suffer a loss of credibility.[24]

While this use of Central America as a symbol of U.S. resolve

---

[22] William H. Luers, "The Soviets and Latin America: A Three Decade U.S. Policy Tangle," *Washington Quarterly* 7 (Winter 1984): 29. Ambassador Luers was a widely respected career diplomat.

[23] Richard L. Millett, "The Best of Times, the Worst of Times: Central American Scenarios—1984," in *The Central American Crisis: Policy Perspectives*, ed. Abraham F. Lowenthal and Samuel F. Wells, Jr., Working Paper 119, Latin American Program, The Wilson Center, Washington, D.C., 1982, p. 81.

[24] On this point see Abraham F. Lowenthal and Samuel F. Wells, Jr., "Introduction," in *The Central American Crisis: Policy Perspectives*, p. 2.

was attractive to many policy makers, it also aroused extraordinary opposition, in part because it resurrected the ghost of an earlier symbol, Vietnam. In March 1983, Senator Daniel Inouye, a moderate Democrat and a widely respected World War II hero, stood on the floor of the Senate and told his colleagues:

> As I have listened to our policy-makers in the White House and in the Department of State, I have become convinced that, wise and honorable though they may be, they are mistaken; there is a tragic flaw in their reasoning. They look toward Central America and the demands for a negotiated end to the bloodletting and they see Laos; they speak of spreading violence, of falling dominoes, and they hear the ominous echo—Vietnam.[25]

The reason many policy makers who, like Senator Inouye, had opposed U.S. policy toward Vietnam also opposed the Reagan administration's policy toward Central America was not because the circumstances were similar in both cases. Rather, it was because the rationale for U.S. participation was the same: to maintain the balance of power. Officials who were reluctant to accept this rationale for a war in an obscure corner of the globe in the 1960s were still reluctant in the 1980s. To these officials all that had changed in the intervening years was that we had selected a different, but only slightly less obscure, corner of the globe as the site to employ our simplifying interpretation of instability.

To miss this parallel between Vietnam and Central America is to misunderstand the nature of the opposition to the Reagan administration's policy toward Central America. In the initial years of the Vietnamese conflict the United States fought for a number of reasons, including the need for access to raw materials and the desire to help the people of that tragic land resist what was believed to be tyranny. At first, U.S. motivations were complex. But as the war evolved, the justifications narrowed, and altruism in particular was rejected. By 1965, an aide to Secretary of Defense McNamara advised his superiors to think only of one thing: the U.S. image in the world.

> It is essential . . . that [the] US emerge as a "good doctor." We must have kept promises, been tough, taken risks, gotten bloodied, and hurt the enemy very badly. We must avoid harmful appearances which will affect judgments by, and pro-

[25] *Congressional Record*, March 14, 1983, p. S2742.

vide pretexts to, other nations regarding how the US will behave in future cases of particular interest to those nations—regarding US policy, power, resolve and competence to deal with their problems. In this connection, the relevant audiences are the Communists (who must feel strong pressures), the South Vietnamese (whose morale must be bouyed), our allies (who must trust us as "underwriter") and the US public (which must support our risk-taking with US lives and prestige).[26]

Ultimately there was only one reason to prevail in Vietnam: the need not to be perceived as weak in order to preserve U.S. credibility. Secretary of State Dean Rusk spoke of Vietnam the exact way that President Reagan spoke of Central America: "The integrity of the American commitment is at the heart of this problem. I believe that the integrity of the American commitment is the principal structure of peace throughout the world. . . . Now, if our allies or, more particularly, if our adversaries should discover that the American commitment is not worth anything, then the world will face dangers of which we have not yet dreamed. And so it is important for us to make good on that American commitment to South Viet-Nam."[27] A month earlier Rusk had argued that "if our commitment to South Viet-Nam did not mean anything, what would you think if you were a Thai and considered what our commitments meant to Thailand? What would you think if you were West Berliners and you found that our assurance on these matters did not amount to very much? Now this is utterly fundamental in maintaining the peace of the world, utterly fundamental."[28]

Like Central America in the 1980s, Vietnam in the 1960s became an utterly fundamental symbol. It was unimportant that the United States had to ally itself with a government in Saigon that failed to command the respect of its people. It was unimportant that Vietnam had no essential raw materials to sell, guarded no strategic sea lines, nor provided any other tangible benefit for U.S.

[26] Memorandum from John T. McNaughton, Assistant Secretary of Defense for International Security Affairs, to Robert S. McNamara, Secretary of Defense, March 24, 1965. Reprinted in *The Pentagon Papers: The Defense Department History of United States Decisionmaking on Vietnam*, 4 vols. (Boston: Beacon Press, 1973), 3: 700-701.

[27] Secretary Rusk made this statement during a CBS television interview, August 9, 1965. It is reprinted in the Department of State *Bulletin*, August 30, 1965, p. 342.

[28] Secretary Rusk made this statement during an ABC television interview, July 11, 1965. It is reprinted in the Department of State *Bulletin*, August 2, 1965, p. 188.

security. What was important is that the Soviet Union not have it. And so we fought an enemy we did not understand, supported a government we thought we could reform, and over the course of two decades we wasted the lives of 58,000 U.S. citizens and untold numbers of Vietnamese.[29]

In the end Vietnam left the U.S. sphere of influence and joined that of the Soviet Union. Because officials such as Dean Rusk had said that preventing this from occurring was utterly fundamental to the maintenance of the balance of power, the United States presumably lost some credibility when it lost Vietnam. U.S. officials, nervous by nature, became increasingly anxious, wondering where the next Soviet challenge would come. As we have seen, many policy makers thought it was Central America, beginning in the late 1970s and carrying over into the 1980s.

Not surprisingly, it was easy for critics to draw parallels between U.S. policy in Vietnam and in Central America. There are many more differences than similarities, of course, but one striking parallel towered above any list of dissimilarities: in the minds of many policy makers, a victory by the United States in both Vietnam and Central America was perceived as essential to U.S. credibility and, therefore, to world peace. Halting instability in both places was perceived as an indicator—a symbol—of the broader struggle to maintain the fragile balance of world power. Because U.S. officials perceived instability in this way, Central America, like Vietnam, could be intrinsically unimportant and yet utterly fundamental to the maintenance of world peace. To these officials in the 1980s Central America was the site where the United States struggled, this time in its own backyard, to maintain its credibility.

THE UNDERLYING BELIEFS ABOUT POWER

The political culture of the United States is based upon the concept of dividing and balancing power—among federal, state, and local governments, among the branches of government, and among the nations of the world. Given this culture, no one in Washington challenges the basic wisdom of the theory that peace is a product of the stable division of power between the United States and the Soviet Union. Even the most vehement opponents of the Reagan administration's policy toward Latin America con-

---

[29] It is possible to argue that the lives were not wasted because the war demonstrated Washington's willingness to expend substantial resources on even relatively minor interests and therefore deterred attacks upon major U.S. interests.

centrated their attacks not upon the validity of the balance of power theory but upon the judgment of officials who attributed global balance of power stakes to what they, the critics, perceived as local conflicts among Central Americans.

Inevitably these attacks raised the question of policy makers' beliefs. The judgment of most policy makers is clearly influenced by a series of beliefs that, taken together, serve to simplify and, in my view, to distort the meaning of instability in Latin America. Four of these beliefs have become particularly important in *predetermining* policy makers' interpretation of instability as an attempt to disturb the global balance of power.

Hegemony is Desirable

There is a difference between the maintenance of a sphere of influence and the maintenance of hegemony, or dominant influence. The difference is best illustrated by comparing U.S. policy toward Western European NATO countries, on the one hand, and U.S. policy toward Latin America in general and the Caribbean region in particular, on the other.

Most policy makers would agree that Washington's NATO allies are within the U.S. sphere of influence, but their agreement would be reluctant and certainly not for attribution. Policy makers would prefer to emphasize the concept of community, with Western Europe and the United States sharing a common heritage, common aspirations, and common foreign policy interests. Differences of opinion, they would say, are primarily tactical—whether to deploy Pershing II and cruise missiles or whether to boycott the Moscow olympics—and even extremely sensitive issues (such as Communist party participation in the governments of France and Italy) are always subject to discussion and compromise. On balance, however, the power of the United States is recognized as the glue that holds NATO together. Washington exerts much more influence upon Europe than Europe exerts upon the United States, and this alliance-based power asymmetry is the essence of what is meant today by the term "sphere of influence."

Hegemony, in sharp contrast, implies domination. In early 1953, the National Security Council produced a report, "U.S. Objectives and Courses of Action with Respect to Latin America," for the incoming Eisenhower administration. The report asserted that "our purpose should be to arrest the development of irresponsibility and extreme nationalism and their belief in their immunity from the exercise of US power."[30] A year later a small group of Guatemalans—trained, financed, and supported by the CIA—overthrew the freely elected government of Jacobo Arbenz, demonstrating clearly that Central America, at least, was not immune from the exercise of U.S. power.[31] This was hegemony.

Seventeen years later, in September 1970, the people of Chile freely elected a socialist, Salvador Allende, to serve as their president. At a secret White House meeting months earlier, on June 27, National Security Adviser Henry Kissinger had remarked, "I don't see why we need to stand by and watch a country go Communist due to the irresponsibility of its own people,"[32] and had attempted through covert action to deny Allende his victory. This was hegemony. At a press conference four years later President Ford was asked about the CIA destabilization program that contributed to Allende's downfall in 1973. "I think this [covert action] is in the best interest of the people of Chile and, certainly, in our own best interest," he responded. "I am not going to pass judgment on whether it is permitted or authorized under international law. It is a recognized fact that historically, as well as presently, such actions are taken in the best interests of the countries involved."[33] This was hegemony.

When a mid-level State Department official refers to Honduras as "our Central American whore," and when an Army Green Beret jokes about "the USS Honduras" in a Tegucigalpa bar, it is hegemony. When President Reagan says that the only way he will accept the continuation of the Sandinista government is if its leaders will

[30] National Security Council Report 144, March 6, 1953, p. 11. This interesting report continues: "The withholding of favors should be limited to as few American states as possible since, if too widely applied it would be divisive. It should be done so as not to arouse solid Latin American opposition."

[31] On the Guatemalan intervention, see Richard M. Immerman, *The CIA in Guatemala: The Foreign Policy of Intervention* (Austin: University of Texas Press, 1982); Stephen C. Schlesinger and Stephen Kinzer, *Bitter Fruit: The Untold Story of the American Coup in Guatemala* (Garden City, N.Y.: Doubleday, 1982).

[32] Kissinger's remark was first printed in the *New York Times*, September 11, 1974, p. 14. The remark was supposed to be the first sentence in Victor Marchetti and John D. Marks, *The CIA and the Cult of Intelligence* (New York: Knopf, 1974), but it was censored. For a fascinating, uncensored version of Marchetti and Marks, see Center for National Security Studies, "The Consequences of 'Pre-Publication Review': A Case Study of CIA Censorship of The CIA and the Cult of Intelligence," CNSS Report No. 109 (Washington, D.C.: Center for National Security Studies, September 1983).

[33] *Public Papers of the Presidents of the United States. Gerald R. Ford, 1974* (Washington, D.C.: GPO, 1975), pp. 151, 156.

"say uncle," it is hegemony.[34] And it is not the way the United States treats its NATO allies.

Hegemony is such a natural feature of inter-American relations that it often goes unrecognized. In fact, those officials who most believe hegemony is desirable also consciously reject the charge that they wish to dominate Latin America. "The U.S. is not asserting 'sphere of influence' arrogance in Latin America," an angry Secretary of State Haig told a Senate committee. The problem, Mr. Haig continued, is that "the Soviet Union, Cuba and other Communist countries have acted arrogantly by attempting to overthrow a friendly government, and shipping arms and other military supplies to guerrillas. The U.S. responded to these interventions of the Soviet Union and others at the request of the Salvadoran government."[35] This seems fair: the United States asserts the right to support legitimate governments if the Soviets support guerrilla insurgents. But the United States asserts the right to support the latter groups as well—in Guatemala in 1954, Cuba in 1960 to 1961, and Nicaragua's Contras since 1981. The unspoken presumption is that the United States has the right to intervene in support of its side, while the Soviet Union has no such right in Latin America.[36] This is hegemony.

In discussing the subject of Soviet military bases with various officials, I often asked if they thought the United States should move some of its forces away from Soviet borders—in Turkey, for example—in return for Moscow's agreement not to move forces into the Caribbean region. The response usually bordered on disbelief that the question had been asked. We "belong" in Turkey; they do not "belong" in Latin America. Why? Well, the Turks invited us. Could the Cubans invite the Soviets? Absolutely no comparison—the Castro government does not need protection; we are not going to invade. More important, the Soviet Union has no legitimate reason to have a base in Latin America. Our bases in Turkey are for the purpose of containing Soviet expansion. The Soviets want bases in Latin America in order to threaten the United

[34] Press conference, February 21, 1985.

[35] U.S. Congress, Senate, Committee on Appropriations, *Foreign Assistance and Related Programs Appropriations, Fiscal Year 1982*, 97th Cong., 1st Sess., 1981, pt. 1, p. 55.

[36] For a criticism of this presumption, see Abraham F. Lowenthal, "The United States and Latin America: Ending the Hegemonic Presumption," *Foreign Affairs* 55 (October 1976): 199-213.

States, to expand further into the region, and thereby tilt the balance of power.

Similarly, officials were asked whether the United States should withdraw all of its military personnel from Central America in return for an agreement from the Cubans and Soviets to do likewise. Most respondents thought the question was absurd, despite Assistant Secretary of State Enders' public suggestion in 1982 that "each country should put a common ceiling on the number of outside military and security advisers and troops, subject to reciprocity and full verification. Why not make it zero?"[37] The dominant position among policy makers is that U.S. military advisors "belong" in Central America while the Soviets do not. Enders' successor, Langhorne Motley, returned State's Bureau of Inter-American Affairs to the mainstream on this issue when he said that "with the end of the Cuban/Soviet military presence, the region would cease being a battlefield in the East-West conflict, a role the region neither wants nor can afford."[38] The idea that the United States, too, should end its military presence is simply not a topic for consideration. In complex Europe, U.S. policy makers will bargain—if the Soviets dismantle their SS-20s, for example, NATO will not deploy Pershing IIs and cruise missiles. In simplified Central America there is nothing about which to bargain. The Soviets should go home, the Sandinistas should say "uncle," and instability will end in Latin America. Pax Americana is hegemony.

Some policy makers criticize the hegemonic presumption. They wince at policy makers' references to Latin America as "our backyard"; they, like Ambassador William Luers, prefer to conceive of Latin America as "our immediate neighborhood."[39] To some this is simply an indication of respect for the sensitivities of Latin Americans, but to most it is a more practical recognition that the United States can no longer sustain its hegemonic role. Virtually everyone who has watched the postwar evolution of inter-American relations (as opposed to watching the balance of East-West power) has noticed the decline in U.S. power. This observation

[37] Speech to the Commonwealth Club of San Francisco, San Francisco, California, August 20, 1982. It is no secret that this statement surprised some Reagan administration officials, particularly Ambassador Jeane Kirkpatrick and CIA Director William Casey who, together with National Security Advisor William Clark, soon forced Enders from his post.

[38] Speech to the Foreign Policy Association, New York, January 19, 1984.

[39] Luers, "The Soviets and Latin America," p. 17.

spans much of the political spectrum, from the Hoover Institution's moderately conservative Robert Wesson ("the influence of the United States in Latin America has tended to wane ever since it reached an exceptional high in the wake of World War II") to Columbia University's moderately liberal Alfred Stepan ("the special conditions that contributed to the U.S. economic, political and military dominance in Latin America after World War II are gone forever").[40] What specialists have observed, however, is at variance with what the U.S. public and many senior policy makers continue to believe. "The problem which most Americans have in thinking about Latin America," writes career diplomat Viron Vaky, "is that they have come to consider the dominant position in the world and the overwhelming hegemony the United States exercised in the Hemisphere in the 20 years following World War II as the normal state of affairs."[41]

Critics of the hegemonic presumption often argue that the loss of hegemony may make policy more complex—we can no longer solve our problems with Cuba by using Senator Lodge's approach of grabbing the Cubans by the scruff of the neck and shaking them until they behave—but the loss is not necessarily undesirable. Unlike an earlier era, they assert, weak client states make poor allies in the late twentieth century. Some years ago an AID official argued that Latin Americans needed to become more autonomous in order to become more reliable as allies:

> The quest for national identity and purpose which is likely to occur over the next decades will probably sustain, if not broaden, anti-American sentiment. But there is some reason to expect that the fulfillment of this quest, with its resultant boost to Latin-American self-confidence, may reverse the tide, just as Latin America's lack of self-confidence has largely caused it to flow. . . . Our goal is something comparable to our relationship with Western Europe, one in which Latin America is no longer our "responsibility," one guided by enlightened self-interest. Latin America's ability to participate in

[40] Robert Wesson, ed., *U.S. Influence in Latin America in the 1980s* (New York: Praeger, 1982), p. 17; Alfred Stepan, "The United States and Latin America: Vital Interests and the Instruments of Power," *Foreign Affairs* 58 (1980): 662. See also David Scott Palmer, "Military Governments and U.S. Policy: General Concerns and Central American Cases," *AEI Foreign Policy and Defense Review* 4 (1982): 28.

[41] Viron P. Vaky, "Hemispheric Relations: 'Everything is Part of Everything Else,' " *Foreign Affairs* 59 (1981): 639.

this kind of community with us depends on its self-confidence.[42]

There was a time when U.S. policy makers did not need to care whether Latin Americans liked their subservient role in a hegemonic system, when the encouragement of Latin American self-confidence was even lower on the list of U.S. priorities than it is today, and when the type of hemispheric community the United States favored was one in which Washington yelled, "Jump," and Latin Americans asked, "How high?" Now, former Carter official Richard Feinberg observes, "countries do not want merely to be clients of the United States"; indeed, they probably never cared for that role. Now, however, what they want makes a difference for, as Jorge Domínguez argues, today "the defense of the United States depends in part on the cooperation of others such as Mexico, France, and West Germany."[43] This type of statement does not sit well with policy makers who think of Mexico as the land of enchiladas and salsa. These officials can understand the need for cooperation with the people who invented quiche and perfected the BMW, but anyone can make a tortilla.

What seems increasingly obvious is that Latin Americans have by themselves developed a considerable amount of self-confidence. To these Latin Americans continued hegemony in the 1980s is as offensive as segregation was to U.S. blacks in the 1950s. Jaime Wheelock, Minister of Agriculture and one of the nine Sandinista comandantes, said it most clearly: "For Nicaragua, to be directed by the United States is now unacceptable. The people will fight against this to the end."[44] To some U.S. policy makers this is merely revolutionary bravado, but to others it is interpreted as a declaration of independence. As one long-time observer of Nicaraguan politics, historian Richard Millett, notes, "the utter bottom line of what it means to be a Sandinista is never to humble yourself in front of the United States. No matter what else it means that is the bottom line. Any policy which thinks you can scare them, push them or get them to publicly do a mea culpa, is totally misinformed."[45] Although a foe of the Sandinistas, Millett, like

[42] Lawrence E. Harrison, "Waking from the Pan-American Dream," *Foreign Policy*, no. 5 (Winter 1971-1972), pp. 177, 179.

[43] Domínguez, *U.S. Interests and Policies in the Caribbean and Central America*, p. 31.

[44] Interview, Managua, Nicaragua, November 3, 1984.

[45] U.S. Congress, House, Committee on Foreign Affairs, Subcommittee on Inter-

most critics of the hegemonic presumption, argues that to insist that the Sandinistas say "uncle" is as likely to be productive as to demand that U.S. blacks take a seat in the back of the bus.

This new self-confidence signals a fundamental change in post-war international relations: continued U.S. hegemony now requires excessively heavy costs. "If the Reagan administration chooses to pursue the hegemonism of a bygone era," writes Robert Leiken, "it will eventually find the sphere of Latin America-U.S. strategic cooperation drastically restricted and Soviet influence in Latin America mounting steadily."[46] In the early and mid-1980s, Mr. Leiken, who greatly feared the rise of Soviet influence, spoke for the mainstream of Washington policy makers, the current of official opinion that only very reluctantly accepted the fact that there was no way to retain hegemony's historical benefits without incurring enormously increased costs. No one in Washington questions the fact that the United States retains great power to bend Latin America to its will, but this mainstream recognizes that the United States can no longer "bend" Mexico, Brazil, Argentina, or Cuba without facing losses equal to those it would suffer if it were to attempt to bend Norway, Italy, or Spain.[47] To many policy makers the costs of exerting hegemony over these nations have now come to surpass any conceivable benefit, and much of the rest of Latin America is not far behind.

The irony is that while many forms of U.S. power have grown enormously, they have almost all become less useful. For example, except in tiny outposts like Grenada, U.S. armed force can only destroy; it can no longer gain objectives.[48] It can destroy the Sandinista government, but, unlike an earlier era, it cannot bring anti-Communist stability to Nicaragua. Similarly, U.S. economic

---

American Affairs, *Honduras and U.S. Policy: An Emerging Dilemma*, 97th Cong., 2d Sess., 1982, p. 86.

[46] Robert S. Leiken, "Eastern Winds in Latin America," *Foreign Policy*, no. 42 (Spring 1981), p. 113.

[47] Bruce Russett argues that while it has decreased in terms of raw power, hegemony in the form of being able to control outcomes has not decreased and that subtle forms of control such as the spread of U.S. cultural norms have made U.S. hegemony stronger as other forms have made it weaker. See Bruce Russett, "The Mysterious Case of Vanishing Hegemony; or, Is Mark Twain Really Dead?" *International Organization* 39 (Spring 1985): 207-231.

[48] On this issue see Barbara Tuchman, "The American People and Military Power in an Historical Perspective," in *America's Security in the 1980s: Part I*, Adelphi Papers No. 173 (London: International Institute for Strategic Studies, 1982), pp. 11-12.

power has welded much of Latin America to the United States, but in the process Latin Americans have gained the ability to determine the solvency of the U.S. banking system. Today, write critics of the hegemonic presumption such as Stanley Hoffmann, "successful power means influencing, not controlling, others, and affecting, not determining outcomes."[49] To what appears to be a rapidly growing number of policy makers, there is reason to question the wisdom of hegemony.

Appearances, however, are often deceiving, particularly when they are obtained during a brief period—the 1980s—when critics have not had the opportunity to exercise power and demonstrate their rejection of hegemony. During an earlier era—the late 1970s—there was less criticism. Many of the critics of the 1980s are on record as believing, for example, that it was appropriate for the United States to toil tirelessly to halt human rights abuses in Pinochet's Chile, Videla's Argentina, and Somoza's Nicaragua. Like policy makers in the late 1970s, they pressured repressive governments to change their policy of abusing human rights.

However desirable it may have been to work forcefully in defense of human rights, the *unilateral* approach of the Carter policy was implicitly hegemonic.[50] It took advantage of Washington's dominant influence over Latin American states and sought to exploit that influence for the sake of a noble cause. No one has ever made this point more clearly than Representative (now Senator) Tom Harkin, one of the most honorable human rights activists of our time. "We have to start being more adamant and more forceful in our relationships with those countries" in Latin America that violate human rights, he told a gathering in 1978:

> We always hear it said, "Well, we don't want to interfere in those countries. We don't want to go in there and mess in their internal affairs." I don't see why not. We have been doing it for over a hundred years anyway. . . . We are going to influence Latin America. We will influence every country there. The question is how. Are we going to keep supporting these dictators down there who violate human rights with some kind

[49] Hoffmann, *Dead Ends*, p. 8.

[50] Although, in my view, it would have been more effective to focus upon strengthening international, multilateral mechanisms for protecting and promoting human rights, the point here is not to argue the merits of the Carter human rights policy. The point, rather, is to suggest that the very officials who most criticized the hegemonic beliefs of the Reagan administration seemed to share these beliefs when they exercised control of United States policy toward Latin America.

of sense of security? Or will we forcefully, once and for all, say "No, we won't put up with it"?[51]

Policy makers like Senator Harkin perceive a difference between the use of U.S. power to protect human rights in Latin America and the more selfish use of U.S. power to maintain a sphere of influence. But in interview after interview, officials of the Reagan administration responded that they, too, have noble causes: anti-Communism, democracy, and freedom were the most frequently cited examples. Everyone has a noble cause. Everyone in Washington wants to do good to Latin America.

My interest here, however, is not to evaluate the purposes to which U.S. hegemony is directed but to determine whether one part of the policy-making spectrum believes in maintaining hegemony while the other believes that hegemony has outlived its usefulness. Some of the officials who criticized the Reagan administration are fairly candid and suggest that hegemony is a "natural" feature of relations between neighboring states of unequal power. To Ambassador Robert White, for example, "realistically, the choices of the governments and peoples of Central America are to some extent limited by the power of the United States. That may be unpalatable to hear. Certainly it is not an ideal situation. It does not give total freedom. But few things are ideal in this world, and this is one of the realities that leaders of small nations have to take into account."[52]

Other critical policy makers tend to avoid admitting to hegemonic designs; indeed, they become angry—often furious—at the suggestion that *their* beliefs are hegemonic. Thus one has to watch what they do or what they suggest doing and then ask, "Is that hegemonic?" It very often is. If I did not know the author of the following quotation, I would guess that it could have been written to defend the policy of funding the Nicaraguan Contras: "The United States should pursue a strategy of identifying and offering moral and, in some cases, material support to groups that share our values and principles, which, after all, are universal— human rights, social justice, economic development, and self-

[51] Tom Quigley et al., *U.S. Policy on Human Rights in Latin America (Southern Cone): A Congressional Conference on Capitol Hill* (New York: Fund for New Priorities in America, 1978), pp. 75-76.

[52] Robert White, "There Is No Military Solution in El Salvador," *The Center Magazine* 14 (July-August 1981): 13.

determination."[53] Similarly, it is difficult not to label as hegemonic the proposal by liberal Ramsey Clark for an activist human rights policy: "Preaching helps, but I believe it is time for meddling, too."[54]

The point is obvious: doing good is not the same thing as nonintervention, and opposing the Reagan administration's policy toward Central America is not the same thing as rejecting hegemony. Only when critics of the Reagan administration regain control of U.S. policy will we know whether the belief in hegemony has truly diminished. Until then, what little data we have suggests that most policy makers continue to view hegemony as a desirable feature of inter-American relations.

U.S. Loss = Soviet Gain

Perhaps the simplest way to visualize the balance of world power is as a playground see-saw, with the two superpowers perched on either end surrounded by their allies. It is not necessary for one participant to move to the opposite side in order to disturb the balance; all that must occur is for a participant to jump off—to leave the game by becoming nonaligned. In this sense Cuba was a double loss for the United States. Its weight was first subtracted from the U.S. side of the balance and then added to the Soviet side. The important point, however, is that even true nonalignment is damaging to policy makers who conceive of the world in bipolar terms. To them nonalignment would represent a serious crack in hemispheric solidarity, a minus for the United States and therefore a plus for the Soviet Union. Thus when Victor Johnson of the House Foreign Affairs Committee staff said that he could live with Central America as a region of "Yugoslavias," many officials thought he had taken leave of his senses. Others, like Senators Leahy and Pell, thought it would be acceptable to "be left with a mini-Yugoslavia," but only because that would be "a lot more preferable than a mini-Cuba."[55]

[53] Robert A. Pastor, "Spheres of Influence: Seal Them or Peel Them?" *SAIS Review* 4 (Winter-Spring 1984): 86.

[54] U.S. Congress, House, Committee on Foreign Affairs, Subcommittee on International Organizations and Movements, *International Protection of Human Rights*, 93rd Cong., 1st Sess., 1973, p. 35.

[55] U.S. Congress, Senate, Committees on Foreign Relations and on Appropriations, *El Salvador: The United States in the Midst of a Maelstrom*, 97th Cong., 2d Sess., 1982, pp. 20-21.

While fewer and fewer policy makers continue to conceive of the world in rigid bipolar terms, the clear majority continues to interpret instability in Latin America primarily in terms of its potential impact upon the U.S.-Soviet rivalry. No one who examines United States policy toward Latin America will conclude that bipolarism is a relic of the past.

U.S. officials, particularly career foreign service officers, seem truly torn, however. On the one hand, most know that the old bipolar days are over. They know they should not mind if Central America were to become, if not a Yugoslavia, then a Switzerland or an Austria. This feeling is communicated in the report of the Kissinger Commission, which asserted in an abstract way that "indigenous reform, even indigenous revolution, is not a security threat to the United States."[56] But, on the other hand, policy makers know that for a Latin American country to have an indigenous revolution, then leave the Rio Treaty and become completely nonaligned, would be to make an unfavorable statement about alliance with the United States. To most officials in Washington there is a cost to losing an ally, a cost that is measured in terms of the global balance of power.

One way to manage the dilemma posed by these conflicting attitudes is to simplify reality by creating a mental world in which indigenous revolution (and indigenous revolutionaries) are acceptable in theory but impossible to find in practice. This creative process, which was the topic of Chapter 3, is widely practiced in Washington. It is what led President Reagan to assert that "the Soviet Union underlies all the unrest that is going on. If they weren't engaged in this game of dominoes, there wouldn't be any hot spots in the world." Policy makers who believe this—or anything close to this—are simply not prepared to perceive indigenous revolution in Latin America. They cannot see countries leave the see-saw; only those that shift sides are visible. Abraham Lowenthal noted the practical application of this policy-making peculiarity in his analysis of the Dominican intervention of 1965: "Given the shared preoccupation of American officials with the "second Cuba" possibility, there was a tendency . . . to err on the side of magnifying the Communist risk, by reporting on all the possible Communist connections of other Dominican actors in the crisis and by passing on to Washington all reports of presumed Com-

[56] U.S. National Bipartisan Commission, "Report of the National Bipartisan Commission on Central America," p. 4.

munist plans and intentions."[57] As we also saw in Chapter 3, virtually every policy maker who criticizes the applicability of the balance of power approach to instability in Latin America focuses upon the error of automatically equating instability with Communist adventurism. Many officials would agree with analysts like William Kane, who wrote that "the basic problem with U.S. policy has been the misidentification of important political groups in Latin America as pawns or allies of the Sino-Soviet powers."[58] This misidentification, critics argue, is the direct result of the classic oversimplification of our time, the belief in bipolarity.

Thus when a specific case of instability arises, many policy makers are willing to accommodate rather than confront the fomenters of instability. They do not perceive a threat. In 1982, former ambassador Robert White stated this view most clearly: "I do not believe that the security of the United States is so finely balanced ... that ... whether El Salvador goes one way or another temporarily is of that great consequence. I think that whatever combination of forces of the left take over in El Salvador, it will be possible for this Government to work with them."[59] The point these officials make is not that a threat is impossible but that it is improbable. "Every country in Central America will be willing to trade with the United States and to supply us with any goods we wish to purchase regardless of the nature of the regime" that emerges from instability, writes Morton Halperin.[60] The fact that a new regime might be socialist should not be taken as an indicator of a shift in the balance of power, argues Richard Barnet, for "socialism in Latin America, especially if it comes about by a popular movement or a democratic election, is not a threat to the United States. Every nationalist regime of the Left that has come to power in the Western Hemisphere has preferred to do business with the United States and to receive aid from the United States rather than relate to the Soviet Union—for obvious reasons."[61]

[57] Abraham F. Lowenthal, *The Dominican Intervention* (Cambridge, Mass.: Harvard University Press, 1972), p. 154.

[58] William Everett Kane, *Civil Strife in Latin America: A Legal History of U.S. Involvement* (Baltimore: The Johns Hopkins University Press, 1972), p. 227.

[59] U.S. Congress, House, Committee on Foreign Affairs, Subcommittee on Inter-American Affairs, *Presidential Certification on El Salvador*, vol. 1, 97th Cong., 2d Sess., 1982, p. 238.

[60] Morton H. Halperin, "U.S. Interests in Central America: Designing a Minimax Strategy," in *The Central American Crisis: Policy Perspectives*, p. 23.

[61] Richard J. Barnet, *Real Security: Restoring American Power in a Dangerous Age* (New York: Simon and Schuster, 1981), p. 115.

But "the crucial question," argues Halperin, has nothing to do with trade, foreign aid, and investment; rather it

> is whether those who are leading the revolutionary struggle in El Salvador (and Guatemala) view themselves as part of an international revolution and, for ideological or other reasons, are determined to subordinate the interests of their country to the foreign-policy goals of the Soviet Union and Cuba. I know of no revolutionary leadership that has taken this position in the past, and there is no reason to believe that the current guerrilla leadership in these countries will do so.[62]

Even if the worst occurred—if a Latin American country were to "go Communist"—Tom Farer points out that many policy makers (those he calls "neo-realists") believe that "most Third World states are only of trivial significance to the East-West strategic balance."[63] In short, a significant minority of policy makers believe that the probability of instability in Latin America leading to a tilt in the balance of power is almost zero. To most members of this group the concept of a balance of power does not apply to Latin America.

But a majority of policy makers rejects this interpretation. Their approach to the Nicaraguan revolution is an excellent example. As we saw in Chapter 1, accommodation was the core of the initial Carter policy toward the Sandinistas. When Assistant Secretary of State Vaky was asked just before the fall of Somoza if he feared a Communist takeover, he responded, "I have no doubt there will be an effort by Communist elements to achieve just that. I think it would be a mistake to assume that that will automatically occur."[64] Other Carter administration officials were not as calm as Mr. Vaky; a near-panic seized many who feared another Cuba.

The important point, however, is that whether they advocated a policy of accommodation or confrontation, from the beginning most policy makers were concerned about (and many were obsessed by) the existence of "Communist elements" in Nicaragua. As Lowenthal found in the Dominican Republic in 1965, this made it extremely difficult to perceive an indigenous revolution,

[62] Halperin, "U.S. Interests in Central America," p. 25.

[63] Tom J. Farer, "Searching for Defeat," *Foreign Policy*, no. 40 (Fall 1980), pp. 157-158.

[64] U.S. Congress, House, Committee on Foreign Affairs, Subcommittee on Inter-American Affairs, *United States Policy towards Nicaragua*, 96th Cong., 1st Sess., 1979, p. 45.

if one existed. All eyes were riveted upon any scrap of evidence regarding Nicaragua's role in the balance of power. With each passing day there were fewer assertions that Nicaragua might be moving away from the United States and seeking its own position between East and West—non-alignment—and more and more hand-wringing over Soviet inroads. Ambassador Lawrence Pezzullo warned, "we have to give them aid. They need it so desperately that they'll be forced to look for it elsewhere if we turn them down."[65] The Carter administration's assistant secretary of state for congressional relations argued that "delays in moving forward with this aid undermine the role of our friends and serve the interests of the elements in Nicaragua who would shed no tears if U.S. relations with Nicaragua were to be harmed."[66] The message was clear: bribe them with aid to stay on our side or lose them to the Soviet Union. By the end of the Carter administration the debate had become one in which Switzerland, Austria, or even Yugoslavia were forgotten.

Once the simplifying assumption of bipolarity had asserted its control over United States policy toward Nicaragua, we lost forever a rare opportunity to test the accuracy of rival beliefs about the applicability of the balance of power theory to Latin America. Once the possibility of nonalignment had been eliminated in the minds of Washington policy makers, every step away from the United States was inevitably perceived as a step toward Moscow. This, in turn, set in motion the dynamic seen earlier in Grenada.[67] Lost in the process was the answer to the most interesting question of all: the course the Nicaraguan revolution would have taken if the United States had stood back and permitted it to run its course. That freedom was exactly what many policy makers could not permit, for as Representative Dan Mica remarked, "I just find it extremely difficult to believe that if we stand up as a great na-

[65] *Newsweek*, September 8, 1980, p. 48.

[66] U.S. Congress, House, Committee on Foreign Affairs, Subcommittee on Inter-American Affairs, *Review of Presidential Certification of Nicaragua's Connection to Terrorism*, 96th Cong., 2d Sess., 1980, p. 25.

[67] To many officials, the issue was not only whether the Nicaraguan revolution would lead to alignment with the Soviet Union but the potential consequences of the linkage between the Nicaraguan revolution and the active insurgency in El Salvador. Thus policy makers' assessments of the Soviet-Nicaraguan relationship were clouded by their simultaneous assessments of the Nicaraguan-El Salvadoran relationship. Had El Salvador not been involved, the Nicaraguan revolution would have provided a better opportunity to examine policy makers' beliefs about bipolarity.

tion and say we will stop our effort, that the Communist-inspired forces on the other side would do the same."[68] It was another victory for bipolarity.

Control of Territory = Control of People

The third simplifying belief underlying the use of the balance of power theory to Latin America is that the balance is calculated largely, if not exclusively, in terms of the geographic territory within the sphere of influence (or hegemonic system) of one or the other superpowers. Latin America is conceived as a place and, of course, it is. But it is other, more complex things as well, including nearly four hundred million humans. In the making of United States policy toward Latin America, however, many policy makers focus upon the territory rather than its inhabitants. Like Mr. Reagan, they say, "If Central America were to fall, . . ." not, in contrast, "If Central Americans were to fall, . . ." They talk about "our backyard," not "our neighbors." They say, "if we cannot defend ourselves *there*, we cannot expect to prevail *elsewhere*." "There" and "elsewhere," like "backyard," "doorstep," and "neighborhood," are places, territory. Policy makers do not say, "If we cannot maintain the allegiance of Central Americans, we cannot expect to maintain the allegiance of our allies elsewhere." No doubt they would respond that that is what they mean, but it is not what they say. Policy makers tend to talk in terms of territory; they seem to assume that if the territory of Latin America is on the side of the United States in the balance of power, then the people who inhabit the territory, like the raw materials and surrounding sea lines, will also be counted on Washington's side.

This territorial interpretation of the balance of power once made good sense from the point of view of U.S. security. Early in the nineteenth century Latin America was not much more than territory—sparsely populated and still largely disorganized, the region had no autonomous weight in international relations. What mattered to leaders like John Quincy Adams is that the territory of Latin America not be used by European powers to threaten U.S. security, which U.S. officials assumed Europeans could and would do if the United States did not prevent it. The result was the orig-

[68] U.S. Congress, House, Committee on Foreign Affairs, *Foreign Assistance Legislation for Fiscal Year 1982*, 97th Cong., 1st Sess., 1981, pt. 7, p. 249.

inal statement of the territory-oriented U.S. policy toward Latin America, the Monroe Doctrine.

A small number of policy makers argue that changing circumstances require rethinking the territorial basis of the Monroe Doctrine. They note today what Bernard Brodie observed two decades ago: "In the field of modern strategy, time tends to deal severely with concepts as well as facts";[69] they agree today with what British Prime Minister Benjamin Disraeli said a century ago:

> The Monroe Doctrine is one which, with great respect to the Government of the United States, is not, in my opinion, suited to the age in which we live. The increase in the means of communication between Europe and America have made one great family of the countries of the world; and that system of government which, instead of enlarging would restrict the relations between those two quarters of the globe, is a system which is not adapted to this age.[70]

The British Prime Minister obviously had in mind the best interests of Great Britain, not Latin America and certainly not the United States, but today there are U.S. policy makers who share his view. In language less refined than that of Mr. Disraeli one congressional aide observed, "we can no longer treat the countries of Latin America the way my dog treats the neighborhood fire hydrants." Almost everything about inter-American relations has changed since 1823, these officials argue, while U.S. policy makers have remained fixated on territory as a measure of strength. Even a near-sacred concept like the Monroe Doctrine sometimes outlives its usefulness, and its continued utilization often makes more difficult the very goal (in this case, security) it was intended to achieve.

Two changes in particular are often cited as specific reasons to rethink the emphasis on territory in discussing Latin America's role in the balance of power. The first is that many people around the world, and especially in Latin America, do not like it. They view as ludicrous the idea that the United States claims control over access to an entire hemisphere by virtue of a unilateral declaration issued in 1823. In this sense the Monroe Doctrine is often perceived as no more legitimate than the 1494 Treaty of Torde-

[69] Bernard Brodie, *Strategy in the Missile Age* (Princeton, N.J.: Princeton University Press, 1965), p. v.

[70] *Hansard's Parliamentary Debates*, 3d Series, volume 142, column 1,511, June 16, 1856.

sillas, in which Spain and Portugal divided the New World between themselves. Critics argue that the realities of the late twentieth century, and in particular the sensitivity of small states to
infringements upon their sovereignty, require the United States to
abandon its claim to control who gets in or out of Latin America.

The second change involves technological innovation and the
increase in Soviet power, including economic power. The original
focus of the Monroe Doctrine—to deter formal colonization—
tends at times to obscure the general thrust of the document,
which was to prevent the use of territory in the Western Hemisphere to threaten U.S. interests. Today, when policy makers perceive the threat primarily in terms of a shift in the balance of
power (and not necessarily the formal colonization of territory), it
is difficult to determine whether the Monroe Doctrine has been
violated. Latin America is now "globalized" in the sense that
nonregional actors like the Soviet Union have a major presence in
the region. Cuba, of course, is the major example, and perhaps Nicaragua as well. But there are many less-dramatic examples. In
1960, there were three Soviet embassies in Latin America, and the
United States was actively interested in reducing the number to
zero. Today there are nineteen, and only a few in Washington talk
about rolling back the number. Gone forever is the era when other
powers stayed out of Latin America except to trade.

As a result it has become ever more difficult to decide whose territory belongs on whose side of the balance of power. Cuba and
Honduras are easy to place, but the very existence of Cuba demonstrates that the territorial imperative of the Monroe Doctrine is
no longer applicable to modern Latin America. And what about a
country like Argentina that traditionally pays little attention to
U.S. wishes? Does Argentina belong on the side of the United
States, to whose banks Argentina is irretrievably indebted for the
foreseeable future, or on the side of the Soviet Union, currently Argentina's principal trading partner and just about the only country
in the world that wants to purchase more than a small quantity of
what Argentina has to sell on the world market?[71]

To officials and analysts who criticize the applicability of the
balance of power theory in Latin America it is silly to ask questions like this. Argentina, like most of Latin America, is on its

[71] For a review of Argentina's recent disregard for the balance of power, see Roberto Russell, "Argentina y la política exterior del régimen autoritario (1976-1983):
una evaluación preliminar," *Estudios Internacionales* 17 (April-June 1984).

own side, and where it sits on any specific issue depends upon how Argentines calculate their own interests, not on how Washington (or Moscow) calculates the global balance of power. But, critics continue, entering into that calculation will be the general attitude of Argentines toward the United States, and, Margaret Thatcher aside, there is nothing that more annoys the Argentines than references to their country as part of the "backyard" of the United States.

As an alternative to policy makers' fixation upon territory, critics suggest that there is a more realistic way to calculate Latin America's historic role in the balance of power. It focuses upon people. Because of the long history of U.S. domination of the region, the United States is often perceived as responsible, at least in part, for what Latin America is today. That, for example, is the message of the literature on dependency, with deep roots in the intellectual community around the world. Hegemony aside, Latin America has been identified as part of the U.S. sphere of influence. Thus when people, including Latin American people, are deciding where to stand on (or whether to care about) the East-West conflict, it is probable that at least some of them will look at Latin America for evidence of the results that come from selecting the side of the United States. In this sense Latin America is a showcase; it is an example of how human beings spend their lives when their territory is included in the U.S. sphere of influence.

This must be a sobering thought for those in Washington who believe that Latin America has a role to play in the balance of power. It is sobering because in most of Latin America the conditions of life for an average person are unpleasant. Some do very well, and there is promise that others will do better in the years ahead, but the sad fact is that most Latin Americans lead lives that are, by contemporary world standards, deprived. Given the painfully obvious evidence that the United States has neither the will nor the capacity to end this deprivation, perhaps nothing more complex than self-interest dictates that the time has come to cease believing Latin America is "our backyard."

Supporting Existing Governments = Protecting U.S. Security

Looking back at the postwar history of United States policy toward Latin America, all policy makers recognize that Washington has committed some serious blunders. Perhaps the most serious of these is the repeated tendency to befriend and support Latin American governments that fail to command the respect of their own people. From the vantage point of time, virtually everyone now agrees, for example, that U.S. support for Fulgencio Batista (and most of his predecessors) helped determine the unfortunate course of postrevolutionary U.S.-Cuban relations. As the revolution was occurring in the late 1950s, however, many in Washington asserted that U.S. security interests required continued support for the Batista government, for no one wanted another Guatemala. It was not until April 1958, less than nine months before Castro's victory, that Secretary of State Dulles declared Batista expendable by announcing an arms embargo: "We allow arms to go to other countries primarily to meet international defense requirements. . . . We don't like to have those go where the purpose is to conduct a civil war."[72] With this short statement at the very end of a press conference that focused on other subjects, the Secretary of State separated the continuation of the Batista government from the protection of U.S. interests. It was too late.

Similarly, as time passes it becomes increasingly difficult to identify Washington policy makers who believe it was wise for the United States to support Anastasio Somoza until close to the bitter end. The Somozas ruled Nicaragua for about forty-five years; U.S. support continued for forty-four years and a few months. Only in September 1978, did the United States indicate, again in the form of an unofficial arms embargo, that U.S. security interests no longer required a continuation of *somocismo*. Once more it was too late.

A few officials like Jeane Kirkpatrick and Jesse Helms still argue that U.S. support for Batista and Somoza never should have ceased, for they were only authoritarians, not totalitarians. But most now agree that Washington waited too long to withdraw its support. This is a debate we cannot resolve. What the record clearly indicates, however, is that in 1958 and 1978 U.S. policy makers felt they had little choice but to abandon these two longtime friends. At the time the perception in Washington was that Batista and Somoza were about to be deposited on the trash heap of history. In both cases the general population of Cuba and Nicaragua had become so disaffected with their governments that there was nothing the United States could do to re-create the domestic support that had disappeared, it seemed, almost overnight. All Washington could do was respond to desperate requests for mili-

[72] *New York Times*, April 9, 1958, p. 10.

tary aid and, in the process, become even further identified with an ever-more-shaky regime. And so Washington pulled the plug. To policy makers as different as John Foster Dulles and Jimmy Carter the protection of U.S. security now required an end to what had been long, cordial friendships.

The tendency for policy makers to identify U.S. interests with the continuation of a specific Latin American government is often criticized, but each successive group of postwar policy makers seems prone to repeat the process. At the time that it happens the tendency seems impossible to resist. By the mid-1980s, for example, the Reagan administration seemed to have no choice but to identify U.S. interests with the continuation of the Duarte government in El Salvador. Push had come to shove in Central America, and Mr. Duarte and his hardy band of Christian Democrats were perceived in Washington as the only alternative to either the far right, which the U.S. Congress would not support, and the far left, which the administration believed it could not permit to reach power for the sake of the balance of world power. Most of the time most administrations feel that they face similar situations; there are rarely many choices, and often all of the choices are far from desirable. In retrospect, few policy makers seem to prefer leaders such as Batista, Somoza, Trujillo, or whatever general happens to be in charge of Guatemala or Honduras at any particular moment.[73] But in the heat of the moment, when instability is the focus of concern, many policy makers believe their continuation in power is important to U.S. security.

Nearly everyone in Washington knows this is a problem. The most common cause of the problem is said to be a lack of foresight. In calculating U.S. policy, critics argue, U.S. officials give too little weight to the long-term costs and too much weight to the short-term benefits of identifying U.S. interests with an existing government. No doubt this short-term/long-term playoff does cause part of the problem, for incumbent policy makers have an understandable incentive to err on the side of short-term benefits and to

[73] There are always some officials who actually like these tyrants. In the 1980s, Senator Jesse Helms was uncommonly fond of the Pinochet dictatorship; in the 1970s, the Somoza lobby included a host of former officials and about a half-dozen members of Congress; in the 1950s, Trujillo had more than one senator in his sugar lobby; and in the 1940s, Ambassador Sumner Welles, who more than any other U.S. citizen was responsible for the Batista dictatorship, wrote in his memoirs that Batista was an "extraordinarily brilliant and able fellow." Sumner Welles, *The Time for Decision* (New York: Harper and Brothers, 1944), p. 197.

downplay the long-term costs that might have to be paid by their successors. But the root of the problem probably lies in the conceptualization of Latin America as part of the global balance of power.

The problem begins when decisions must be made on how to implement the policy of keeping the region on the side of the West in the East-West conflict. There are a variety of ways to implement this policy, ranging from the armed occupation of the entire hemisphere to the winning of hearts and minds of its inhabitants, but all of these methods require the cooperation of at least some Latin Americans. That means that policy makers must first identify groups who will help maintain Latin America on the U.S. side of the balance of power. That is not an easy task, for Latin American political groups (parties, labor unions, interest groups, even the military) do not divide neatly upon Washington's view of U.S. security interests; or, stated differently, it is extremely rare to find a significant political group that belongs to "the West" or to "the East" in the sense that its political activity exists in order to maintain or to disrupt the balance of world power. Latin American political actors have their own quite different goals; as in the United States, most often these goals focus upon the maintainence or attainment of power.

But these differing goals do not necessarily conflict with Washington's goal of maintaining Latin America within the U.S. sphere of influence. In the early days of the Cold War, the United States identified the Latin American political actors that were overtly anti-Communist, and Washington agreed to support them in their internal political power struggles; these anti-Communist actors agreed, in turn, to support the United States in its international struggle. It often was (and is) a marriage of convenience. As it happened, the early postwar period was a particularly propitious time for the United States to strike this bargain, for U.S. power was at its apogee and a variety of political actors sought our friendship. In most cases—Guatemala was a notable exception—the United States decided to support those who held political power at the time and especially those who had demonstrated their friendship in the battle against the Axis.[74] Incumbent governments agreed to support the United States, symbolizing their support by signing

[74] For an interesting illustration of this tendency, see G. Pope Atkins and Larman C. Wilson, *The United States and the Trujillo Regime* (New Brunswick, N.J.: Rutgers University Press, 1971).

the Rio Treaty, and the United States agreed to support them with various forms of aid, not the least of which was military assistance.

Then, as the Cold War heated up, it became ever more important that the United States suffer no defections in Latin America. Not being fools, our allies in Latin America quickly realized their key role in Washington's perceptions of the balance of power. All these allies needed to do to obtain aid from the United States was to pursue an anti-Communist foreign policy and to maintain domestic stability. Too often this stability was achieved by stifling domestic political dissent with weapons provided by the United States, provided on the understanding that they would be used to fight Communist adventurism. The domestic opposition to our friends reacted with dismay at being shot with bullets provided by the U.S. military assistance program; those who survived became increasingly radicalized, and they came to perceive the United States as hostile to them and to their principles, which they invariably defined as justice and freedom. If the United States stands against the cause of justice and freedom, then it must be that the United States is the enemy.

What happened, in short, is that given Latin America's position after World War II, and given Washington's simplifying perception of the world as a bipolar struggle for power, the U.S. unwittingly chose sides in internal political struggles in which it had no interest. Containment "was initially a sound doctrine," write Jordan and Taylor, but it led to "a widely accepted idea that any status quo was good, any revolution bad."[75] This simplified view of Latin American politics explains why Washington had such difficulty leaving Somoza's side. To many policy makers his friendship became the way the United States ensured Nicaragua's position alongside the United States on the balance of power. U.S. security interests became equated, willy-nilly, with the continuation of *somocismo*.

It is commonplace to criticize policy makers for forgetting that the United States has interests to protect in Latin America, not friends. Officials responsible for the day-to-day implementation of U.S. policy never talk this way, however, because they know that the United States cannot protect its interests in the region without the help of friends. Unwilling to occupy the region militarily in order to maintain the balance of power, and unwilling to accept

[75] Amos A. Jordan and William J. Taylor, Jr., *American National Security: Policy and Process* (Baltimore: The Johns Hopkins University Press, 1981), p. 284.

the risks involved in letting Latin Americans choose their side (or no side) without significant input from Washington, U.S. policy makers use friends to protect U.S. interests. To many officials no other alternative seems available.

The question that is debated in Washington, then, is not whether the United States is wise to identify and support allies but rather whether the single-minded commitment to keeping Latin America on the U.S. side of the balance of power has caused the United States to become involved in Latin American political struggles that have nothing to do with global issues but simply reflect the complex, continuously evolving domestic politics of each Latin American country. It is here, critics argue, that the application of the balance of power concept to Latin America does its most severe damage. It creates chasms of hostility between the United States, on the one hand, and emerging Latin American political groups, on the other. These are mental chasms—intellectual and emotional—that U.S. policy makers cannot leap because they are weighted down by their Cold War beliefs about the balance of power.

Specifically, once policy makers identify a friend as the champion of U.S. interests, then the belief in a bipolar balance of power makes it effortless to identify that friend's domestic opposition as an enemy of the United States. The friend is sitting next to the United States on a precarious see-saw, and the opposition is trying to knock him off. The natural reaction is to help the friend hold on. Given a bipolar belief about the nature of the balance of power, many policy makers in Washington are psychologically unprepared to think of other political groups as potential friends.

Because this belief orients U.S. officials to see the domestic adversaries of our friends as our own enemies, we act accordingly. Thus by the time the United States *wants* to switch its friendship away from a crumbling government, it is often too late. By then the imperatives of maintaining the balance of power have created a situation in which opposition groups, so long a target of U.S. attacks, no longer want to be our friend. By that time there are too many wounds. Many of the participants and innocent bystanders have had their lives destroyed—parents, children, or spouses killed—and they now have intangible needs that even the most sincere apology and promise of accommodation cannot possibly satisfy, at least in the short term. Readers should imagine a situation in which their family has been destroyed by a Nicaraguan National Guard trained and supplied by the United States, and then ask themselves if they would accept the Carter administration's economic aid package and let bygones be bygones. Imagine the attitude of a survivor of the National Guard's attack on Chinandega in 1978, who told an OAS investigating team her story:

It was Thursday, September 14, when the airplanes began to strafe our houses in Barrio La Libertad. My husband, my 5 year old daughter and I were crouched in a corner of our house, crying and thinking that we would die right then and there because the bullets and shrapnel were destroying our small wooden house. We decided to go out and seek shelter in a safe place; we left by the kitchen, my husband with our daughter in his arms. A plane flew very low, it seemed as if it was coming straight at us, and fired some rockets which hit my daughter's shoulder and my husband who was carrying her. Everywhere I looked I could see the heart and intestines of my child; she was in pieces, destroyed. My husband, who had already lost his arms, took about thirty steps, with blood spouting everywhere, until he fell dead. He had a wound to the chest; he had a part of a still-smoking rocket stuck in his leg. The left leg was bare to the bone. I wanted to lift my child, but she was in pieces; I didn't know what to do. I ran and I got her little arm and tried to put it on her, I tried to put everything that was coming out of her back in but she was already dead. She was my only daughter, and I had a difficult time having her; and I used to dress her up and spoil her. I don't know what I'm going to do, I'm going to go crazy.[76]

This type of experience must effect the existential calculus of the victims who survive. There are no data, but I imagine that some survivors join the armed opposition. They do so not because they seek to reduce poverty through socioeconomic reforms or because they want to tilt the balance of power, but because they hate.

Once with the opposition, they will learn, if they do not already know, who supplied the rocket and the plane and the pilot training that killed their families. There they will learn that the opposition groups to which they now belong have turned to our ideological adversaries for support and by doing so have fulfilled Washington's expectations and provided the United States with further justification for supplying weapons to a government that clings to power by bombing its own citizens. There they will learn, perhaps never

[76] Organization of American States, Inter-American Commission on Human Rights, "Report on the Situation of Human Rights in Nicaragua," mimeographed (Washington, D.C.: OAS General Secretariat, October 1978), p. 33. This document is essential reading for an understanding of the Nicaraguan revolution.

to forget, a song describing the United States as the enemy of humanity.

THE simplifying concept of the balance of power is at the very core of the heated debate over United States policy toward Latin America. It is difficult to communicate how threatened many policy makers feel by the prospect that instability will lead to the loss of one or more Latin American countries from the U.S. sphere of influence. In the 1980s, virtually all Reagan policy makers expressed a deep and abiding fear that instability in Latin America would cause a tilt in the global balance of power. However much these fears may or may not be justified, they are deeply and honestly felt.

At the same time it is equally difficult to communicate how profoundly disturbed many other policy makers have become over the use of what they perceive as an irrelevant aspect of the Cold War rivalry to determine the perceived consequences of instability in Latin America. Virtually none of these critical officials have worked in positions of prominence in the executive branch during the Reagan administration. Rather, they are members of Congress and their staffs, or former officials, waiting in Washington think-tanks and law offices for a change of administrations and the opportunity to return to public service. At times an interviewer can almost touch their sense of despair over the policies of the Reagan administration.

When to the emotional content of the subject is added its intangible nature—we are, after all, discussing policy makers' perceptions of an unmeasurable concept, the balance of power—the potential for continual disagreement becomes painfully apparent. On some aspects of inter-American relations policy makers can sit down together, discuss various points of view, agree on a few facts, and sometimes even reach a conclusion about the nature of U.S. interests and the appropriate way to protect them. On the balance of world power, however, no agreement seems possible. Every fact is capable of a worst-case and a best-case interpretation, and policy makers in the 1980s cluster almost unerringly at one of the two poles.[77] As we enter the final years of the twentieth century, the gap between policy makers seems unbridgeable. In Washington today, the cost of simplification is the polarization of the policy-making process.

[77] For an interesting examination of this polarization, see I. M. Destler, Leslie H. Gelb, and Anthony Lake, *Our Own Worst Enemy: The Unmaking of American Foreign Policy* (New York: Simon and Schuster, 1984).

CONCLUSION

THE BATTLES that have been occurring in Washington over United States policy toward Latin America are the product of policy makers' differing beliefs about the causes and consequences of instability.

In the 1980s, the policy makers who believe Communist adventurism is the cause of instability in Central America based their belief on what they consider irrefutable facts: Nicaragua had fallen to people who called themselves Marxists, El Salvador was falling to people who called themselves Marxists, and Guatemala was besieged by two guerrilla groups with suspicious names: the Guerrilla Army of the Poor and the Revolutionary Organization of the People in Arms. But other policy makers responded with the same kind of answer—they believe poverty caused the instability in Central America, and they, too, based their belief on what they consider to be an irrefutable fact: the awakening of Central America's poverty-stricken masses.

Since these beliefs about the causes of instability in Central America are so different, it is tempting to conclude that both groups are correct but that each sees only part of the big picture. This was the approach of the Kissinger Commission, which asserted that "the roots of the crisis are both indigenous and foreign."[1] To do this, however, the commission had to gloss over fundamentally *contradictory* beliefs by simply asserting that they are both correct. This was also the approach of the Kennedy administration a generation ago, and it provided the basis for an uneasy truce. At that time, however, policy makers' beliefs about poverty were much less firm, perhaps because they were held by officials who understood intuitively the destabilizing capacity of poverty mixed with social change but who had not experienced it firsthand. These Kennedy officials—men like Arthur Schlesinger, Jr., and Richard Goodwin—were neither intellectually equipped nor emotionally inclined to challenge their rivals' claims that the poor could be physically separated from the fomenters of instability.

[1] U.S. National Bipartisan Commission on Central America (Kissinger Commission), "Report of the National Bipartisan Commission on Central America," mimeographed [Washington, D.C.: Department of State(?) January 1984], p. 4.

And so they accepted a compromise in which the "antipoverty" policy makers would provide economic aid while the "anti-Communist" policy makers would provide military and police assistance for counterinsurgency units. These units were supposed to go into the countryside and urban slums and destroy only Communist insurgents, leaving the poor alone. In fact, of course, they often shot everything that moved, but that is not the point to be made here. The point is that the Kennedy compromise was predicated upon one fundamental assumption: that the poor in Latin America could be physically separated from the "Communists" who were causing instability.

Today the Kennedy compromise is not possible, for the many policy makers who have lived and worked among the poor in Latin America believe that this physical separation is impossible. To these policy makers it makes no sense to argue, as Secretary of State Haig did, that Cuba "is the platform, the instigator, and the operative leadership behind the situation in El Salvador," because these policy makers believe that it is just not so. Similarly, to the many policy makers who, like President Reagan, believe that Soviet Communism is "the focus of evil in the modern world" and that Moscow "underlies all the unrest that is going on," it makes no sense to argue, as Walter LaFeber did, that "the overwhelming number of Central Americans were in rebellion because their children starved."[2] The rival beliefs about the causes of instability are mutually exclusive; they cannot be accommodated by merging them together.

On the surface the disputes among policy makers over the consequences of instability appear to be largely technical and factual, but at root they depend upon policy makers' differing beliefs about such abstract concepts as the intentions of Soviet leaders, the role of weak states in the balance of power, and the nature of aggression in international relations.

These differences surface primarily during discussions of strategic denial. On the subject of strategic access, the differences among policy makers are not profound. On *raw materials*, policy makers will probably never reach agreement on a single answer to the question, "How dependent is the United States upon access to supplies from Latin America?" But that is probably not the most important question. Rather, the key to the issue of raw materials

[2] Walter LaFeber, "Inevitable Revolutions," *Atlantic Monthly* 249 (June 1982): 83.

seems to be policy makers' answer to the question, "Who cares?" The answer is that most officials care a little bit, but nobody cares very much. On *military bases and military support*, most policy makers believe that access to Latin America is not of major significance to U.S. security. The United States neither needs nor desires additional military bases in Latin America, and the value of existing bases has declined with the development of new technologies such as satellite communications. There is a particularly strong agreement that the United States does not need the assistance of Latin American militaries, except in those cases where a multinational force is appropriate for political reasons. On Latin American *sea lines of communication*, policy makers have not had a particularly difficult time reaching a rough agreement. Most have agreed to focus upon only one of the four major Latin American SLOCs, the Caribbean. Except for an occasional speech that mentions the Panama Canal, no one in Washington worries much about the other three. Of the Caribbean, no one disregards the presence of Soviet warships and no one likes the fact that the Cuban navy now includes submarines, but most officials believe that the existence of these hostile forces in the Caribbean poses no direct threat that is not already posed by Soviet warships elsewhere.

In brief, on the question of strategic access to Latin America, most policy makers believe that the security of the United States would not be adversely affected if Latin America were, say, to sink to the bottom of the sea. In an abstract sense that avoids the calculation of human losses, these officials would be relieved if Latin America did not exist. Give these officials the opportunity to choose between the continued physical presence of our neighboring republics to the south, on the one hand, and a vast ocean broken only by the ice packs of Antarctica, on the other, and the probability is quite high that they would select the latter option. Why? Because if the region were to disappear, then there would no longer be a need to worry about Latin America's vulnerability to Soviet and Soviet/Cuban adventurism. And that is the true significance of Latin America to U.S. security: it is not that policy makers want much of anything from Latin America but rather that they do not want the Soviet Union to have it.

Although a clear majority of policy makers believes that Soviet military bases in the Caribbean region would pose a security threat to the United States, a much more important belief is that the establishment of a Soviet base would send a signal to the rest of the world that the Soviet Union is capable of making further inroads into the U.S. sphere of influence in Latin America. A base, like other forms of an increased Soviet presence, would therefore symbolize a shift in the global balance of power. This is unquestionably *the* crucial consequence that dominates debates about instability in Latin America. Many officials believe that the "loss" of a Latin American country from the U.S. sphere of influence would send shock waves around the globe, undermining U.S. leadership and damaging the fragile structure of world peace.

The opposition to this position does not question the balance of power theory but rather the theory's application to instability in Latin America. One part of this opposition believes that because instability stems from poverty and injustice—indigenous causes—the consequences are therefore unrelated to the global balance of power. Another part is far less certain; members of this part agree that poverty is the cause of instability, but they worry about the possibility that the Soviet Union might take advantage of it. Here, then, is the struggle. When instability flares in Latin America, policy makers who believe that instability is caused by Communist adventurism—the conservatives in the upper-right of Figure C.1—attempt to form a fragile coalition with those who disagree with them on the cause of instability but agree that Communists are likely to take advantage of an unstable situation—the moderates in the upper-left of the figure. At the same time, a third group of policy makers—the liberals in the lower-left—also attempts to form a coalition with the moderates, a coalition based upon their common belief that poverty is the cause of instability.

THE CONSERVATIVES: WINNING THE BATTLES . . .

The evidence conservatives offered in the 1980s to support the belief that Communist adventurism caused instability in Central America did not seem convincing. The Soviet Union did not invade Central America as it did Afghanistan. There were no Soviet advisors fighting with the Sandinistas against Somoza, nor were any Soviets found with the guerrillas in El Salvador and Guatemala. The most common conservative response was that the Soviets were invisible because the Cubans acted as their surrogate. Deputy Assistant Secretary of State James Michel was strongly impressed by "the Soviet Union's ability to act through others, particularly through personnel or organizations acting in the name of small developing countries . . . like Grenada, Nicaragua,

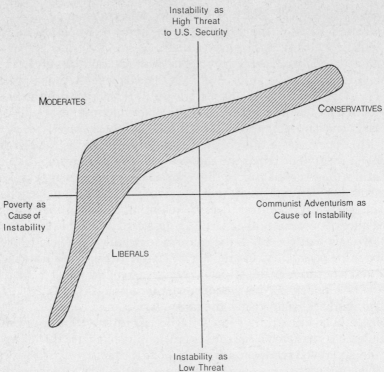

Instability as
High Threat
to U.S. Security

MODERATES

CONSERVATIVES

Poverty as
Cause of
Instability

Communist Adventurism as
Cause of Instability

LIBERALS

Instability as
Low Threat
to U.S. Security

FIGURE C.1. Distribution of Policy Makers' Beliefs
about Instability in Latin America, circa 1980

and even Cuba."[3] But the Cubans, too, were difficult to find—so
difficult, in fact, that no one ever unearthed a single one of them
fighting with the Sandinistas, the FMLN, or any of the Guatemalan guerrilla groups.[4] Nor, with the single exception of Orlando
Tardencillas, were there any Nicaraguans to be found fighting in
El Salvador or Guatemala. To most policy makers who believe
that poverty causes instability the Reagan administration officials

[3] Prepared statement before the House Foreign Affairs Subcommittee on Inter-
American Affairs, February 28, 1985.

[4] Leiken asserts that "Cuban military advisers . . . accompanied FSLN forces into
battle" a few months before Somoza's overthrow, but his supporting evidence from
news clippings is far from authoritative. Robert S. Leiken, "The USSR and Central
America: Great Expectations Dampened?" in *Central America and the Western
Alliance*, ed. Joseph Cirincione (New York: Holmes and Meier, 1985), p. 159.

who searched for evidence of Communist adventurism in every unstable situation in Central America resembled nothing so much as Périgord pigs during truffle season, nosing around the tropical shrubbery in search of the elusive prize, a Soviet instigator.

Most of this evidence was presented in reports produced by the Department of State's Office of Public Diplomacy for Latin America and the Caribbean, whose director, Otto Reich, left no shrubbery unturned in his singleminded quest for evidence of Communist expansionism.[5] Indeed, even a bit of distortion was considered appropriate. For example, the quotation in the title of one document, "*Revolution Beyond Our Borders*," comes from a July 1981 speech by Nicaraguan Interior Minister Tomás Borge, but it is a mistranslation. Borge did not say, as the Department of State's truffle hunters alleged, that "this revolution goes beyond our borders." Rather, he said that "this revolution transcends national boundaries," and he did so in a context that clearly means ideological transcendence, not physical support for revolution. Borge said: "This does not mean we export our revolution. It is enough— and we couldn't do otherwise—for us to export our example." For reasons that seem obvious the State Department decided not to quote this part of Borge's address.

It would be easy to dismiss this problem of proof as evidence of how incorrect policy makers are who believe that Communist adventurism causes instability in Latin America. To let the matter go at that, however, would be to miss completely the basis of the belief. To ask for *physical* evidence that Communism causes instability is to miss the fact that this belief is not based upon evidence that can be seen. Policy makers who believe that poverty

[5] For excellent examples of this type of effort, see the various reports produced by the Department of State in the mid-1980s: U.S. Department of State and Department of Defense, *News Briefing on Intelligence Information on External Support of the Guerrillas in El Salvador* (Washington, D.C.: Office of Public Diplomacy for Latin America and the Caribbean, Department of State, August 8, 1984); U.S. Department of State and Department of Defense, *Background Paper: Nicaragua's Military Build-Up and Support for Central American Subversion* (Washington, D.C.: Department of State and Department of Defense, July 18, 1984); U.S. Department of State and Department of Defense, *The Soviet-Cuban Connection in Central America and the Caribbean* (Washington, D.C.: Department of State and Department of Defense, March 1985); U.S. Department of State and Department of Defense, *The Sandinista Military Build-up* (Washington, D.C.: Department of State and Department of Defense, revised May 1985); U.S. Department of State, *"Revolution Beyond Our Borders": Sandinista Intervention in Central America*, Special Report No. 132 (Washington, D.C.: Department of State, September 1985).

causes instability can take skeptics to Central America, show them extraordinary amounts of poverty, and say, "See, there's your proof."[6] In stark contrast, policy makers who believe that Communism causes instability are unable to take skeptics to Central America, show them a large number of Communists, and say, "See, there's your proof." But, in their own way, these policy makers have ample evidence to support their perceptions. They have four beliefs that, unlike those cited by their adversaries, are firmly embedded in a different reality, the reality of postwar international relations.

As an illustration of the strength of this position imagine a nationally televised debate in which one contestant is required to explain Latin American instability in terms of poverty while the other contestant is limited to using Communist adventurism.

Most viewers will find the first explanation strange, for they have no frame of reference to help understand Latin America's structural poverty. The misery caused by earthquakes, hurricanes, and similar natural disasters can be seen on the evening news, but how many U.S. citizens are familiar with the suffering caused by unjust land tenure patterns or the conversion of tropical agricultural land to production for export markets? The Juan Valdéz that we all see on television seems well-fed and reasonably happy as he carefully selects each coffee bean for U.S. consumers. Social mobilization, political mobilization, elite intransigence—few viewers will have ever heard of these terms, or of Christian grass-roots communities, or of death squads. On the other hand, it is true that many U.S. citizens are appalled by descriptions of abject poverty, and a capable liberal or moderate with sufficient time can probably describe Latin America in such a way that many viewers will agree that poverty is the cause of at least some of the instability in the region.

Then a conservative stands up and begins to talk in the language of the people, beginning with a discussion of the need to contain Soviet Communism—the very heart and soul of postwar United States foreign policy. The speaker turns next to Latin America and taps perhaps the deepest root of U.S. foreign policy—the Monroe

---

[6] As we have seen, however, that is not proof of causality but of covariation. An appropriate belief system is needed to cement the causal linkage between poverty and instability. Given the obvious existence of covariation, however, there is a natural predisposition to entertain the notion of causality. United States citizens who would find Central American poverty unbearable are predisposed to believe that Central Americans would not stand for it any longer, either.

Doctrine. With that said, the contest is over—poverty and social science gibberish are no match for anti-Communism and the Monroe Doctrine. But just to score an additional point or two, the speaker adds a few well-understood words about Latin American political culture. The speaker admits that there are human rights abuses in places like El Salvador, but "what we have to remember, however, is that the situation in El Salvador has, for virtually all of El Salvador's history, been one in which the local commander, even the local commander of a small unit who might only be a sergeant or a corporal, has had a very large degree of autonomy. That has been their system, and it is going to take a good deal of restructuring and reeducation to change that system."[7] Fortunately there are officials in the Department of State who are eager to take up where Teddy Roosevelt left off in civilizing Latin America. In the meantime we cannot forget that there are radicals in Latin America who accept millions of dollars of U.S. aid, then turn around and sing songs about us being "the enemy of humanity." These Latin American radicals are the same people who are always talking about poverty and injustice, trying to milk sympathy from dewy-eyed liberals and misinformed moderates who are too naive to recognize a Communist threat in America's own backyard.

The debate is over, the voters select the winners, and the victors go to Washington and take control of United States policy toward Latin America. There a winner like conservative Representative Bill Young will come face-to-face with a liberal like Heather Foote from the Washington Office on Latin America. There they will try to communicate but, in the end, they will only illustrate the inability of these two groups of policy makers to discuss—let alone agree on—almost any issue involving instability in Latin America:

> MR. YOUNG: Do you share a concern that there is a Communist movement in that part of the world?
> MS. FOOTE: Are you referring to Central America?
> MR. YOUNG: Yes.
> MS. FOOTE: Central America as a region has unfortunately suffered some of the most serious problems of all of Latin America, as regards concentration of wealth and land. And it is a region that has not changed in an evolutionary fashion. . . .

[7] The statement is by the State Department's John Bushnell in U.S. Congress, Senate, Committee on Foreign Relations, *Foreign Assistance Authorization for Fiscal Year 1982*, 97th Cong., 1st Sess., 1981, p. 437.

Many people have reached out to attempt a solution. There is no question that there is participation by Communists in some of these efforts. I think, though, that the problems in Central America are very much home grown. I don't think the Central Americans, and particularly those in El Salvador, needed any outside instigation to become indignant and insistent about their living conditions.

MR. YOUNG: Do you believe there is any outside instigation from Cuban or Communist organizations in El Salvador.

Ms. FOOTE: I think Cuba is very interested in a move away from oligarchic rule.

MR. YOUNG: Into what form of rule? There is going to be some kind of rule there, isn't there?

Ms. FOOTE: It is hard to answer that question any more.[8]

Heather Foote wanted to discuss what she believed to be the cause of instability: poverty and injustice; Bill Young preferred to focus upon what he believed were the consequences of Communist adventurism: a threat to U.S. security. Neither wanted to hear what the other had to say.

Who, then, was listening? Certainly it was not the public at large, the citizens who elected Mr. Young and his colleagues to represent them in Congress. Virtually all analyses continue to demonstrate that the U.S. public remains indifferent to issues of inter-American relations. Immediately after President Reagan's April 1983 speech on Central America to a joint session of Congress, telecast live during prime time, presidential pollster Richard Wirthlin discovered that seven out of ten citizens *who had viewed the speech* did not know that Mr. Reagan had mentioned Nicaragua.[9] This speech, in which the President used extremely harsh language to lash out at the Sandinistas, apparently fell on deaf ears. Similarly, a CBS News/*New York Times* poll taken in mid-1983 indicated that only 8 percent of the American public could identify which side (government or insurgent) the United States supported in both El Salvador and Nicaragua. Twenty-five percent of the respondents knew (or guessed correctly) that the U.S. supported the government in El Salvador; 13 percent knew the U.S. sided with the insurgents in Nicaragua; 8 percent knew

[8] U.S. Congress, House, Committee on Appropriations, Subcommittee on Foreign Operations and Related Agencies, *Foreign Assistance and Related Programs Appropriations for 1981*, 96th Cong., 2d Sess., 1980, pt. 1, p. 405.

[9] *International Herald Tribune*, August 11, 1983, p. 2.

both.[10] Nonetheless, of the 92 percent of the respondents who did *not* know what side the U.S. supported, 86 percent had an opinion, pro or con, on U.S. policies toward the two countries. Stated differently, *almost nine out of every ten citizens who do not know what U.S. policy is nevertheless have an opinion of that policy.*

These data confirm what we have known for years: since knowledge of inter-American relations is minimal, the public's opinions about U.S. policy toward Latin America are formed on the basis of other criteria—ideology, policy traditions, recent experience, and, perhaps most important, the general opinion of the incumbent U.S. administration. In the early and mid-1980s, respondents who felt positive toward the Reagan administration were inclined to favor the Reagan policy toward Central America, despite the fact that most of them did not know what that policy was.

The real audience for the debate between the conservatives and the liberals is the moderate policy makers and the members of the attentive public who fit into the upper-left cell of Figure C.1. As anyone who has ever attempted to lobby on an issue of inter-American relations can attest, these moderates have an intellectual fissure: they agree with liberals on the cause of instability, but they agree with conservatives on its effect. When instability flares in Latin America, the basic issue in Washington becomes which of the two moderate beliefs will dominate their thinking and therefore determine their policy position? So long as instability is not linked in moderates' minds with Communism, they will form a coalition with liberals. If conservatives can forge a link between Communism and instability—and that, as we have seen, is not a particularly difficult task—moderates will reconsider; slowly they will stop emphasizing the cause of instability and begin to focus on its potential consequences for U.S. security.

In this tug-of-war for control over the hearts and minds of moderate policy makers, liberals initially have a major advantage: moderates agree with them on the cause of instability. But they also have a significant disadvantage: moderates do not believe that liberals place a sufficiently high priority upon protecting U.S. security interests in Latin America. As instability grows, moderates become increasingly wary of their liberal colleagues.

This does not happen all at once, of course, for some moderates worry more than others. Some are oriented toward East-West relations; some toward North-South relations. Some face tough re-

[10] *New York Times,* July 1, 1983, pp. A1-A2.

election campaigns from right-wing challengers; others face a daily barrage of mail from left-wing constituents. But eventually the scale tips. While moderates do not forget their beliefs about poverty causing instability, their discussions of poverty become "background" material that lacks animation. Poverty causes instability, they seem to be saying, but that was long ago. Once the level of instability becomes acute, poverty becomes just one more underlying cause of unrest, like low sugar prices, large foreign debts, and dependency upon the industrialized West. As instability increases, moderates make fewer references to the revolution of rising expectations among the poor and more references to Cuba and the Soviet Union.[11]

Gradually, conservatives win the tug-of-war. United States policy toward the Nicaraguan revolution provides the most recent example of this slow defection of moderate policy makers. The turning point occurred in 1985, when the House of Representatives voted twice on the administration's request for aid to the Nicaraguan Contras. In April the liberal-moderate poverty coalition was barely able to prevail, and the House rejected Mr. Reagan's request by a slim margin of 215 to 213. Then Nicaraguan president Daniel Ortega made a much-publicized trip to Moscow, and an additional handful of moderates began to worry more about the consequences of the Nicaraguan revolution and less about the poverty that caused the instability. In June the House reconsidered and voted 232 to 196 to provide the Contras with $27 million in humanitarian aid. Of the twenty-two Representatives who switched their vote, nineteen had an ADA rating in the moderate range of 35 to 65, and all twenty-two were in the 20 to 80 range.[12] The Rea-

[11] A good illustration of this shift is provided by the Council of the Americas, a business organization dominated by moderates. In 1982, the council's Washington representative argued that "at the root of political instability are poverty, illiteracy, frustrated aspirations, unequal income distribution, and inadequate economic opportunities." By 1983, however, the organization's president did not disagree, but he was not interested in the issue of causality. Instead he argued that "there is a high level of consensus about the need to provide military support to the Salvador government, [about the need] for interdiction of weapons supplies to the Salvadoran rebels, [and] about the need to defend Honduras." Moreover, he observed that U.S. business leaders would be "perfectly happy" if the Sandinista government were overthrown. *Los Angeles Times*, September 25, 1983, pt. 1, p. 2.

[12] Each year the Americans for Democratic Action (ADA) rates each member of Congress on a scale of 0 to 100, with a low score indicating a conservative voting record and a high score indicating a liberal voting record.

The number of switchers does not seem to balance with the number of votes that changed, for although 428 representatives voted each time, they were not exactly

gan administration won the battle to reassert the prominence of conservative beliefs about instability in United States policy toward Latin America.

. . . Losing the War?

In a sense, it was a remarkable victory. But it must be remembered that the Reagan administration had entered office determined to make Central America a symbol of the U.S. resolve to halt the spread of Communism and maintain the balance of power. After years of continuous effort the Contra aid votes of 1985 and 1986 indicated division rather than resolve. The votes demonstrated that the United States remains bitterly divided over issues of inter-American relations. Despite the remarkable personal popularity of Mr. Reagan and the overall strength of his administration, there was no hint of the creation of a policy-making consensus. The divergent beliefs of U.S. policy makers did not change. What changed in the 1980s is that one set of beliefs that had been in decline experienced a resurgence; the divisions among policy makers were thereby exacerbated. Nothing was resolved. The administration narrowly won battle after battle over specific issues of inter-American relations, but it lost the war to provide a coherent vision of United States policy toward Latin America upon which policy makers could agree and therefore act. As we enter the final years of the century, profound dissensus remains the overarching reality of United States policy toward Latin America.

It is not obvious that any administration can ever win this war. It may be that the United States will simply muddle through the coming decades, having a human rights policy one year, a Contra war the next; an aid package here, an invasion there; tough talk, soft talk; usually managing to cope, but just barely.

This dissensual policy-making process and the vacillation it produces have become luxuries the United States can no longer afford. For most of the years since 1823, it really did not matter

---

the same 428. A total of 25 representatives switched their vote to favor the administration, but 3 were freshmen and therefore as yet had no ADA rating. The mean ADA rating of the 22 representatives was 50.

Several votes were cast on each of the two occasions. The ones selected here for comparision were the key votes: on April 24, the Michel Amendment (rejected 213 to 215, with a "yea" vote supporting the administration's position); on June 12, the amendment to continue indefinitely the Boland Amendment (rejected 196 to 232, with a "nay" vote supporting the administration's position).

much whether the United States had its ducks in order in its policy toward Latin America. Far from the site of world conflicts and tucked securely into the soft underbelly of the United States, until very recently Latin America remained isolated in the quiet backwater of international relations. The fact that much of it sat rotting under the control of U.S.-supported tyrants did not bother many officials whose foreign policy interests lay elsewhere. No less a humanitarian than Franklin Roosevelt could dismiss the problems of tyranny in Nicaragua with a facile remark that the first Somoza may be an SOB, but at least he's *our* SOB.

A lot of good that does us today. The third and last Somoza is dead, his Mercedes blown up by a bazooka in Paraguay, and down in Managua we are no longer consulted by his successors. The current generation of Nicaraguans has apparently decided not to permit the United States to select its leaders, as Washington selected Somoza, nor to dictate the nature of Nicaraguan public policy. And, unlike an earlier era, this generation of Nicaraguans appears to be able to resist U.S. hostility. The United States is reaping the bitter harvest of years of disregard for the rights and the well-being of the people of Nicaragua.

A wiser policy in the past might have avoided the problems of the present. A wiser policy would have separated the protection of U.S. interests from the continuation of the Somoza dictatorship. A wiser policy would have expressed a more serious concern for the abject poverty of the vast majority of Nicaraguans. But a wiser policy would have required Washington to change many of its time-honored beliefs about Latin American stability and about Cold War security. Unable to effect this change, U.S. policy makers watched their influence decline to an all-time low in Nicaragua. By the mid-1980s, it rested upon the fighting ability of a rag-tag group of Honduran-based guerrillas whose efforts would not continue for a month without U.S. financial support. In a few short years the United States moved from a position of extraordinary influence in Nicaragua to a position where our influence had to be exercised through a hopeless band of blood-thirsty mercenaries. That the Reagan administration insisted on referring to these minatory misfits as "freedom fighters" is an indication not of ignorance but of desperation—no one else in Nicaragua would be our ally. In the two-century history of inter-American relations it is difficult to identify a sorrier spectacle than United States policy toward postrevolutionary Nicaragua. We blew it completely.

Perhaps the saddest aspect of the debacle in Nicaragua is that all this could have been avoided if, much earlier, U.S. policy makers had been willing to adjust their beliefs about Latin American instability and about Cold War security to conform to a changing reality in Latin America. But here there is, perhaps, some reason for optimism for the future. As we have seen in the preceding chapters, there is considerable evidence among liberal and moderate officials that these beliefs about instability and security are, in fact, susceptible to change.

Changing Beliefs

Latin American Stability

The concept of stability, long considered the primary means of ensuring U.S. security in Latin America, is undergoing a profound reassessment that focuses upon three specific problems that have arisen repeatedly in recent years.

The first problem is that *an undiscriminating commitment to stability damages U.S. security.* Policy makers have always known that an ethical dilemma sometimes accompanies the U.S. emphasis upon stability in Latin America: for the sake of stability Washington has at times become allied with repugnant dictatorships, authoritarian systems directed by men like Batista, Pinochet, Somoza, Stroessner, and Trujillo. The alliance is not accidental, as no less a realist than Hans Morgenthau noted, for "tyranny becomes the last resort of a policy committed to stability as its ultimate standard."[13]

The moral reservations that would normally accompany U.S. support for tyranny have always been rationalized by the argument that these tyrannical governments provide anti-Communist stability—a higher value—and might, over time, moderate their repression. If, as Jeane Kirkpatrick and others suggest, the choice is often between authoritarianism and totalitarianism, then there is a certain morality in opting for the former because it is clearly the lesser of the two evils.

But Kirkpatrick clearly misidentified the problem. Recent history demonstrates that U.S.-backed authoritarian regimes cannot maintain stability forever; they *invariably* collapse, and when they fall they tend to be replaced by new regimes whose leaders are not inclined to forgive and forget the years of U.S. support for tyranny. Washington's problems in Iran and Nicaragua after 1979 and

[13] *The New Republic*, October 11, 1975, p. 17.

Cuba after 1959 are the direct result of U.S. support for the preceding authoritarian regimes. Thus the alleged tradeoff of accepting authoritarianism to avoid totalitarianism, which may or may not be an accurate portrayal of the available alternatives, is irrelevant to U.S. security. For U.S. foreign policy the *real* tradeoff occurs when short-sighted officials support authoritarian stability today and ignore the growing evidence that this support leads to threatening instability in the future. The lesson of Iran, Cuba, and now Nicaragua is clear: the chickens always come home to roost.

This is more than a moral issue. At question is the *instrumental* value of an unquestioning commitment to stability as a guarantor of U.S. security in Latin America. Liberals and most moderates have concluded, as Stanley Hoffmann did, that "there can be no stability where there is profound dissatisfaction with the political, economic, or social situation."[14] Agreement with this assertion implies a level of discrimination that has often not existed in U.S. policy, the avoidance of macrocategories that lead the United States to embrace "authoritarianism" in order to avoid "totalitarianism." And it implies a focus upon poverty and injustice. No one knows where thoughts like this will lead U.S. policy, but the Reagan administration's decision to distance itself from the Duvalier authoritarianism in Haiti and the Marcos authoritarianism in the Philippines indicates how broadly this type of thinking has permeated the policy-making process. In the 1970s, respected analysts such as Hans Morgenthau and Stanley Hoffmann challenged policy makers to rethink the value of an undiscriminating commitment to stability; by the mid-1980s, increasing numbers of Washington policy makers were prepared to accept the challenge.

The second problem is that, for better or for worse, *the United States no longer has the strength to enforce its concept of stability in Latin America.* Cuba and now Nicaragua are two sobering experiences. Short of an armed invasion by the U.S. Marines, Washington policy makers have done everything they can to rid the hemisphere of the Castro and Sandinista governments. Yet both governments persist. If they were alive today, policy makers of Theodore Roosevelt's generation would be startled to observe that the United States can no longer dictate the behavior of the Cuban government.

Other officials would be particularly surprised by the ability of the Nicaraguan government to absorb more than five long years of

[14] Stanley Hoffmann, *Primacy or World Order: American Foreign Policy since the Cold War* (New York: McGraw-Hill, 1978), p. 30.

322

punishment from the Reagan administration, including both economic sanctions and a guerrilla war, and to show no sign of weakening. In an interview in December 1985 with a senior career diplomat, a moderate assigned to ARA's Office of Central American Affairs, I asked for an opinion regarding the future of Nicaragua. "They've learned to operate their helicopters," he responded. "They've won. The rest of the decade we'll be working on damage control." Barring an invasion, the response was accurate in the sense that access to weapons was obviously the *sine qua non* of survival, and the Nicaraguans, like the Cubans, felt fortunate that Washington's rivals were willing to help. But as the United States discovered in Vietnam, access to sophisticated weapons is only a necessary, but not a sufficient, criterion of success. The willingness of the Nicaraguan population to fight in defense of their revolution is what turned back the Contra attacks and gave pause to policy makers contemplating an invasion. Such an invasion may lie in the future, but it is doubtful that the marines can restore U.S.-style stability to Nicaragua. Too many Nicaraguans would rather die than submit. Just yesterday a quintessential banana republic, Nicaragua has taken charge of its own future.

It is common today to talk about declining hegemony and the limits of U.S. power, but examples are often ambiguous and difficult to interpret. Nicaragua is different. It is obvious what has happened—the United States has lost its power to dictate at low or even moderate cost the nature of stability in Nicaragua. Not long ago one of the most perceptive observers of inter-American relations, Laurence Whitehead, wrote that "America's enormous power in relation to her near neighbors . . . leaves her free to pursue a regional policy of over-insurance." This, Whitehead noted, is "a luxury almost unknown to the foreign-policy makers of other nations."[15] The lesson of Cuba and Nicaragua (and Afghanistan) is that this luxury has become increasingly expensive; Third World nations now sometimes fight back. In Guatemala in 1954, the costs of over-insurance were so modest as to be inconsequential—no one even bothered to calculate them. Today the costs are measured not only in thousands of lives but in the way they jeopardize other, more important goals. A U.S. invasion anywhere in Central America would be seriously damaging to a broad range of U.S. foreign policy goals, not the least of which is the goal of asserting U.S. leadership in the world. No nation that aspires to lead the West

[15] Laurence Whitehead, "Explaining Washington's Central American Policies," *Journal of Latin American Studies* 15 (November 1983): 362.

can afford the image of a neighborhood bully, of a crude nation that solves its problems by force. Rambo is a movie, not a foreign policy. And over-insurance, however reassuring, is a frivolous expenditure now beyond our means.

The third problem is that *a different type of instability is beginning to emerge in Latin America.* A generation of dictatorships has now been replaced by democratically elected governments in Latin America. Our generation's twilight of the tyrants is occurring.[16] Only a few years ago General Pinochet was the norm; today he, along with Paraguay's Stroessner, is the anomaly, the exception to the rule. Today guns are not as valuable as they were in determining the leaders of many Latin American nations, including all of the most important ones; today aspiring leaders need votes.

This transition to democracy has implications for the nature of instability in the years immediately ahead. In most countries the survival of popularly elected governments is dependent upon their ability to satisfy the demands of an electorate organized into fragile multiclass coalitions. All the perennial conflicts over social policy, particularly the mix of government expenditures for consumption and investment, have reappeared with increased vigor. Indeed, today's conflicts are probably more acute than ever, for now there is the need to manage Latin America's deep involvement in the international economy, especially the large foreign debt that overhangs virtually every country in the region. We are, in short, at the beginning of an era of heightened political conflict—instability—as fragile democratic governments attempt once more to answer the perennial questions of who gets what.

No one knows the form instability will take in each country, nor can anyone hope to predict its intensity. All that is certain is that the democratic genie is now out of the bottle in most of Latin America, along with its capacity for generating instability. This is a type of instability that differs in one clear and irrefutable way from the Central American instability that has mesmerized Washington during the 1980s: it has nothing to do with the Cold War.

Cold War Security

For outside observers concerned about such issues as human rights, poverty, and injustice in Latin America, the patent nobility

---

[16] The original "twilight" was identified by Tad Szulc in his *Twilight of the Tyrants* (New York: Holt, 1959).

of their concerns seems so obvious that they often wonder why policy makers refuse to give these issues the priority they deserve. Observers are almost always poised to accuse policy makers of being insensitive, of neglecting principle, of compromising with the generals in Guatemala and Chile.

My impression is that policy makers are deeply (and in most cases justifiably) offended by this criticism. Most care as much as any of us about the protection and promotion of these values. But they, unlike us, have been assigned the responsibility for protecting our security. Because the penalty for failure is so high, officials act with extreme caution, invariably preferring to err on the side of safety. And when they perceive a conflict between the protection of U.S. security and the promotion of values such as human rights, security invariably wins.

The impact of this national security mentality upon United States policy toward Latin America is seen whenever instability is addressed: the discussion generally gravitates to an assessment of the intentions of the only power that can cause an unmitigated disaster, the Soviet Union. Latin America becomes inert, a passive object of no intrinsic value, a place where the United States and the Soviet Union play out the drama of international politics. It is difficult to overemphasize how policy makers' dominant concern for security leads to this conception of Latin America. The effect is to convert the debate over U.S. policy toward Latin America into a debate over Soviet intentions.

The perennial problem of this debate is that there is no objective procedure to determine what Soviet intentions are. Even our best efforts are little more than informed guesses. In his analysis of the host of studies about the Cuban missile crisis, Ned Lebow concludes that all of them "are, at best, clever speculations about Soviet behavior consistent with the few established facts." More to the point,

American analysts' lack of knowledge extends to just about every other aspect of Soviet decision-making in the Cuban missile crisis. If this is true of Cuba, an event that took place almost twenty-years ago and that has been extensively studied, how much truer it is of more recent Soviet initiatives, whose understanding benefits from neither the meager documentary evidence available on the missile crisis nor from the greater analytical detachment that develops in the fullness of time. In effect, imaginative analysts can readily devise an explanation consonant with their own intellectual or ideological orientation.[17]

Clearly there are severe intellectual limitations on the accuracy of any assessment of Soviet intentions. But because there is no way to determine for certain Moscow's "real" intentions, many policy makers feel obliged by the high costs of error to assume the worst. For Latin America, policy makers assume that instability is a manifestation of Soviet intentions to snatch a country from the U.S. sphere of influence.

But today not everyone in Washington is willing to accept automatically a worst-case interpretation of Latin American instability. As Ambassador William Luers notes,

> there is a debate over Soviet intentions that affects U.S. thinking about Latin America. Do they seek world conquest? To communize the world? Or only secure borders? Are they only desirous of equal status and equal influence as a superpower? Or more ambitiously to change the "correlation of forces." These are legitimate and difficult questions. One can almost position a foreign policy expert on the political spectrum between conservative and liberal . . . as to how he or she answers these questions. Conservatives stress Soviet expansionism, world strategy and conquest. "Realists" [our moderates] and liberals tend toward a somewhat more benign view of Soviet motives—secure borders and equal status as a superpower.[18]

Until fairly recently only a few policy makers spoke with this ambivalence about the intentions of the Soviet Union; most simply assumed that Moscow was out to change the "correlation of forces"—the global balance of power. Now, as uncertainty replaces orthodoxy in attributing intent, there is room for policy makers to reconsider their beliefs about instability in Latin America.

This reconsideration can be seen, first, in officials' attitudes toward the larger countries of Latin America. The autonomy of Argentina, Brazil, Chile, Colombia, Mexico, Peru, and Venezuela is now accepted throughout Washington as a fact. These are no longer perceived as inert hunks of territory that the Soviet Union

[17] Richard Ned Lebow, "The Cuban Missile Crisis: Reading the Lessons Correctly," *Political Science Quarterly* 98 (Fall 1983): 456.

[18] William H. Luers, "The Soviets and Latin America: A Three Decade U.S. Policy Tangle," *Washington Quarterly* 7 (Winter 1984): 7.

occasionally attempts to grab through subversion in order to tilt the global balance of power. These are real countries with real people and very real problems all their own; they are anything but inert. To the extent that this belief takes hold of the consciousness of U.S. policy makers, the focus on Cold War beliefs about Soviet intentions will necessarily weaken. The United States will be confronted with problems—foreign debt, protectionism, drug trafficking, immigration—that cannot be sloughed off as Communist adventurism.

Second, the reconsideration can be seen in policy makers' slow loss of innocence. What appeared to be a total victory in 1945 soon led to the realization that victories are never complete in international relations. The simple blacks and whites soon began to blur into every shade of gray.

Policy makers learned other lessons about complexity as well. In Latin America they learned that development was not easy. They learned, as Herbert Goldhamer observed, that

> societies are highly resistant to conscious attempts to control their direction and rate of change. Changes may indeed be effected but they are not often of the magnitude or the character intended. The attempt of the United States in the sixties to reshape the Latin societies, to alter their rate of development, and to oppose changes of which it disapproved, discounted the intractability of societies, especially to outside control.[19]

A more recent lesson, this time from Southeast Asia, was that Washington's tendency to militarize issues—to solve problems by shooting people—is often counterproductive.[20] By the early 1980s, a large number of policy makers believed that brute force would never resolve U.S. problems in Central America. Of El Salvador, Senators Leahy and Pell argued that "the resolution of the present conflict cannot, and will not, be a military one"; Ambassador Robert White asserted that "there is no military solution to the problem of El Salvador"; and shortly before he was fired, Assistant Secretary of State Thomas Enders expressed his belief that "U.S. troops are no solution now" to the problems of instability in Cen-

---

[19] Herbert Goldhamer, *The Foreign Powers in Latin America* (Princeton, N.J.: Princeton University Press, 1972), p. 260.

[20] For a discussion of this tendency, see Stanley Hoffmann, *Dead Ends: American Foreign Policy in the New Cold War* (Cambridge, Mass.: Ballinger Publishing Company, 1983), especially p. 206.

tral America.[21] Alfred Stepan wrote that "military intervention may finally be obsolete as a weapon in the U.S. hemispheric policy arsenal."[22]

That may or may not be correct. It is always difficult to determine the precise moment when something as complex as a tendency toward military domination ends. But one thing is certain: to all liberals and most moderates the days of winning a clear victory over a clear foe called Communism in Latin America, of "developing" Latin America with foreign aid, of sending in the marines to quiet things down, these days are fading—not gone, but fading. Conservative officials obviously hold on to these ideas, preparing for the day when the United States might once again restore a Pax Americana to the Caribbean basin and put an end, once and for all, to the instability that threatens U.S. security. Certainly that was one of the lessons of Grenada. But to focus upon the resurgence of Cold War internationalism in the 1980s is to miss the important evidence that this resurgence may have been a last gasp.

Third, and finally, the reconsideration can be seen in a growing malaise among many officials. A policy focus upon stopping Soviet expansion into Latin America has the singular defect of being inalterably negative. It provides an answer to the question of what the United States does *not* want Latin America to be but no satisfactory, realistic vision of the type of relationship the United States should attempt to develop with the nations of the region. Representative Benjamin Rosenthal illustrated this uneasiness in an exchange with Secretary of State Haig:

> The whole tenor of your statement and testimony, I sense, is that the foreign policy of the United States today is based essentially on a reaction to the Soviets, a reaction to international terrorism, a reaction to international unrest, and violent change. There does not seem to be, and I say this respectfully, a positive tone to the statement. I cannot figure out from it what the positive fundamental program of U.S. foreign policy is. What are the things we stand for? Are we

[21] The statements were made, respectively, in U.S. Congress, Senate, Committees on Foreign Relations and on Appropriations, *El Salvador: The United States in the Midst of a Maelstrom,* 97th Cong., 2d Sess., 1982, p. 7; Robert White, "There Is No Military Solution in El Salvador," *The Center Magazine* 14 (July-August 1981): 9; and Thomas O. Enders, speech before the Commonwealth Club of San Francisco, August 20, 1982.

[22] Alfred Stepan, "United States and Latin America: Vital Interests and the Instruments of Power," *Foreign Affairs* 58 (1980): 661.

only reacting to Soviet adventurism, terrorism, unrest, things of that nature?[23]

In Washington today it is difficult to find a career foreign policy official who takes pride in U.S. leadership in inter-American affairs. Many are profoundly embarrassed by U.S. involvement in dirty little wars, in gross embarrassments like the decision to withdraw from the jurisdiction of the World Court, in cheap victories like Grenada. There is no vision, no *élan vital*. Of course there are times like World War II when policy must focus upon the negative, upon stopping something that must be stopped. But the United States has never been comfortable with this role. As a nation we pride ourselves on our innovation, our creativity, our progress. We like to spend our time solving problems, meeting challenges, moving ahead. As it is practiced in our policy toward Latin America, containment is not well-suited to our temperment. It chafes our nature.

Many policy makers and, I think, particularly our brightest policy makers, do not want their tombstone to read, "He spent his life stopping Communism in Latin America." They want it instead to read, "She spent her life creating a better hemisphere." The policy of containment provides these policy makers with no guidance for this type of creative life. When moderate Republicans like Margaret Daly Hayes write that "a better strategy for dealing with Cuba would be simply to ignore it—write it out of our vocabulary and concentrate on positive, constructive activities that would help minimize Cuba's revolutionary appeal," they are asking for a positive, constructive vision of the future, a vision whose achievement is worth the investment of their lives.[24] For all its shortcomings the Carter human rights policy provided at least a portion of this vision, a goal of helping to create a hemisphere free of torture; it was something to work *for*. With containment there is only something to stop.

Quite obviously not everyone worries about this existential problem. Like most professors, most policy makers focus on meeting their next deadline, on paying their mortgages, on getting their kids' teeth straightened, on managing the hundred-and-one daily demands that eventually combine to serve as an acceptable ex-

[23] U.S. Congress, House, Committee on Foreign Affairs, *Foreign Assistance Legislation for Fiscal Year 1982*, 97th Cong., 1st Sess., 1981, pt. 1, p. 166.
[24] Margaret Daly Hayes, "The Stakes in Central America and U.S. Policy Responses," *AEI Foreign Policy and Defense Review* 4 (1982): 17-18.

ample of a full career. But every so often a researcher sees a hint that on those occasional quiet moments when policy makers must confront the substance of their lives, many find that containment is unfulfilling.

WHAT is occurring, in short, is that in the minds of many policy makers Latin America is becoming more than a place where East fights West. Fundamental questions about security will remain forever at the center of policy debates, and the assessment of an adversary's intentions will always be a part of U.S. policy toward Latin America, but the size of the part is decreasing. It is not decreasing because some policy makers are less concerned than they should be about U.S. Security but because they recognize that Latin America is quickly outgrowing the Cold War.

Over the short span of time since World War II the United States has been forced to learn many lessons about international relations that other nations took centuries to absorb. All the lessons were difficult, and some, like Vietnam, were traumatic. In many respects the lessons from Latin America have been among the most difficult, for we are not accustomed to thinking that there is much to learn from the region. But U.S. policy makers tend to be bright, adaptive people, fully capable of learning what they need to learn in order to protect our security. Unquestionably, much of this learning is not reflected in recent United States policy toward Latin America, and that is lamentable. It is tragic to prolong needlessly and thereby make more traumatic the period of adjustment that inevitably lies ahead. But just because we seem determined to do everything the hard way in places like Nicaragua does not mean that the adjustment will not occur. Immediately below the surface of contemporary policy is a quiet understanding among a significant and, I think, a growing number of policy makers that the Cold War is not what needs to be fought today in Latin America, that the struggle now is against the indigenous foes of poverty and injustice, and that the entire hemisphere would be a more pleasant and a more secure place to live if all of us were to cease fighting the battles of the past and instead focus our energies upon the future.

BIBLIOGRAPHY

Aaron, Harold R. "The Inter-American Military Force." *Military Review* 45 (June 1965): 63-68.

Abel, Christopher A. "A Breach in the Ramparts." *Proceedings of the U.S. Naval Institute* 106 (July 1980): 46-50.

Aberbach, Joel D., and Rockman, Bert A. "Clashing Beliefs Within the Executive Branch: The Nixon Administration Bureaucracy." *American Political Science Review* 70 (June 1976): 456-468.

Adams, Charles Francis, ed. *Memoirs of John Quincy Adams, Comprising Portions of His Diary from 1795 to 1848.* 12 vols. Philadelphia, Pa.: J.B. Lippincott, 1874-1875.

Adams, Michael R. "Coast Guarding the Caribbean." *Proceedings of the U.S. Naval Institute* 108 (August 1982): 61-65.

Agee, Philip. *White Paper Whitewash: Interviews with Philip Agee on the CIA and El Salvador,* edited by Warner Poelchau. New York: Deep Cover Books, 1981.

Altimir, Oscar. "The Extent of Poverty in Latin America." World Bank Staff Working Paper No. 522. Washington, D.C.: World Bank, 1982.

American Petroleum Institute. *Basic Petroleum Data Book: Petroleum Industry Statistics.* Washington, D.C.: American Petroleum Institute, quarterly issues, June 1977 through May 1986.

Americas Watch. *Human Rights in Nicaragua.* New York: Americas Watch, 1984.

———, Helsinki Watch, and Lawyers Committee for International Human Rights. *The Reagan Administration's Human Rights Policy: A Mid-Term Review.* New York: Americas Watch, December 10, 1982.

Anderson, Thomas D. *Geopolitics of the Caribbean: Ministates in a Wider World.* New York: Praeger, 1984.

Armacost, Michael H. "Strength and Diplomacy: Toward a New Consensus?" Speech to the Boston World Affairs Council, Boston, January 25, 1985.

Arnson, Cynthia. *Background Information on El Salvador and U.S. Military Assistance to Central America.* Washington, D.C.: Institute for Policy Studies, March 1982.

———. *El Salvador: A Revolution Confronts the United States.* Washington, D.C.: Institute for Policy Studies, 1982.

Ascher, William. *Scheming for the Poor: The Politics of Redistribution in Latin America.* Cambridge, Mass.: Harvard University Press, 1984.

Atkins, G. Pope. *Latin America in the International Political System.* New York: The Free Press, 1977.

Atkins, G. Pope, and Wilson, Larman C. *The United States and the Trujillo Regime*. New Brunswick, N.J.: Rutgers University Press, 1971.

Axelrod, Robert, ed. *Structure of Decision: The Cognitive Maps of Political Elites*. Princeton, N.J.: Princeton University Press, 1976.

Bagley, Worth H. *Sea Power and Western Security: The Next Decade*. Adelphi Papers No. 139. London: International Institute for Strategic Studies, 1977.

Bailey, Thomas A. *A Diplomatic History of the American People*. 7th ed. New York: Appleton-Century-Crofts, 1984.

Baily, Samuel L. *The United States and the Development of South America, 1945-1975*. New York: New Viewpoints, 1976.

Baloyra, Enrique A. *El Salvador in Transition*. Chapel Hill: University of North Carolina Press, 1983.

Barber, Williard F., and Ronning, C. Neale. *Internal Security and Military Power: Counterinsurgency and Civic Action in Latin America*. Columbus: Ohio State University Press, 1966.

Barnet, Richard J. *Intervention and Revolution*. New York: Mentor, 1972.

————. *Real Security: Restoring American Power in a Dangerous Age*. New York: Simon and Schuster, 1981.

————. *Roots of War*. New York: Atheneum, 1972.

————, and Falk, Richard A. "Cracking the Consensus: America's New Role in the World." *Working Papers for a New Society* 6 (March/April 1978): 41-48.

Barth, John. *The Floating Island*. New York: Avon Books, 1956.

Beaulac, Willard L. *A Diplomat Looks at Aid to Latin America*. Carbondale: Southern Illinois University Press, 1970.

Beckerman, W. *In Defense of Economic Growth*. London: Jonathan Cape, 1974.

Beirne, Charles J., S.J. "Jesuit Education for Justice: The Colegio in El Salvador, 1968-1984." *Harvard Educational Review* 55 (February 1985): 1-19.

Bemis, Samuel Flagg. "A Way to Stop the Reds in Latin America." *U.S. News and World Report*, December 28, 1959, pp. 77-80.

Bennett, Douglas C., and Sharpe, Kenneth E. "Capitalism, Bureaucratic Authoritarianism, and Prospects for Democracy in the United States." *International Organization* 36 (Summer 1982): 633-663.

Bennett, James T., and Williams, Walter E. *Strategic Minerals: The Economic Impact of Supply Disruptions*. Washington, D.C.: Heritage Foundation, 1981.

Bergsman, Joel. "Income Distribution and Poverty in Mexico." World Bank Staff Working Paper No. 395. Washington, D.C.: World Bank, June 1980.

Bernstein, Alvin H., and Waghelstein, John D. "How to Win in El Salvador." *Policy Review*, no. 47 (Winter 1984), pp. 50-52.

Berrios, Rubén. "Economic Relations between Nicaragua and the Socialist Countries." Working Paper No. 166, Latin American Program, The Wilson Center, Washington, D.C., 1984.

Berryman, Phillip. "El Salvador: From Evangelization to Insurrection." In *Religion and Political Conflict in Latin America*, edited by Daniel H. Levine, pp. 58-78. Chapel Hill: University of North Carolina Press, 1986.

———. *The Religious Roots of Rebellion: Christians in Central American Revolutions*. Maryknoll, N.Y.: Orbis Books, 1984.

Bishop, George F.; Oldendick, Robert W.; Tuchfarber, Alfred J.; and Bennett, Stephen E. "Pseudo-Opinions on Public Affairs." *Public Opinion Quarterly* 44 (Summer 1980): 198-209.

Bitar, Sergio. "United States-Latin American Relations: Shifts in Economic Power and Implications for the Future." *Journal of Inter-American Studies and World Affairs* 26 (February 1984): 3-31.

Blachman, Morris J.; LeoGrande, William M.; and Sharpe, Kenneth. *Confronting Revolution: Security through Diplomacy in Central America*. New York: Pantheon, 1986.

Black, Jan Knippers. *United States Penetration of Brazil*. Manchester, Eng.: Manchester University Press, 1977.

Blasier, Cole. "The Cuban-U.S.-Soviet Triangle: Changing Angles." *Cuban Studies* 8 (1978): 1-9.

———. *The Giant's Rival: The USSR and Latin America*. Pittsburgh: University of Pittsburgh Press, 1983.

Blechman, Barry M., and Kaplan, Stephen S. *Force Without War: U.S. Armed Forces as a Political Instrument*. Washington, D.C.: Brookings Institution, 1978.

———, and Levinson, Stephanie E. "Soviet Submarine Visits to Cuba." *Proceedings of the U.S. Naval Institute* 101 (September 1975): 31-39.

———, with Berman, Robert P.; Binkin, Martin; and Weinland, Robert G. "Toward a New Consensus in U.S. Defense Policy." In *Setting National Priorities: The Next Ten Years*, edited by Henry Owen and Charles L. Schultze, pp. 59-128. Washington, D.C.: Brookings Institution, 1976.

Bonner, Raymond. "The Thousand Small Escalations of Our Latin War." *New York Times*, September 16, 1984, p. 31.

———. *Weakness and Deceit: U.S. Policy and El Salvador*. New York: Times Books, 1984.

Bouchey, L. Francis. "Reagan Policy: Global Chess or Local Crap Shooting?" *Caribbean Review* 11 (Spring 1982): 20-23.

Braden, Spruille. *Diplomats and Demogogues: The Memoirs of Spruille Braden*. New Rochelle, N.Y.: Arlington House, 1971.

Brodie, Bernard. *Strategy in the Missile Age*. Princeton, N.J.: Princeton University Press, 1965.

Brodin, Katarina. "Belief Systems, Doctrines, and Foreign Policy." *Conflict and Cooperation* (Stockholm), vol. 8 (1972): 97-112.

Brown, Cynthia, ed. *With Friends Like These: The Americas Watch Re-*

port on *Human Rights and U.S. Policy in Latin America.* New York: Pantheon, 1985.

Brown, Stephen Dechman. "The Power of Influence in United States-Chilean Relations." Ph.D. dissertation, University of Wisconsin—Madison, 1983.

Browning, David. "Agrarian Reform in El Salvador." *Journal of Latin American Studies* 15 (November 1983): 399-426.

————. *El Salvador: Landscape and Society.* Oxford: Clarendon Press, 1971.

Brzezinski, Zbigniew. "Strategic Implications of the Central American Crisis." In *Central America and the Western Alliance,* edited by Joseph Cirincione, pp.105-110. New York: Holmes and Meier, 1985.

Buckley, Tom. *Violent Neighbors: El Salvador, Central America, and the United States.* New York: Times Books, 1984.

Buckley, William F., Jr., and Reagan, Ronald. "On Voting Yes or No on the Panama Canal Treaties." *National Review* 30 (February 17, 1978): 210-217.

Buell, Raymond Leslie. *Isolated America.* 2d ed. New York: Alfred A. Knopf, 1940.

Buergenthal, Thomas. "The Inter-American Court of Human Rights." *American Journal of International Law* 76 (1982): 231-245.

Burden, William A. M. *The Struggle for Airways in Latin America.* New York: Council on Foreign Relations, 1943.

Burke, W. Scott. "Human Rights and El Salvador." *Strategic Review* 11 (Spring 1983): 62-67.

Calicott, Wilfred Hardy. *The Western Hemisphere: Its Influence on United States Policies to the End of World War II.* Austin: University of Texas Press, 1968.

Carlen, Claudia, ed. *The Papal Encyclicals, 1958-1981.* 5 vols. Wilminton, N.C.: McGrath Publishing Company, 1981.

Carlin, James F. "Bismuth." *Mineral Commodity Profiles Series.* Washington, D.C.: Bureau of Mines, 1979.

Carrigan, Ana. *Salvador Witness: The Life and Calling of Jean Donovan.* New York: Simon and Schuster, 1984.

Carter, Jimmy. *Keeping Faith: Memoirs of a President.* New York: Bantam Books, 1982.

Center for National Security Studies. "The Consequences of 'Pre-Publication Review': A Case Study of CIA Censorship of The CIA and the Cult of Intelligence." CNSS Report No. 109. Washington, D.C.: Center for National Security Studies, September 1983.

Center for Strategic and International Studies. *Russia and the Caribbean: Part One and Part Two.* Special Report Series No. 13. Washington, D.C.: Georgetown University, 1973.

Centro de Documentación de INIES. *Cronología de las relaciones Esta-*

dos Unidos-Nicaragua, 1979-1984. Managua: Editorial de Ciencias Sociales, 1985.

Chester, Edward W. *The United States and Six Atlantic Outposts: The Military and Economic Considerations*. Port Washington, N.Y.: Kennikat Press, 1980.

Child, John. "Estados Unidos y latinoamérica: conceptos estratégicos militares." *Estratégia*, no. 63 (March-April 1980), pp. 71-90.

———. "From 'Color' to 'Rainbow': U.S. Strategic Planning for Latin America, 1919-1945." *Journal of Inter-American Studies and World Affairs* 21 (May 1979): 233-259.

———. "Issues for U.S. Policy in the Caribbean Basin in the 1980s: Security." In *Western Interests and U.S. Policy Options in the Caribbean Basin*, Report of the Atlantic Council's Working Group on the Caribbean Basin, pp. 139-185. Boston: Oelgeschlager, Gunn and Hain, 1984.

———. "Military Aspects of the Panama Canal Issue." *Proceedings of the U.S. Naval Institute* 106 (January 1980): 46-51.

———. "Strategic Concepts of Latin America: An Update." *Inter-American Economic Affairs* 34 (Summer 1980): 61-82.

———. *Unequal Alliance: The Inter-American Military System, 1938-1978*. Boulder, Colo.: Westview Press, 1980.

Christmann, Timothy J. "TacAir in Grenada." *Naval Aviation*, November-December 1985, pp. 6-9.

Clark, J. Ruben. *Memorandum on the Monroe Doctrine*. Washington, D.C.: Government Printing Office, 1930. Memorandum is dated December 17, 1928.

Clements, Charles. *Witness to War: An American Doctor in El Salvador*. New York: Bantam Books, 1984.

Cochrane, James D. "U.S. Policy toward Recognition of Governments and Promotion of Democracy in Latin America since 1963." *Journal of Latin American Studies* 4 (November 1972): 275-291.

Cohen, Stephen B. "Conditioning U.S. Security Assistance on Human Rights Practices." *American Journal of International Law* 76 (1982): 246-279.

Collier, David. "Industrial Modernization and Political Change: A Latin American Perspective." *World Politics* 30 (July 1978): 593-614.

Collins, John M. *U.S.-Soviet Military Balance: Concepts and Capabilities, 1960-1980*. New York: McGraw-Hill, 1980.

Commager, Henry Steele. "Outmoded Assumptions." *Atlantic Monthly* 249 (March 1982): 12-22.

Conn, Stetson, gen. ed. *The Western Hemisphere: The Framework of Hemispheric Defense*, vol. 12, pt. 1 of *The United States Army in World War II*. Washington, D.C.: Office of the Chief of Military History, Department of the Army, 1964.

———. *The Western Hemisphere: Guarding the United States and Its Outposts*, vol. 12, pt. 2 of *The United States Army in World War II*.

Washington, D.C.: Office of the Chief of Military History, Department of the Army, 1964.

Connell-Smith, Gordon. "The United States and the Caribbean: Colonial Patterns, Old and New." *Journal of Latin American Studies* 4 (May 1972): 113-122.

Cook, Fred J. "Juggernaut: The Warfare State." *The Nation*, October 28, 1961, pp. 276-337.

————. *Maverick: Fifty Years of Investigative Reporting*. New York: Putnam, 1984.

Cooper, Marc, and Goldin, Greg. "Playboy Interview with José Napoleón Duarte." *Playboy* 31 (November 1984): 63-75.

Corbett, Charles D. "Toward a U.S. Defense Policy: Latin America." *Military Review* 55 (June 1975): 11-18.

Crassweller, Robert D. *The Caribbean Community: Changing Societies and U.S. Policy*. New York: Praeger, 1972.

Cravero, Kathleen Ann. "Food and Politics: Domestic Sources of U.S. Food Aid Policies, 1949-1979." Ph.D. dissertation, Fordham University, 1982.

Crespo, Horacio, and Nudelman, Ricardo. "The Narrow Path: Democracy in Central America." In *Central America and the Western Alliance*, edited by Joseph Cirincione, pp. 199-211. New York: Holmes and Meier, 1985.

Cronan, D. S. *Underwater Minerals*. London: Academic Press, 1980.

Cuello, José, and Domínguez, Asdrubal. "The Dominican Republic: Two Years After." *World Marxist Review*, March 1968, pp. 38-39.

Dallek, Robert. *The American Style of Foreign Policy: Cultural Politics and Foreign Affairs*. New York: Knopf, 1983.

Dam, Kenneth W. "The Larger Importance of Grenada." Speech to the Associated Press Managing Editors' Conference, Louisville, Kentucky, November 4, 1983.

Daniel, Donald C., and Hansel, Howard M. "Navy." *Soviet Armed Forces Review Annual* 7 (1982-1983): 214-236.

Dassin, Joan, ed. *Torture in Brazil: A Report by the Archdiocese of Sao Paulo*. New York: Random House, 1986.

Davies, John Paton. *Foreign and Other Affairs*. New York: W.W. Norton, 1964.

Davis, Douglas W. "The Inter-American Defense College: An Assessment of Its Activities." Ph.D. dissertation, University of Maryland, 1968.

Davis, Lynn E. "United States Policy toward Central America." In *The Central American Crisis: Policy Perspectives*, edited by Abraham F. Lowenthal and Samuel F. Wells, Jr., pp. 7-16. Working Paper No. 119, Latin American Program, The Wilson Center, Washington, D.C., 1982.

Davis, Nathaniel. *The Last Two Years of Salvador Allende*. Ithaca, N.Y.: Cornell University Press, 1985.

—————. "U.S. Covert Actions in Chile, 1971-1973." *Foreign Service Journal* 55 (November 1978): 10-14, 38-39 and (December 1978): 11-13, 43.

Davis, Shelton H., and Hodson, Julie. *Witness to Political Violence in Guatemala: The Suppression of a Rural Development Movement*. Boston: Oxfam America, 1982.

Dealy, Glen C. "Pipe Dreams: The Pluralistic Latins." *Foreign Policy*, no. 57 (Winter 1984-1985): pp. 108-127.

—————. *The Public Man: An Interpretation of Latin American and Other Catholic Countries*. Amherst: University of Massachusetts Press, 1977.

Debray, Régis. *Strategy for Revolution*. New York: Monthly Review Press, 1971.

De Conte, Alexandre. *Herbert Hoover's Latin American Policy*. Stanford: Stanford University Press, 1951.

Deere, Carmen Diana, and Diskin, Martin. "Rural Poverty in El Salvador: Dimensions, Trends, and Causes." Geneva: International Labor Organization, January 1983. Mimeographed.

Destler, I. M. "The Elusive Consensus: Congress and Central America." In *Central America: Anatomy of Conflict*, edited by Robert S. Leiken., pp. 319-335. New York: Pergamon Press, 1984.

—————; Gelb, Leslie, H.; and Lake, Anthony. *Our Own Worst Enemy: The Unmaking of American Foreign Policy*. New York: Simon and Schuster, 1984.

Deutsch, Karl W. "Social Mobilization and Political Development." *American Political Science Review* 55 (September 1961): 493-514.

Devine, Frank J. *El Salvador: Embassy Under Attack*. New York: Vantage Press, 1981.

Dickey, Christopher. "Central America: From Quagmire to Cauldron?" *Foreign Affairs* 62 (1984): 659-694.

DiGiovanni, Cleto, Jr. "U.S. Policy and the Marxist Threat to Central America." *Heritage Foundation Backgrounder*, no. 128, October 15, 1980.

Diskin, Martin. "Agrarian Reform in El Salvador." In *Latin American Agriculture: Structure and Reform*, edited by William Thiesenhusen, pp. 107-157. Winchester, Mass.: Allen and Unwin, 1987.

Dix, Robert H. *Colombia: The Political Dimensions of Change*. New Haven: Yale University Press, 1967.

Dodson, Michael. "Democratic Ideals and Contemporary Central American Politics." Paper presented at the Annual Meeting of the International Studies Association, Atlanta, Georgia, March 1984.

Doenitz, Karl. *Memoirs*. Translated by R. H. Stevens. London: Weidenfeld and Nicolson, 1959.

Domínguez, Jorge I. "The United States and Its Regional Security Interests: The Caribbean, Central, and South America." *Daedalus* 109 (Fall 1980): 115-133.

337

Domínguez, Jorge I. *U.S. Interests and Policies in the Caribbean and Central America*. Washington, D.C.: American Enterprise Institute, 1982.

Domínguez, Virginia R., and Domínguez, Jorge I. *The Caribbean: Its Implications for the United States*. Headline Series No. 253. New York: Foreign Policy Association, 1981.

Duncan, W. Raymond. "Soviet Interests in Latin America: New Opportunities and Old Constraints." *Journal of Inter-American Studies and World Affairs* 26 (May 1984): 163-198.

Dupuy, Trevor N.; Hayes, Grace P.; and Andrews, John A. C. *The Almanac of World Military Power*. 3d ed. New York: R.R. Bowker, 1974.

Durham, William H. *Scarcity and Survival in Central America: Ecological Origins of the Soccer War*. Stanford: Stanford University Press, 1979.

Eckes, Alfred E., Jr. *The United States and the Global Struggle for Minerals*. Austin: University of Texas Press, 1979.

Eckstein, Susan. "Cuban Internationalism." In *Cuba: Twenty-Five Years of Revolution*, edited by Sandor Halebsky and John M. Kirk, pp. 372-390. New York: Praeger, 1985.

Edwards, Mickey. "Soviet Expansion and Control of the Sea-Lanes." *Proceedings of the U.S. Naval Institute* 106 (September 1980): 46-51.

Eisenhower, Milton S. *United States-Latin American Relations: Report to the President*. Washington, D.C.: Department of State, 1953.

————. *The Wine is Bitter: The United States and Latin America*. Garden City, N.Y.: Doubleday, 1963.

Enders, Thomas O. "Building the Peace in Central America." Speech to the Commonwealth Club of California, San Francisco, August 20, 1982. Mimeographed.

————. "The Central American Challenge." *AEI Foreign Policy and Defense Review* 4 (1982): 8-12.

Erickson, Stephen. "The Panama Canal: Potential Problems in the Years Ahead." *The Fletcher Forum* 5 (Winter 1981): 119-133.

Erisman, H. Michael, ed. *The Caribbean Challenge: U.S. Policy in a Volatile Region*. Boulder, Colo.: Westview Press, 1984.

Escude, Carlos Andrés. "The Argentine Eclipse: The International Factor in Argentina's Post World War II Decline." Ph.D. dissertation, Yale University, 1981.

Etheredge, Lloyd S. *Can Governments Learn? American Foreign Policy and Central American Revolutions*. New York: Pergamon Press, 1985.

————. *A World of Men: The Private Sources of American Foreign Policy*. Cambridge, Mass.: MIT Press, 1978.

Fagen, Richard R. "The Real Clear and Present Danger." *Caribbean Review* 11 (Spring 1982): 18-19, 52-53.

————, and Pellicer, Olga, eds. *The Future of Central America: Policy*

Choices for the U.S. and Mexico. Stanford: Stanford University Press, 1983.

Falcoff, Mark. "The El Salvador White Paper and Its Critics." *AEI Foreign Policy and Defense Review* 4 (1982): 18-24.

————, and Royal, Robert, eds. *Crisis and Opportunity: U.S. Policy in Central America and the Caribbean.* Washington, D.C.: Ethics and Public Policy Center, 1984.

Farer, Tom J. "Reagan's Latin America." *New York Review of Books,* March 19, 1981.

————, "Searching for Defeat." *Foreign Policy,* no. 40 (Fall 1980), pp. 155-174.

Feierabend, Ivo K., and Feierabend, Rosalind L. "Aggressive Behaviors within Polities, 1948-1962: A Cross-National Study." *Journal of Conflict Resolution* 10 (September 1966): 249-271.

Feinberg, Richard E. *The Intemperate Zone: The Third World Challenge to U.S. Foreign Policy.* New York: W.W. Norton, 1983.

————, ed. *Central America: International Dimensions of the Crisis.* New York: Holmes and Meier, 1982.

Felix, David. "Income Distribution and the Quality of Life in Latin America: Patterns, Trends, and Policy Implications." *Latin American Research Review* 18, no. 2 (1983): 3-33.

Fenlon, Leslie K., Jr. "The Umpteenth Cuban Confrontation." *Proceedings of the U.S. Naval Institute* 106 (July 1980): 40-45.

Ferber, Robert, ed. *Consumption and Income Distribution in Latin America.* Washington, D.C.: Organization of American States, 1980.

Ferguson, Yale H. "The Ideological Dimension in United States Policies toward Latin America, 1945-1976." In *Terms of Conflict: Ideology in Latin American Politics,* edited by Morris J. Blachman and Ronald G. Hallman, pp. 193-235. Philadelphia, Pa.: ISHI, 1977.

————. "The Reagan Administration's Latin American Policies: An Assessment." Paper presented at the International Congress of the Latin American Studies Association, Mexico City, September 30, 1983. Mimeographed.

————. "Reflections on the Inter-American Principle of Nonintervention: A Search for Meaning in Ambiguity." *Journal of Politics* 32 (August 1970): 628-654.

Finan, John J., and Child, John. *Latin America, International Relations: A Guide to Information Sources.* Detroit: Gale Research, 1981.

Fontaine, Roger; DiGiovanni, Cleto, Jr.; and Kruger, Alexander. "Castro's Specter." *Washington Quarterly* 3 (Autumn 1980): 3-27.

Forché, Carolyn, and Wheaton, Philip. *History and Motivations of U.S. Involvement in the Control of the Peasant Movement in El Salvador: The Role of AIFLD in the Agrarian Reform Process, 1970-1980.* Washington, D.C.: Ecumenical Program for Interamerican Communication and Action (EPICA), 1980.

Foster, Gregory D. "On Selective Intervention." *Strategic Review* 11 (Fall 1983): 48-63.

Frechette, Myles R. R. "Letter to the Editor." *Foreign Policy*, no. 48 (Fall 1982), pp. 175-179.

Gabriel, Richard. "Scenes from an Invasion." *Washington Monthly*, February 1986, pp. 34-41.

Gaddis, John Lewis. *Strategies of Containment: A Critical Appraisal of Postwar American National Security Policy*. New York: Oxford University Press, 1982.

Galbraith, John Kenneth. *The Affluent Society*. Boston: Houghton Mifflin, 1958.

Gantenbein, James W., ed. *The Evolution of Our Latin-American Policy: A Documentary Record*. New York: Columbia University Press, 1950.

Garrett, William B. "Arms Transfers, Congress, and Foreign Policy: The Case of Latin America, 1967-1976." Ph.D. dissertation, Johns Hopkins University, 1982.

Garthoff, Raymond L. *Détente and Confrontation: American-Soviet Relations from Nixon to Reagan*. Washington, D.C.: The Brookings Institution, 1985.

Gedda, George. "A Dangerous Region." *Foreign Service Journal* 60 (February 1983): 18-21, 34.

George, Alexander L. "The 'Operational Code': A Neglected Approach to the Study of Political Leaders and Decision-Making." *International Studies Quarterly* 13 (June 1969): 190-222.

Goetze, Richard B., Jr. "Transgovernmental Interaction within the Inter-American System: The System of Cooperation among the Air Forces of the Americas." Ph.D. dissertation, American University, 1973.

Goldhamer, Herbert. *The Foreign Powers in Latin America*. Princeton, N.J.: Princeton University Press, 1972.

González, Edward. "Cuba: The Impasse." In *U.S. Influence in Latin America in the 1980s*, edited by Robert Wesson, pp. 198-216. New York: Praeger, 1982.

————; Jenkins, Brian Michael; Ronfeldt, David; and Sereseres, Caesar. Report No. R-3150-RC. *U.S. Policy for Central America: A Briefing*. Santa Monica, Ca.: RAND Corporation, March 1984.

Gorman, Stephen M. "The High Stakes of Geopolitics in Tierra del Fuego." *Parameters* 8 (June 1978): 45-56.

Goure, Leon, and Rothenberg, Morris. *Soviet Penetration of Latin America*. Miami: Center for Advanced International Studies, University of Miami, 1975.

Govett, M. H. "The Geographic Concentration of World Mineral Supplies." *Resources Policy* 1 (December 1975): 357-370.

————. "Geographic Concentration of World Mineral Supplies, Production,and Consumption." In *World Mineral Supplies: Assessment and*

Perspective, edited by G.J.S. Govett and M. H. Govett, pp. 99-145. Amsterdam: Elsevier, 1976.

——, and Govett, G.J.S. "The New Economic Order and World Mineral Production and Trade." *Resources Policy* 4 (December 1978): 230-241.

Graves, Ernest. "U.S. Policy toward Central America and the Caribbean." In *The Central American Crisis: Policy Perspectives*, edited by Abraham F. Lowenthal and Samuel F. Wells, Jr., pp. 17-22. Working Paper No. 119, Latin American Program, The Wilson Center, Washington, D.C., 1982.

——, and Hildreth, Steven A., eds. *U.S. Security Assistance: The Political Process*. Lexington, Mass.: D.C. Heath, 1985.

Gray, Anthony Whitford, Jr. "The Evolution of United States Naval Policy in Latin America." Ph.D. dissertation, American University, 1982.

Grayson, George W. *The Politics of Mexican Oil*. Pittsburgh: University of Pittsburgh Press, 1980.

Green, David. *The Containment of Latin America: A History of the Myths and Realities of the Good Neighbor Policy*. Chicago: Quadrangle Books, 1971.

Griffin, Charles. "On the Present Discontents in Latin America." *Vassar Alumnae Magazine* 48 (1963): 12-16.

Griffin, Keith. *The Political Economy of Agrarian Change: An Essay on the Green Revolution*. Cambridge, Mass.: Harvard University Press, 1974.

Gutman, Roy. "America's Diplomatic Charade." *Foreign Policy*, no. 56 (Fall 1984), pp. 3-23.

Haglund, David G. " 'Grey Areas' and Raw Materials: Latin American Resources and International Politics in the Pre-World War II Years." *Inter-American Economic Affairs* 36 (Winter 1982): 23-51.

Halle, Louis J. *American Foreign Policy: Theory and Reality*. London: George Allen and Unwin, 1960.

Hallowell, A. Irving. "Sociopsychological Aspects of Acculturation." In *The Science of Man in the World Crises*, edited by Ralph Linton, pp. 171-200. New York: Columbia University Press, 1945.

Halperin, Morton H. "Limited War: The Nature of the Limiting Process." In *American Defense Policy*, edited by John E. Endicott and Roy W. Stafford, Jr., pp. 156-160. 4th ed. Baltimore: The Johns Hopkins University Press, 1977.

——. "U.S. Interests in Central America: Designing a Minimax Strategy." In *The Central American Crisis: Policy Perspectives*, edited by Abraham F. Lowenthal and Samuel F. Wells, Jr., pp. 23-26. Working Paper No. 119, Latin American Program, The Wilson Center, Washington, D.C., 1982.

Hanley, Paul Tompkins. "The Inter-American Defense Board." Ph.D. dissertation, Stanford University, 1964.

Hansard's Parliamentary Debates. 3d Series.

Harrison, Donald F. "U.S.-Mexican Military Collaboration during World War II." Ph.D. dissertation, Georgetown University, 1976.

Harrison, Lawrence E. *Underdevelopment Is a State of Mind: The Latin American Case.* Lanham, Md.: University Press of America, 1985.

———. "Waking from the Pan-American Dream." *Foreign Policy,* no. 5 (Winter 1971-1972), pp. 163-181.

Hart, Thomas G. *The Cognitive World of Swedish Security Elites.* Stockholm: Scandinavian University Books, 1976.

Hartz, Louis. *The Liberal Tradition in America: An Interpretation of American Political Thought since the Revolution.* New York: Harcourt, Brace and World, 1955.

Hayes, Margaret Daly. *Latin America and the U.S. National Interest: A Basis for U.S. Foreign Policy.* Boulder, Colo.: Westview Press, 1984.

———. "Promoting U.S. Security Interests in Central America." *AEI Foreign Policy and Defense Review* 5 (1984): 46-54.

———. "Security to the South: U.S. Interests in Latin America." *International Security* 5 (Summer 1980): 130-151.

———. "The Stakes in Central America and U.S. Policy Responses." *AEI Foreign Policy and Defense Review* 4 (1982): 12-18.

———. "United States Security Interests in Central America in Global Perspective." In *Central America: International Dimensions of the Crisis,* edited by Richard E. Feinberg, pp. 85-102. New York: Holmes and Meier, 1982.

Henrickson, Alan K. "East-West Rivalry in Latin America: 'Between the Eagle and the Bear.' " In *East-West Rivalry in the Third World,* edited by Robert W. Clawson, pp. 261-290. Wilmington, Del.: Scholarly Resources, 1986.

Heritage Foundation. "The Soviet Military Buildup in Cuba." *Backgrounder,* no. 189, June 11, 1982.

Hernández, Ernesto. *Colombia en Corea.* Bogotá: Imprenta de las Fuerzas Armadas, 1953.

Herring, Hubert. *Good Neighbors: Argentina, Brazil, Chile and Seventeen Other Countries.* New Haven: Yale University Press, 1941.

Herrmann, Richard K. "Perceptions and Foreign Policy Analysis." In *Foreign Policy Decision Making: Perception, Cognition, and Artificial Intelligence,* edited by Donald A. Sylvan and Steve Chan, pp. 25-52. New York: Praeger, 1984.

Herzog, Arthur. *The War-Peace Establishment.* New York: Harper and Row, 1963.

Hill, Clarence A., Jr. "United States Strategic Interests in the Southern Hemisphere and the Need for a Common Maritime Defense." In *Argentine-United States Relations and South Atlantic Security,* edited by Z. Michael Szaz, pp. 1-13. Washington, D.C.: American Foreign Policy Institute, 1980.

Himmelstein, Jerome L., and Kimmel, Michael S. "States and Revolu-

tions: The Implications and Limits of Skocpol's Structural Model."
American Journal of Sociology 86 (March 1981): 1,145-1,154.

Hoffmann, Stanley. *Dead Ends: American Foreign Policy in the New Cold War*. Cambridge, Mass.: Ballinger Publishing Company, 1983.

———. *Primacy or World Order: American Foreign Policy since the Cold War*. New York: McGraw-Hill, 1978.

Hogan, James Michael. "The 'Great Debate' Over Panama: An Analysis of Controversy over the Carter-Torrijos Treaties of 1977." Ph.D. dissertation, University of Wisconsin—Madison, 1983.

Holsti, Ole R. "Foreign Policy Viewed Cognitively." In *The Structure of Decision: The Cognitive Maps of Political Elites*, edited by Robert Axlerod, pp. 18-54. Princeton, N.J.: Princeton University Press, 1976.

———. "Individual Differences in 'Definition of the Situation.' " *Journal of Conflict Resolution* 14 (September 1970): 303-310.

———. "The Study of International Politics Makes Strange Bedfellows: Theories of the Radical Right and the Radical Left." *American Political Science Review* 68 (March 1974): 217-242.

———. "The Three-Headed Eagle: The United States and System Change." *International Studies Quarterly* 23 (September 1979): 339-359.

———, and Rosenau, James N. *American Leadership in World Affairs: Vietnam and the Breakdown of Consensus*. Winchester, Mass.: George Allen and Unwin, 1984.

———. "Vietnam, Consensus, and the Belief Systems of American Leaders." *World Politics* 32 (October 1979): 1-56.

Holt, W. Stull. "The United States and the Defense of the Western Hemisphere, 1815-1940." *Pacific Historical Review* 10 (March 1941): 29-38.

Hopkins, Gary. "Guantanamo Bay: Big Mission in the Caribbean." *All Hands*, September 1983, pp. 20-30.

Howard, Harry N. *Turkey, the Straits, and U.S. Policy*. Baltimore: The Johns Hopkins University Press, 1974.

Hughes, Charles Evan. "Observations on the Monroe Doctrine." *American Journal of International Law* 17 (1923): 611-628.

Hull, Cordell. "Memorandum of Conversation with Luis Fernando Guachalla, Bolivian Minister to the United States, April 11, 1939." Department of State unpublished documents, 824.6363 St2/336.

Humphreys, R. A. *Latin America and the Second World War*. 2 vols. London: Athlone, 1981.

Hurwitz, Leon. "Contemporary Approaches to Political Stability." *Comparative Politics* 5 (April 1973): 449-463.

Iglesias, Enrique. "Development and Equity: The Challenge of the 1980s." *CEPAL Review* 15 (December 1981): 7-46.

Immerman, Richard M. *The CIA in Guatemala: The Foreign Policy of Intervention*. Austin: University of Texas Press, 1982.

International Labor Organization. *Employment, Growth and Basic*

Needs, A One-World Problem. Geneva: International Labor Organization, 1976.

Ireland, Gordon. *Boundaries, Possessions, and Conflicts in Central and North America and the Caribbean*. Cambridge, Mass.: Harvard University Press, 1941.

————. *Boundaries, Possessions and Conflicts in South America*. Cambridge, Mass.: Harvard University Press, 1938.

Jacobsen, C. G. "Soviet Attitudes toward Aid to and Contacts with Central American Revolutionaries." Washington, D.C.: Department of State, June 1984. Mimeographed.

Janis, Irving. *Victims of Groupthink: A Psychological Study of Foreign-Policy Decisions and Fiascoes*. 2d ed. Boston: Houghton Mifflin, 1982.

Jenkins, Brian; Sereseres, César; and Einaudi, Luigi. *U.S. Military Aid and Guatemalan Politics*. Santa Monica (?), Ca.: California Arms Control and Foreign Policy Seminar, March 1974.

Jervis, Robert. "Hypotheses on Misperception." *World Politics* 20 (April 1968): 454-479.

————. *Perception and Misperception in International Politics*. Princeton, N.J.: Princeton University Press, 1976.

Jhabvala, Farrokh, ed. *Maritime Issues in the Caribbean*. Gainesville: University Presses of Florida, 1983.

Johansen, Robert C. "Toward an Alternative Security System." World Policy Paper No. 24. New York: World Policy Institute, 1983.

Johnson, Robert H. "Exaggerating America's Stakes in Third World Conflicts." *International Security* 10 (Winter 1985-1986): 32-68.

Johnson, Victor C. "Discussion Paper on Central America." In *The Central American Crisis: Policy Perspectives*, edited by Abraham F. Lowenthal and Samuel F. Wells, Jr., pp. 45-52. Working Paper No. 119, Latin American Program, The Wilson Center, Washington, D.C., 1982.

Jordan, Amos A., and Taylor, William J., Jr. *American National Security: Policy and Process*. Baltimore: The Johns Hopkins University Press, 1981.

Jordan, David C. "U.S. Options—and Illusions—in Central America." *Strategic Review* 12 (Spring 1984): 53-62.

Jorden, William J. *Panama Odyssey*. Austin: University of Texas Press, 1984.

Josephson, Matthew. *Empire of the Air: Juan Trippe and the Struggle for World Airways*. New York: Harcourt, Brace, 1944.

Kane, William Everett. *Civil Strife in Latin America: A Legal History of U.S. Involvement*. Baltimore: The Johns Hopkins University Press, 1972.

Kaplan, Morton A. "American Policy toward the Caribbean Region: One Aspect of American Global Policy." In *Issues in Caribbean International Relations*, edited by Basil A. Ince et al., pp. 51-60. Lanham, Md.: University Press of America, 1983.

344

Katz, Mark N. *The Third World in Soviet Military Thought*. Baltimore: The Johns Hopkins University Press, 1982.

Kaufman, Edy; Shapria, Yoram; and Barromi, Joel. *Israel-Latin American Relations*. New Brunswick, N.J.: Transaction Books, 1979.

Keagle, James Martin. "Toward an Understanding of U.S. Latin American Policy." Ph.D. dissertation, Princeton University, 1982.

Kegley, Charles W., Jr., and Wittkopf, Eugene R. *American Foreign Policy: Pattern and Process*. 2d ed. New York: St. Martin's Press, 1982.

————. "Beyond Consensus: The Domestic Context of American Foreign Policy." *International Journal* 38 (Winter 1982-1983): 77-106.

————. "The Reagan Administration's World View." *Orbis* 26 (Spring 1982): 223-244.

Keisling, Phil. "The Tallest Gun in Foggy Bottom." *The Washington Monthly*, November 1982, pp. 50-56.

Kennan, George F. "Foreign Policy and the Professional Diplomat." *Wilson Quarterly* 1 (Winter 1977): 148-157.

————. *Memoirs, 1925-1950*. Boston: Little, Brown, 1967.

Kennedy, Robert, and Marcella, Gabriel. "U.S. Security on the Southern Flank: Interests, Challenges, Responses." In *Western Interests and U.S. Policy Options in the Caribbean Basin*. Report of the Atlantic Council's Working Group on the Caribbean Basin, pp. 187-241. Boston: Oelgeschlager, Gunn and Hain, 1984.

Kenworthy, Eldon. "Why the United States Is in Central America." *Bulletin of the Atomic Scientists*, October 1983, pp. 14-19.

Kirkpatrick, Jeane J. "Dictators and Double Standards." *Commentary* 68 (November 1979): 34-45.

————. "Doctrine of Moral Equivalence." Speech to the Royal Institute of International Affairs, London, April 9, 1984. Mimeographed.

————. "U.S. Security and Latin America." *Commentary* 71 (January 1981): 29-40.

————. "U.S. Security and Latin America." In *Rift and Revolution: The Central American Imbroglio*, edited by Howard J. Wiarda, pp. 329-359. Washington, D.C.: American Enterprise Institute, 1984.

Kissinger, Henry A. "Domestic Structure and Foreign Policy." In *International Politics and Foreign Policy*, revised edition, edited by James N. Rosenau, pp. 261-275. New York: The Free Press, 1969.

Kling, Merle. "Violence and Politics in Latin America." In *Sociological Review Monographs*, No. 11, edited by Paul Halmos, pp. 119-132. Keele: University of Keele, 1967.

Knight, H. Gary. "The Law of the Sea and Naval Missions." *Proceedings of the U.S. Naval Institute* 103 (June 1977): 32-39.

Knight, Peter T., and Plank, John N. "United States Policy toward Latin America." In *The Next Phase in Foreign Policy*, edited by Henry Owen, pp. 88-112. Washington, D.C.: Brookings Institution, 1973.

Komorowski, Raymond A. "Latin America—An Assessment of U.S. Strategic Interests." *Proceedings of the U.S. Naval Institute* 99 (May 1973): 148-171.

Krasner, Stephen D. "Oil Is the Exception." *Foreign Policy*, no. 14 (Spring 1974), pp. 68-83.

Lacroix, Richard L. J. "Integrated Rural Development in Latin America." World Bank Staff Working Paper No. 716. Washington, D.C.: World Bank, 1985.

LaFeber, Walter. "Inevitable Revolutions." *Atlantic Monthly* 249 (June 1982): 74-83.

———. *Inevitable Revolutions: The United States and Central America.* New York: W.W. Norton, 1984.

———. "The Reagan Administration and Revolutions in Central America." *Political Science Quarterly* 99 (Spring 1984): 1-25.

Lamb, Chris. "Belief Systems and Decision Making in the *Mayaguez* Crisis." *Political Science Quarterly* 99 (Winter 1984-1985): 681-702.

Langley, Lester D. *The United States and the Caribbean in the Twentieth Century.* Athens: University of Georgia Press, 1982.

Larson, Deborah Welch. *Origins of Containment: A Psychological Explanation.* Princeton, N.J.: Princeton University Press, 1985.

Lebow, Richard Ned. "The Cuban Missile Crisis: Reading the Lessons Correctly." *Political Science Quarterly* 98 (Fall 1983): 431-458.

Leiken, Robert S. "Eastern Winds in Latin America." *Foreign Policy*, no. 42 (Spring 1981), pp. 94-113.

———. "Fantasies and Facts: The Soviet Union and Nicaragua." *Current History* 83 (October 1984): 314-317.

———. *Soviet Strategy in Latin America.* New York: Praeger, 1982.

———. "The USSR and Central America: Great Expectations Dampened?" In *Central America and the Western Alliance*, edited by Joseph Cirincione, pp. 155-176. New York: Holmes and Meier, 1985.

———, ed. *Central America: Anatomy of Conflict.* New York: Pergamon Press, 1984.

Leith, C. K.; Furness, J. E.; and Lewis, Cleona. *World Minerals and World Peace.* Washington, D.C.: Brookings Institution, 1943.

LeoGrande, William M. "The Author Replies." *Foreign Policy*, no. 48 (Fall 1982), pp. 181-184.

———. *Central America and the Polls.* Washington, D.C.: Washington Office on Latin America, May 1984.

———. "Cuba and Nicaragua, from the Somozas to the Sandinistas." *Caribbean Review* 9 (Winter 1980): 11-14.

———. "A Splendid Little War: Drawing the Line in El Salvador." *International Security* 6 (Summer 1981): 27-52.

Lernoux, Penny. "Notes on a Revolutionary Church: Human Rights in Latin America." Alicia Patterson Foundation, February 1978. Mimeographed.

Levine, Barry B., ed. *The New Cuban Presence in the Caribbean*. Boulder, Colo.: Westview Press, 1983.

Levine, Daniel H. "Whose Heart Could Be So Staunch?" *Christianity and Crisis*, July 22, 1985, pp. 311-312.

————, ed. *Religion and Political Conflict in Latin America*. Chapel Hill: University of North Carolina Press, 1986.

Levine, Robert A. *The Arms Debate*. Cambridge, Mass.: Harvard University Press, 1963.

Levy, Marion J., Jr. *Modernization and the Structure of Society: A Setting for International Affairs*. 2 vols. Princeton, N.J.: Princeton University Press, 1966.

Lewis, Vaughn A. "The Bahamas in International Politics: Issues Arising for an Archipelago State." *Journal of Inter-American Studies and World Affairs* 16 (May 1974): 131-152.

Lippmann, Walter. *Public Opinion*. New York: The Free Press, 1965.

Liss, Sheldon B. *Marxist Thought in Latin America*. Berkeley: University of California Press, 1984.

Lodge, Henry Cabot, ed. *Selections from the Correspondence of Theodore Roosevelt and Henry Cabot Lodge, 1884-1918*. 2 vols. New York: Charles Scribner's Sons, 1925.

Logan, John A., Jr. *No Transfer, An American Security Principle*. New Haven: Yale University Press, 1961.

Lowenthal, Abraham F. *The Dominican Intervention*. Cambridge, Mass.: Harvard University Press, 1972.

————. "The United States and Latin America: Ending the Hegemonic Presumption." *Foreign Affairs* 55 (October 1976): 199-213.

————, and Wells, Samuel F., Jr. "Introduction." In *The Central American Crisis: Policy Perspectives*, edited by Abraham F. Lowenthal and Samuel F. Wells, Jr., pp. 1-5. Working Paper 119, Latin American Program, The Wilson Center, Washington, D.C., 1982.

Luers, William H. "The Soviets and Latin America: A Three Decade U.S. Policy Tangle." *Washington Quarterly* 7 (Winter 1984): 3-32.

Luttwak, Edward N. "The Nature of the Crisis." In *Central America and the Western Alliance*, edited by Joseph Cirincione, pp. 71-79. New York: Holmes and Meier, 1985.

McCann, Thomas P. *An American Company: The Tragedy of United Fruit*. New York: Crown Publishers, 1976.

MccGwire, Michael K. "Foreign-Port Visits by Soviet Naval Units." In *Soviet Naval Policy: Objectives and Constraints*, edited by Michael MccGwire, Ken Booth, and John McDonnell, pp. 387-418. New York: Praeger, 1975.

————. "Soviet Maritime Strategy, Capabilities, and Intentions in the Caribbean." In *Soviet Seapower in the Caribbean: Political and Strategic Implications*, edited by James D. Theberge, pp. 39-58. New York: Praeger, 1972.

McCreery, David. "Debt Servitude in Rural Guatemala, 1876-1936." *Hispanic American Historical Review* 63 (November 1983): 735-759.

———. "Development and the State in Reform Guatemala, 1871-1885." Papers in International Studies, Latin America Series No. 10. Athens: Ohio University Center for International Studies, 1983.

McGinnis, Charles I. "A New Look at the Panama Canal." *Military Engineer* 66 (July-August 1974): 219-222.

McIntire, Alexander Haywood, Jr. "Revolution and Intervention in Grenada: Strategic and Geopolitical Implications." Ph.D. dissertation, University of Miami, 1984.

Mack, Andrew. "Why Big Nations Lose Small Wars: The Politics of Asymmetric Conflict." *World Politics* 27 (January 1973): 175-200.

McLean, Epharaim R., Jr. "The Caribbean—An American Lake." *Proceedings of the U.S. Naval Institute* 67 (July 1941): 947-960.

Maggiotto, Michael A., and Wittkopf, Eugene R. "American Public Attitudes toward Foreign Policy." *International Studies Quarterly* 25 (December 1981): 601-632.

Mahan, Alfred Thayer. *The Influence of Sea Power Upon History, 1660-1783.* 12th ed. Boston: Little, Brown, 1918.

———. *The Interest of America in Sea Power: Present and Future.* Boston: Little, Brown, 1897.

Major, John. "Wasting Asset: The U.S. Re-Assessment of the Panama Canal, 1945-1949." *Journal of Strategic Studies* 3 (May 1980): 123-145.

Mandelbaum, Michael, and Schneider, William. "The New Internationalisms: Public Opinion and Foreign Policy." In *Eagle Entangled: U.S. Foreign Policy in a Complex World,* edited by Kenneth A. Oye, Donald Rothschild, and Robert J. Leiber, pp. 34-88. New York: Longman, 1979.

Marchetti, Victor, and Marks, John D. *The CIA and the Cult of Intelligence.* New York: Knopf, 1974.

Maris, Gary L. "International Law and Guantanamo." *Journal of Politics* 29 (May 1967): 261-286.

Martin, John Bartlow. *U.S. Policy in the Caribbean.* Boulder, Colo.: Westview Press, 1978.

Mason, Theodore K. *Across the Cactus Curtain: The Story of Guantánamo Bay.* New York: Dodd, Mead, 1984.

May, Ernest R. *The Making of the Monroe Doctrine.* Cambridge, Mass.: Harvard University Press, 1975.

Mayer, Jean. "The Dimensions of Human Hunger." *Scientific American* 235 (September 1976): 40-49.

Menges, Constantine C. "Central America and the United States." *SAIS Review* 1 (Summer 1981): 13-33.

———. "The United States and Latin America in the 1980s." In *The National Interests of the United States in Foreign Policy,* edited by Prosser Gifford, pp. 53-72. Washington, D.C.: University Press of America, 1982.

Michel, James. "Defending Democracy." In *Central America and the Western Alliance*, edited by Joseph Cirincione, pp. 49-55. New York: Holmes and Meier, 1985.

Midlarsky, Manus I. "Scarcity and Inequality: Prologue to the Onset of Mass Revolution." *Journal of Conflict Resolution* 26 (March 1982): 3-38.

———, and Roberts, Kenneth. "Class, State, and Revolution in Central America: Nicaragua and El Salvador Compared." *Journal of Conflict Resolution* 29 (June 1985): 163-193.

Migdal, Joel S. *Peasants, Politics, and Revolution: Pressures toward Political and Social Change in the Third World*. Princeton, N.J.: Princeton University Press, 1974.

Millett, Richard L. "The Best of Times, the Worst of Times: Central American Scenarios—1984." In *The Central American Crisis: Policy Perspectives*, edited by Abraham F. Lowenthal and Samuel F. Wells, Jr., pp. 81-89. Working Paper 119, Latin American Program, The Wilson Center, Washington, D.C., 1982.

Mills, C. Wright. *The Marxists*. New York: Dell Publishing Company, 1962.

Mitchell, Donald W. "Strategic Significance of Soviet Naval Power in Cuban Waters." In *Soviet Seapower in the Caribbean: Political and Strategic Implications*, edited by James D. Theberge, pp. 27-37. New York: Praeger, 1972.

Moffett, George D., III. *The Limits of Victory: The Ratification of the Panama Canal Treaties*. Ithaca, N.Y.: Cornell University Press, 1985.

Mohan, Rakesh. "An Anatomy of the Distribution of Urban Income: A Tale of Two Cities in Colombia." World Bank Staff Working Paper No. 650. Washington, D.C.: World Bank, July 1984.

Moncivaiz, Adolfo. "U.S. National Security Concern for the 1980s: Oil Export Potential in Latin America." Ph.D. dissertation, University of Miami, 1981.

Moore, Barrington, Jr. *Injustice: The Social Bases of Obedience and Revolt*. White Plains, N.Y.: M. E. Sharpe, 1978.

Morgenthau, Hans J. "John Foster Dulles." In *An Uncertain Tradition: American Secretaries of State in the Twentieth Century*, edited by Norman A. Graebner, pp. 289-308. New York: McGraw-Hill, 1961.

———. *Politics Among Nations: The Struggle for Power and Peace*. New York: Knopf, 1978.

———. "To Intervene or Not to Intervene." *Foreign Affairs* 45 (April 1967): 425-436.

Morison, Samuel Eliot. *History of United States Naval Operations in World War II*, vol. 1, *The Battle of the Atlantic, September 1939–May 1943*. Boston: Little, Brown, 1947.

Mosk, Sanford A. "The Coffee Economy of Guatemala, 1850-1918: Development and Signs of Instability." *Inter-American Economic Affairs* 9 (Winter 1955): 6-20.

Mudge, Arthur W. "A Case Study in Human Rights and Development Assistance: Nicaragua." *Universal Human Rights* 1 (October-November 1979): 93-102.

Nolan, David. *The Ideology of the Sandinistas and the Nicaraguan Revolution.* Coral Gables: Graduate School of International Studies, University of Miami, 1984.

Nolta, Frank. "Passage through International Straits: Free or Innocent? The Interests at Stake." *San Diego Law Review* 11 (May 1974): 815-833.

Nowels, Larry Q. "Central America and U.S. Foreign Assistance: Issues for Congress in 1984." Issue Brief No. IB84075. Washington, D.C.: Congressional Research Service, Library of Congress, updated June 12, 1984.

Organization of American States. *Conferencias internacionales americanas, segundo suplemento, 1945-1954.* Washington, D.C.: Departamento Jurídico, Unión Panamericana, 1956.

————. "Report on the Situation of Human Rights of a Segment of the Nicaraguan Population of Miskito Origin and Resolution of the Friendly Settlement Procedure regarding the Situation of Human Rights of a Segment of the Nicaraguan Population of Miskito Origin." OAS document number OEA/Ser. P AC/CP/doc.355/84. Washington, D.C.: Organization of American States, 1984. Mimeographed.

————, Inter-American Commission on Human Rights. "Report on the Situation of Human Rights in Nicaragua." Washington, D.C.: OAS General Secretariat, October 1978. Mimeographed.

————. "Report on the Situation of Human Rights in the Republic of Nicaragua." Washington, D.C.: OAS General Secretariat, 1981. Mimeographed.

Osgood, Robert E. "Central America in U.S. Containment Policy." In *The Central American Crisis: Policy Perspectives,* edited by Abraham F. Lowenthal and Samuel F. Wells, Jr., pp. 91-96. Working Paper No. 119, Latin American Program, The Wilson Center, Washington, D.C., 1982.

Packenham, Robert A. *Liberal America and the Third World: Political Development Ideas in Foreign Aid and Social Science.* Princeton, N.J.: Princeton University Press, 1973.

Padelford, Norman J., and Gibbs, Stephen R. *Maritime Commerce and the Future of the Panama Canal.* Cambridge, Md.: Cornell Maritime Press, 1975.

Palmer, David Scott. "Military Governments and U.S. Policy: General Concerns and Central American Cases." *AEI Foreign Policy and Defense Review* 4 (1982): 24-29.

————. "Rebellion in Rural Peru: The Origins and Evolution of Sendero Luminoso." *Comparative Politics* 18 (January 1986): 127-147.

————. "U.S. Policy Issues in the Caribbean Basin in the 1980s: Economic, Social, and Political Aspects." In *Western Interests and U.S. Policy Options in the Caribbean Basin*. Report of the Atlantic Council's Working Group on the Caribbean Basin, pp. 99-137. Boston: Oelgeschlager, Gunn and Hain, 1984.

Parker, Phyllis R. *Brazil and the Quiet Intervention, 1964*. Austin: University of Texas Press, 1979.

Parsons, Talcott. *The Social System*. Glencoe, Ill.: The Free Press, 1951.

Pastor, Robert A. "Continuity and Change in U.S. Foreign Policy: Carter and Reagan on El Salvador." *Journal of Policy Analysis and Management* 3 (Winter 1984): 175-190.

————. "Cuba and the Soviet Union: Does Cuba Act Alone?" in *The New Cuban Presence in the Caribbean* edited by Barry B. Levine, pp. 191-209. Boulder, Colo.: Westview Press, 1983.

————. "Our Real Interests in Central America." *Atlantic Monthly*, July 1982, pp. 27-39.

————. "Redefining the Strategic Challenge in Central America." In *Central America and the Western Alliance*, edited by Joseph Cirincione, pp. 81-89. New York: Holmes and Meier, 1985.

————. "Sinking in the Caribbean Basin." *Foreign Affairs* 60 (Summer 1982): 1,038-1,058.

————. "Spheres of Influence: Seal Them or Peel Them?" *SAIS Review* 4 (Winter-Spring 1984): 77-90.

————. "The Target and the Source: El Salvador and Nicaragua." *Washington Quarterly* 5 (Summer 1982): 116-127.

————. "Three Perspectives on El Salvador." *SAIS Review* 1 (Summer 1981): 35-48.

————. "The United States and the Grenada Revolution: Who Pushed First and Why?" Documento de Trabajo No. 26. Centro de Investigaciones del Caribe y América Latina, Universidad Interamericana de Puerto Rico, 1986.

Pate, James L. "CMA in Central America: The Private Sector Suffers Two KIA." *Soldier of Fortune* 10 (January 1985): 75-86.

The Pentagon Papers: The Defense Department History of United States Decisionmaking on Vietnam. 4 vols. Boston: Beacon Press, 1973.

Perkins, Dexter. *A History of the Monroe Doctrine*. New ed. (new title). Boston: Little, Brown, 1955.

————. *The Monroe Doctrine, 1823-1826*. Cambridge, Mass.: Harvard University Press, 1927.

Pfeffermann, Guy Pierre. "The Distribution of Income in Brazil." World Bank Staff Working Paper No. 356. Washington, D.C.: World Bank, September 1979.

Pike, Fredrick B. "Guatemala, the United States, and Communism in the Americas." *Review of Politics* 17 (1955): 232-261.

Policy Alternatives for the Caribbean and Central America (PACCA).

Changing Course: Blueprint for Peace in Central America and the Caribbean. Washington, D.C.: Institute for Policy Studies, 1984.

Portes, Alejandro. "Latin American Class Structures: Their Composition and Change during the Last Decades." *Latin American Research Review* 20, no. 3 (1985): 7-39.

Pratt, W. P., and Brobst, D. A. "Mineral Resources: Potentials and-Problems." U.S. Geological Survey Circular No. 698. Washington, D.C.: U.S. Geological Survey, 1974. Mimeographed.

The President's Committee to Study the United States Military Assistance Program (Draper Committee). "Conclusions Concerning the Mutual Security Program." Washington, D.C.: Department of State (?), August 17, 1959. Mimeographed.

Prina, L. Edgar. "The Merchant Marine and National Defense." *Seapower* 27 (August 1984): 67, 70.

Prosterman, Roy L.; Riedinger, Jeffrey M.; and Temple, Mary N. "Land Reform and the El Salvador Crisis." *International Security* 6 (Summer 1981): 53-74.

Public Papers of the Presidents of the United States. Dwight D. Eisenhower, 1954. Washington, D.C.: Government Printing Office, 1960.

Public Papers of the Presidents of the United States. Dwight D. Eisenhower, 1960-61. Washington, D.C.: Government Printing Office, 1961.

Public Papers of the Presidents of the United States. Gerald R. Ford, 1974. Washington, D.C.: Government Printing Office, 1975.

Public Papers of the Presidents of the United States. John F. Kennedy, 1961. Washington, D.C.: Government Printing Office, 1962.

Public Papers of the Presidents of the United States. John F. Kennedy, 1963. Washington, D.C.: Government Printing Office, 1964.

Public Papers of the Presidents of the United States. Lyndon B. Johnson, 1963-64. Washington, D.C.: Government Printing Office, 1965.

Public Papers of the Presidents of the United States. Lyndon B. Johnson, 1965. 2 vols. Washington, D.C.: Government Printing Office, 1966.

Public Papers of the Presidents of the United States. Richard M. Nixon, 1969. Washington, D.C.: Government Printing Office, 1971.

Public Papers of the Presidents of the United States. Richard M. Nixon, 1970. Washington, D.C.: Government Printing Office, 1971.

Public Papers of the Presidents of the United States. Richard M. Nixon, 1971. Washington, D.C.: Government Printing Office, 1972.

Public Papers of the Presidents of the United States. Ronald Reagan, 1981. Washington, D.C.: Government Printing Office, 1982.

Puffer, Ruth R.; Serrano, C. V.; and Dillon, Ann. *The Inter-American Investigation of Mortality in Childhood.* Washington, D.C.: Pan American Union, 1971.

Quigley, Tom, et al. *U.S. Policy on Human Rights in Latin America (Southern Cone): A Congressional Conference on Capitol Hill.* New York: Fund for New Priorities in America, 1978.

Reagan, Ronald. "The Canal as Opportunity: A New Relationship with Latin America." *Orbis* 21 (Fall 1977): 547-563.

Republic of Cuba, Ministry of Foreign Relations, Information Office. *Guantánamo: Yankee Naval Base of Crime and Provocation*. Havana: Instituto del Libro, 1970.

Richelson, Jeffrey. *The U.S. Intelligence Community*. Cambridge, Mass.: Ballinger Publishing Company, 1985.

Roberts, Jack L. "The Growing Soviet Naval Presence in the Caribbean: Its Politico-Military Impact upon the United States." *Naval War College Review* 23 (June 1971): 31-42.

Rogers, William D. "The United States and Latin America." *Foreign Affairs* 63 (Winter 1984-1985): 560-580.

————, and Meyers, Jeffrey A. "The Reagan Administration and Latin America: An Uneasy Beginning." *Caribbean Review* 11 (Spring 1982): 14-17.

Ronfeldt, David. *Geopolitics, Security, and U.S. Strategy in the Caribbean Basin*. Santa Monica, Ca.: RAND, November 1983.

Root, Elihu. "The Real Monroe Doctrine." *Proceedings of the American Society of International Law*, n.v. (1914), pp. 6-22.

Rosati, Jerel A. "The Impact of Beliefs on Behavior: The Foreign Policy of the Carter Administration." In *Foreign Policy Decision Making: Perception, Cognition, and Artificial Intelligence*, edited by Donald A. Sylvan and Steve Chan, pp. 158-191. New York: Praeger, 1984.

Rosson, W. B. "U.S. Southern Command in Latin America." *Commanders Digest* 16 (October 18, 1973): 2-11.

Rostow, Eugene V. "Re-Arm America." *Foreign Policy*, no. 39 (Summer 1980), pp. 6-8.

Rubin, Barry. *Secrets of State: The State Department and the Struggle over U.S. Foreign Policy*. New York: Oxford University Press, 1985.

Russell, Roberto. "Argentina y la política del régimen autoritario (1976-1983): una evaluación preliminar." *Estudios Internacionales* 17 (April-June 1984).

Russett, Bruce. "The Mysterious Case of Vanishing Hegemony; or, Is Mark Twain Really Dead?" *International Organization* 39 (Spring 1985): 207-231.

Ryan, Paul B. "Canal Diplomacy and U.S. Interests." *Proceedings of the U.S. Naval Institute* 103 (January 1977): 43-53.

St. John, Jeffrey. *The Panama Canal and Soviet Imperialism: War for the World Waterways*. Washington, D.C.: Heritage Foundation, 1978.

Sánchez, Nestor D. "The Communist Threat." *Foreign Policy*, no. 53 (Fall 1983). pp. 43-50.

————. "Revolutionary Change and the Nicaraguan People." *Strategic Review* 12 (Summer 1984): 17-22.

Sanford, Jonathan E. "U.S. Foreign Assistance to Central America." Report No. 84-34 F. Washington, D.C.: Congressional Research Service, Library of Congress, March 2, 1984. Mimeographed.

Sanjuan, Pedro A. "Why We Don't Have a Latin America Policy." *Washington Quarterly* 3 (Autumn 1980): 28-39.

Satterthwaite, Ridgway. "Campesino Agriculture and Hacienda Modernization in Coastal El Salvador: 1949 to 1969." Ph.D. dissertation, University of Wisconsin, 1971.

Schacht, Hjalmar. "Germany's Colonial Demands." *Foreign Affairs* 15 (January 1937): 223-234.

Scheina, Robert L. "Latin American Naval Purpose." *Proceedings of the U.S. Naval Institute* 103 (September 1977): 116-119.

———. "Latin American Navies." *Proceedings of the U.S. Naval Institute* 110 (March 1984): 30-35.

———. "The Malvinas Campaign." *Proceedings of the U.S. Naval Institute* 109 (May 1983): 98-117.

———. "South American Navies: Who Needs Them?" *Proceedings of the U.S. Naval Institute* 104 (February 1978): 61-66.

———. "Where Were Those Argentine Subs?" *Proceedings of the U.S. Naval Institute* 110 (March 1984): 115-120.

Schlesinger, Arthur M., Jr. *A Thousand Days: John F. Kennedy in the White House*. Boston: Houghton Mifflin, 1965.

Schlesinger, Stephen C., and Kinzer, Stephen. *Bitter Fruit: The Untold Story of the American Coup in Guatemala*. Garden City, N.Y.: Doubleday, 1982.

Schoultz, Lars. *Human Rights and United States Policy toward Latin America*. Princeton, N.J.: Princeton University Press, 1981.

Schreiber, Anna P. "Economic Coercion as an Instrument of Foreign Policy: U.S. Economic Measures against Cuba and the Dominican Republic." *World Politics* 25 (April 1973): 387-413.

Scott, Harriet Fast, and Scott, William F. *The Armed Forces of the USSR*. 2d ed. Boulder, Colo.: Westview Press, 1981.

Scott, William A. "Psychological and Social Correlates of International Images." In *International Behavior: A Social Psychological Analysis*, edited by Herbert C. Kelman, pp. 70-103. New York: Holt, Rinehart and Winston, 1965.

Scoville, Herbert, Jr. "Missile Submarines and National Security." *Scientific American* 226 (June 1972): 15-27.

Seabury, Paul, and McDougall, Walter A. *The Grenada Papers*. San Francisco: ICS Press, 1984.

Serafino, Nina. "Nicaragua: A Selected Chronology." Washington, D.C.: Congressional Research Service, Library of Congress, March 1984. Mimeographed.

Sereseres, Caesar D. "Inter-American Security Relations: The Future of U.S. Military Diplomacy in the Hemisphere." *Parameters* 7 (1977): 46-56.

Shafer, Michael. "Mineral Myths." *Foreign Policy*, no. 47 (Summer 1982): 154-171.

Shapiro, Michael J., and Bonham, G. Matthew. "Cognitive Process and Foreign Policy Decision-Making." *International Studies Quarterly* 17 (June 1973): 147-174.

Shue, Henry. *Basic Rights: Subsistence, Affluence, and U.S. Foreign Policy*. Princeton, N.J.: Princeton University Press, 1980.

Shultz, George. "America and the Struggle for Freedom." Speech to the Commonwealth Club of California, San Francisco, February 22, 1985. Mimeographed.

———. "Human Rights and the Moral Dimension of U.S. Foreign Policy." Speech to the Creve Coeur Club of Illinois, Peoria, Illinois, February 22, 1984. Mimeographed.

———. "Morality and Realism in American Foreign Policy." Speech to the National Committee on American Foreign Policy, New York, October 2, 1985. Mimeographed.

———. "Power and Diplomacy." Speech to the Veterans of Foreign Wars, Chicago, August 20, 1984. Mimeographed.

———. "Power and Diplomacy in the 1980s." Speech to the Trilateral Commission, Washington, D.C., April 3, 1984. Mimeographed.

———. "Refections Among Neighbors." Speech to the General Assembly of the Organization of American States, Washington, D.C., November 17, 1982. Mimeographed.

———. "Struggle for Democracy in Central America." Speech to the Dallas World Affairs Council and the Chamber of Commerce, Dallas, April 15, 1983. Mimeographed.

Simon, Laurence R.; Stephens, James C., Jr.; and Diskin, Martin. *El Salvador Land Reform, 1980-1981: Impact Audit with 1982 Supplement*. Boston: Oxfam America, 1982.

Skocpol, Theda. *States and Social Revolutions: A Comparative Analysis of France, Russia, and China*. London: Cambridge University Press, 1979.

———. "What Makes Peasants Revolutionary?" *Comparative Politics* 14 (April 1982): 351-375.

Skoug, Kenneth N., Jr. "The United States and Cuba." Speech to the Face to Face Program of the Carnegie Endowment for International Peace, Washington, D.C., December 17, 1984. Mimeographed.

Slater, Jerome N. "The Dominican Republic, 1961-66." In *Force Without War: U.S. Armed Forces as a Political Instrument*, edited by Barry M. Blechman and Stephen S. Kaplan, pp. 289-342. Washington, D.C.: Brookings Institution, 1978.

———. "The Role of the Organization of American States in United States Foreign Policy, 1947-1963." Ph.D. dissertation, Princeton University, 1965.

Smith, Wayne S. "Dateline Havana: Myopic Diplomacy." *Foreign Policy*, no. 48 (Fall 1982), pp. 157-174.

Snyder, Glenn, and Diesing, Paul. *Conflict Among Nations: Bargaining, Decision Making, and System Structure in International Crises*. Princeton, N.J.: Princeton University Press, 1977.

Somoza, Anastasio (as told to Jack Cox). *Nicaragua Betrayed*. Belmont, Mass.: Western Islands, 1981.

"Soviet Penetration of the Caribbean." Heritage Foundation National Security Record No. 22, June 1980. Mimeographed.

Spanier, John. *American Foreign Policy Since World War II*. 9th ed. New York: Holt, Rinehart and Winston, 1982.

Sprout, Harold, and Sprout, Margaret. *The Rise of American Naval Power, 1776-1918*. Princeton, N.J.: Princeton University Press, 1942.

Stebbins, Richard P., and the Research Staff of the Council on Foreign Relations. *The United States in World Affairs, 1954*. New York: Harper and Brothers, 1956.

Stein, Stanley J., and Stein, Barbara H. *The Colonial Heritage of Latin America: Essays on Economic Dependence in Perspective*. New York: Oxford University Press, 1970.

Steinbruner, John D. *The Cybernetic Theory of Decision*. Princeton, N.J.: Princeton University Press, 1974.

Stepan, Alfred. "The United States and Latin America: Vital Interests and the Instruments of Power." *Foreign Affairs* 58 (1980): 659-692.

Stilkind, Jerry. *Guatemala's Unstable and Uncertain Future*. Washington, D.C.: Center for Development Policy, 1982.

Stockholm International Peace Research Institute. *World Armaments and Disarmament: SIPRI Yearbook, 1982*. Cambridge, Mass.: Oelgeschlager, Gunn and Haig, 1982.

Stodder, Joseph H. "Commonality of Military Doctrine in the Inter-American Region." *Air University Review* 34 (November-December 1982): 70-75.

————, and McCarthy, Kevin. "Profiles of the Caribbean Basin in 1960/1980: Changing Geopolitical and Geostrategic Dimensions." Report No. N-2058-AF. Santa Monica, Ca.: RAND, December 1983.

Stokes, William F. "Violence as a Power Factor in Latin American Politics." *Western Political Quarterly* 5 (September 1952): 445-468.

Strasser, Joseph Charles. "Uncooperative Neighbors Become Close Allies: United States–Mexican Relations, 1941-1945." Ph.D. dissertation, Fletcher School of Law and Diplomacy, Tufts University, 1971.

"The Strategic and Economic Importance of the Caribbean Sea Lanes." *White House Digest*, April 4, 1984. Mimeographed.

Streeten, Paul. "From Growth to Basic Needs." *Finance and Development* 16 (September 1979): 28-31.

Swing, John Temple. "Law of the Sea." *Bulletin of the Atomic Scientists* 39 (May 1983): 14-19.

Szulc, Tad. *The Illusion of Peace: Foreign Policy in the Nixon Years.* New York: Viking Press, 1978.

———. *Twilight of the Tyrants.* New York: Holt, 1959.

Tetlock, Philip E. "Content and Structure in Political Belief Systems." In *Foreign Policy Decision Making: Perception, Cognition, and Artificial Intelligence,* edited by Donald A. Sylvan and Steve Chan, pp. 107-128. New York: Praeger, 1984.

Theberge, James D. "A Minerals Raw Material Action Program." *Foreign Policy,* no. 17 (Winter 1974-1975): 75-79.

———. "Soviet Policy in the Caribbean." In *Soviet Seapower in the Caribbean: Political and Strategic Implications,* edited by James D. Theberge, pp. 3-12. New York: Praeger, 1972.

———. *The Soviet Presence in Latin America.* New York: Crane, Russak and Company, 1974.

Thomas, Vinod. "Differences in Income, Nutrition, and Poverty within Brazil." World Bank Staff Working Paper No. 505. Washington, D.C.: World Bank, February 1982.

Thompson, James. "How Could Vietnam Happen?" *Atlantic Monthly* 221 (April 1968): 47-53.

Thompson, W. Scott. "Choosing to Win." *Foreign Policy,* no. 43 (Summer 1981), pp. 78-83.

Tilly, Charles. "Does Modernization Breed Revolution?" *Comparative Politics* 5 (April 1973): 425-447.

———. *From Mobilization to Revolution.* Reading, Mass.: Addison-Wesley, 1978.

Tocqueville, Alexis de. *Democracy in America,* edited by J. P. Mayer and Max Lerner; translated by George Lawrence. New York: Harper and Row, 1966.

Tomasek, Robert D. "The Deterioration of Relations between Costa Rica and the Sandinistas." Occasional Paper No. 9, Center for Hemispheric Studies, American Enterprise Institute, Washington, D.C., September 1984.

Trask, Roger R. "The Impact of the Cold War on United States-Latin American Relations, 1945-1949." *Diplomatic History* 1 (Summer 1977): 271-284.

Trippe, Juan T. "The Business Future—Southward." *Survey Graphic* 30 (March 1941): 136-139.

Tuchman, Barbara. "The American People and Military Power in an Historical Perspective." In *America's Security in the 1980s: Part I.* Adelphi Papers No. 173, pp. 5-13. London: International Institute for Strategic Studies, 1982.

Tucker, Robert W. *The Purposes of American Power: An Essay on National Security.* New York: Praeger, 1981.

Tulchin, Joseph S. "Two to Tango." *Foreign Service Journal* 59 (October 1982): 18-23.

Turner, Stansfield. *Secrecy and Democracy: The CIA in Transition.* Boston: Houghton Mifflin, 1985.

————. "The Stupidity of Intelligence." *Washington Monthly,* February 1986, pp. 29-33.

Ullman, Richard H. "At War with Nicaragua." *Foreign Affairs* 62 (Fall 1983): 39-58.

————. "Paths to Reconciliation: The United States and the International System of the Late 1980s." In *Estrangement: America and the World,* edited by Sanford J. Ungar, pp. 277-305. New York: Oxford University Press, 1985.

United Nations, Security Council. *Military and Paramilitary Activities In and Against Nicaragua.* Document S/16564. New York: United Nations,1984.

U.S. Agency for International Development. "Agrarian Reform in El Salvador: A Report on Its Status." Audit Report No. 1-519-34-2. Washington, D.C.: AID, January 13, 1984. Mimeographed.

————. *Congressional Presentation, Fiscal Year 1983.* Washington, D.C.: AID, 1982.

————. *U.S. Overseas Loans and Grants and Assistance from International Organizations, July 1, 1945-September 30, 1985.* Washington, D.C.: AID, 1986.

U.S. Atlantic-Pacific Interoceanic Canal Study Commission. *Interoceanic Canal Studies 1970.* Washington, D.C.: Government Printing Office, 1971.

U.S. Central Intelligence Agency. *International Energy Statistical Review.* Washington, D.C.: CIA, February 22, 1983, and June 26, 1984.

————. *Major Petroleum Refining Centers for Export.* Report No. ER 77-10140. Washington, D.C.: CIA, April 1977.

U.S. Congress, House, Committee on Appropriations, Subcommittee on the Department of Defense. *Department of Defense Appropriations for 1982.* 97th Cong., 1st Sess., 1981.

U.S. Congress, House, Committee on Appropriations, Subcommittee on the Department of Transportation and Related Agencies Appropriations. *Department of Transportation and Related Agencies Appropriations for 1976.* 94th Cong., 1st Sess., 1975.

U.S. Congress, House, Committee on Appropriations, Subcommittee on Foreign Operations and Related Agencies. *Foreign Assistance and Related Agencies Appropriations for 1976.* 94th Cong., 1st Sess., 1976.

————. *Foreign Assistance and Related Agencies Appropriations for 1977.* 94th Cong., 2d Sess., 1976.

————. *Foreign Assistance and Related Agencies Appropriations for 1978.* 95th Cong., 1st Sess., 1977.

————. *Foreign Assistance and Related Agencies Appropriations for 1979.* 95th Cong., 2d Sess., 1978.

——. *Foreign Assistance and Related Programs Appropriations for 1981.* 96th Cong., 2d Sess., 1980.

——. *Foreign Assistance and Related Programs Appropriations for 1983.* 97th Cong., 2d Sess., 1982.

——. *Supplemental Appropriations for 1982.* 97th Cong., 2d Sess., 1982.

U.S. Congress, House, Committee on Armed Services. *Report on the Inspection of Military Facilities in Panama and Bermuda.* 97th Cong., 1st Sess., 1981.

U.S. Congress, House, Committee on Banking, Finance, and Urban Affairs, Subcommittee on Economic Stabilization. *A Congressional Handbook on U.S. Materials Import Dependency/Vulnerability.* 97th Cong., 1st Sess., 1981.

——. *U.S. Economic Dependence on Six Imported Strategic Non-Fuel Minerals.* 97th Cong., 2d Sess., 1982.

U.S. Congress, House, Committee on Banking, Finance, and Urban Affairs, Subcommittee on International Development Institutions and Finance. *Human Rights and U.S. Policy in the Multilateral Development Banks.* 97th Cong., 1st Sess., 1981.

——. *Inter-American Development Bank Loan to Guatemala.* 97th Cong., 2d Sess., 1982.

U.S. Congress, House, Committee on Foreign Affairs. *Concerning U.S. Military and Paramilitary Operations in Nicaragua.* 98th Cong., 1st Sess., 1983.

——. *Foreign Assistance Legislation for Fiscal Year 1981.* 96th Cong., 2d Sess., 1980.

——. *Foreign Assistance Legislation for Fiscal Year 1982.* 97th Cong., 1st Sess., 1981.

——. *Foreign Assistance Legislation for Fiscal Year 1983.* 97th Cong., 2d Sess., 1982.

——. *Foreign Assistance Legislation for Fiscal Year 1985.* 98th Cong., 2d Sess., 1984.

——. *Foreign Assistance Legislation for Fiscal Years 1986-87.* 99th Cong., 1st Sess., 1985.

——. *Implementation of Congressionally Mandated Human Rights Provisions.* Vol. 2. 97th Cong., 1st and 2d Sess., 1981-1982.

——. *Inter-American Relations: A Collection of Documents, Legislation, Descriptions of Inter-American Organizations, and Other Material Pertaining to Inter-American Affairs.* 93d Cong., 1st Sess., 1973.

——. *The Mining of Nicaraguan Ports and Harbors.* 98th Cong., 2d Sess., 1984.

U.S. Congress, House, Committee on Foreign Affairs, Subcommittee on Human Rights and International Organizations. *Implementation of Congressionally Mandated Human Rights Provisions.* Vol. 1. 97th Cong., 1st Sess., 1981.

359

U.S. Congress, House, Committee on Foreign Affairs, Subcommittee on Human Rights and International Organizations. *Review of the 37th Session and Upcoming 38th Session of the U.N. Commission on Human Rights.* 97th Cong., 1st Sess., 1981.

U.S. Congress, House, Committee on Foreign Affairs, Subcommittee on Inter-American Affairs. *Honduras and U.S. Policy: An Emerging Dilemma.* 97th Cong., 2d Sess., 1982.

————. *Impact of Cuban-Soviet Ties in the Western Hemisphere, Spring 1980.* 96th Cong., 2d Sess., 1980.

————. *Presidential Certification on El Salvador.* Vol. 1. 97th Cong., 2d Sess., 1982.

————. *Review of Presidential Certification of Nicaragua's Connection to Terrorism.* 96th Cong., 2d Sess., 1980.

————. *United States-Brazilian Relations.* 97th Cong., 2d Sess., 1982.

————. *United States-Mexican Relations: An Update.* 97th Cong., 1st Sess., 1981.

————. *United States Policy toward Grenada.* 97th Cong., 2d Sess., 1982.

————. *United States Policy towards Nicaragua.* 96th Cong., 1st Sess., 1979.

————. *U.S. Policy Options in El Salvador.* 97th Cong., 1st Sess., 1981.

————. *U.S. Policy toward El Salvador.* 97th Cong., 1st Sess., 1981.

U.S. Congress, House, Committee on Foreign Affairs, Subcommittee on Inter-American Affairs, and Committee on Agriculture, Subcommittee on Department Operations, Research, and Foreign Agriculture. *Agricultural Development in the Caribbean and Central America.* 97th Cong., 2d Sess., 1982.

U.S. Congress, House, Committee on Foreign Affairs, Subcommittees on International Economic Policy and Trade and on Inter-American-Relations. *U.S. Economic Sanctions Against Chile.* 97th Cong., 1st Sess., 1981.

U.S. Congress, House, Committee on Foreign Affairs, Subcommittee on International Organizations and Movements. *International Protection of Human Rights.* 93rd Cong., 1st Sess., 1973.

U.S. Congress, House, Committee on Foreign Affairs, Subcommittees on International Security and Scientific Affairs, on Human Rights and International Organizations, and on Inter-American Affairs. *Proposed Transfer of Arms to Uruguay.* 97th Cong., 1st Sess., 1981.

U.S. Congress, House, Committee on Foreign Affairs, Subcommittees on International Security and Scientific Affairs and on Inter-American Affairs. *U.S. Arms Transfer Policy in Latin America.* 97th Cong., 1st Sess., 1981.

U.S. Congress, House, Committee on Foreign Affairs, Subcommittee on Western Hemisphere Affairs. *Central America: The Ends and Means of U.S. Policy.* 98th Cong., 2d Sess., 1984.

————. *Developments in Guatemala and U.S. Options.* 99th Cong., 1st Sess., 1985.

————. *United States Policy toward Guatemala.* 98th Cong., 1st Sess., 1983.

————. *U.S. Policy on Latin America—1985.* 99th Cong., 1st Sess., 1985.

————. *U.S. Policy toward Argentina.* 98th Cong., 1st Sess., 1983.

————. *U.S. Relations with Honduras—1985.* 99th Cong., 1st Sess., 1985.

U.S. Congress, House, Committee on International Relations, Subcommittee on Inter-American Affairs. *Arms Trade in the Western Hemisphere.* 95th Cong., 2d Sess., 1978.

U.S. Congress, House, Committee on International Relations, Subcommittee on International Political and Military Affairs. *Soviet Activities in Cuba—Parts VI and VII: Communist Influence in the Western Hemisphere.* 94th Cong., 2d Sess., 1976.

U.S. Congress, House, Committee on Merchant Marine and Fisheries, Subcommittee on the Panama Canal. *Sea-Level Canal Studies.* 95th Cong., 2d Sess., 1978.

U.S. Congress, House, Permanent Select Committee on Intelligence, Subcommittee on Oversight and Evaluation. *U.S. Intelligence Performance on Central America: Achievements and Selected Instances of Concern.* 97th Cong., 2d Sess., 1982.

U.S. Congress, Library of Congress, Congressional Research Service. *Current Information on the Republic of Panama, the Panama Canal, and the Panama Canal Zone.* Washington, D.C.: CRS, December 1977.

U.S. Congress, Senate, Committee on Appropriations. *Foreign Assistance and Related Programs Appropriations, Fiscal Year 1982.* 97th Cong., 1st Sess., 1981.

————. *Foreign Assistance and Related Programs Appropriations, Fiscal Year 1983.* 97th Cong., 2d Sess., 1982.

————. *Foreign Assistance and Related Programs Appropriations, Fiscal Year 1984.* 98th Cong., 1st Sess., 1983.

————. *Foreign Assistance and Related Programs Appropriations, Fiscal Year 1986.* 99th Cong., 1st Sess., 1985.

U.S. Congress, Senate, Committee on Appropriations, Subcommittee on Military Construction. *Military Construction Appropriations for Fiscal Year 1979.* 95th Cong., 2d Sess., 1978.

U.S. Congress, Senate, Committee on Banking, Housing, and Urban Affairs, Subcommittee on International Finance. *U.S. Export Control Policy and Extension of the Export Administration Act, Part III.* 96th Cong., 1st Sess., 1979.

U.S. Congress, Senate, Committee on Energy and Natural Resources. *Sohio Crude Oil Pipeline.* 96th Cong., 1st Sess., 1979.

U.S. Congress, Senate, Committee on Foreign Relations. *Central America: Treading Dangerous Waters.* 98th Cong., 1st Sess., 1983.

U.S. Congress, Senate, Committee on Foreign Relations. *East-West Relations*. 97th Cong., 2d Sess., 1982.

—————. *Executive Sessions of the Senate Foreign Relations Committee (Historical Series)*. Vol. 10. 85th Cong., 2d Sess., 1958. Made public November 1980.

—————. *Executive Sessions of the Senate Foreign Relations Committee (Historical Series)*. Vol. 11. 86th Cong., 1st Sess., 1959. Made public 1982.

—————. *Foreign Assistance Authorization for Fiscal Year 1982*. 97th Cong., 1st Sess., 1981.

—————. *Nomination of Alexander M. Haig, Jr.* Parts 1 and 2. 97th Cong., 1st Sess., 1981.

—————. *Nomination of Elliott Abrams*. 97th Cong., 1st Sess., 1981.

—————. *Nomination of Robert E. White*. 96th Cong., 2d Sess., 1980.

—————. *The Situation in El Salvador*. 97th Cong., 1st Sess., 1981.

—————. *The Situation in Grenada*. 98th Cong., 1st Sess., 1983.

—————. *United States Foreign Policy Objectives and Overseas Military Installations*. 96th Cong., 1st Sess., 1979.

—————. *U.S. Policy in the Western Hemisphere*. 97th Cong., 2d Sess., 1982.

—————. *U.S. Policy toward Nicaragua and Central America*. 98th Cong., 1st Sess., 1983.

U.S. Congress, Senate, Committee on Foreign Relations, Subcommittee on East Asian and Pacific Affairs. *Export of Alaskan Crude Oil*. 98th Cong., 1st Sess., 1983.

U.S. Congress, Senate, Committee on Foreign Relations, Subcommittee on Western Hemisphere Affairs. *Human Rights in Nicaragua*. 97th Cong., 2d Sess., 1982.

—————. *Latin America*. 95th Cong., 2d Sess., 1978.

U.S. Congress, Senate, Committees on Foreign Relations and on Appropriations. *El Salvador: The United States in the Midst of a Maelstrom*. 97th Cong., 2d Sess., 1982.

U.S. Congress, Senate, Committee on Interior and Insular Affairs. *The Trans-Alaska Pipeline and West Coast Petroleum Supply, 1977-1982*. 93rd Cong., 2d Sess., 1974.

U.S. Congress, Senate, Committee on the Judiciary, Subcommittee on Immigration and Refugee Policy. *Refugee Problems in Central America*. 98th Cong, 1st Sess., 1984.

U.S. Congress, Senate, Committee on the Judiciary, Subcommittee to Investigate the Administration of the Internal Security Act and Other Internal Security Laws. *Organization of American States Combined Reports on Communist Subversion*. 89th Cong., 1st Sess., 1965.

U.S. Department of Commerce, Bureau of the Census. *U.S. Imports for Consumption and General Imports: TSUSA, Commodity by Country*

362

of Origin, Annual 1980. Washington, D.C.: Government Printing Office, 1981.

U.S. Department of Defense. *Annual Report to the Congress, Caspar W. Weinberger, Secretary of Defense, Fiscal Year 1983.* Washington, D.C.: Government Printing Office, 1982.

U.S. Department of Energy, Energy Information Administration. *1981 International Energy Annual.* Washington, D.C.: Department of Energy, September 1982.

U.S. Department of the Interior, Bureau of Mines. *Minerals Yearbook 1980.* 3 vols. Washington, D.C.: Government Printing Office, 1981.

U.S. Department of State. *Congressional Presentation, Security Assistance Programs, FY1983.* Washington, D.C.: Department of State, 1982.

———. *Grenada Documents: An Overview and Selection.* Washington, D.C.: Department of State, September 1984.

———. *"Revolution Beyond Our Borders": Sandinista Intervention in Central America.* Special Report No. 132. Washington, D.C.: Department of State, September 1985.

U.S. Department of State, Bureau of Public Affairs. *Cuban Armed Forces and the Soviet Military Presence.* Special Report No. 103. Washington, D.C.: Department of State, August 1982.

———. *Democracy in Latin America and the Caribbean.* Current Policy No. 604. Washington, D.C.: Department of State, August 1984.

———. *Foreign Assistance Program: FY1986 Budget and 1985 Supplemental Request.* Special Report No. 128. Washington, D.C.: Department of State, May 1985.

U.S. Department of State and Department of Defense. *Background Paper: Nicaragua's Military Build-Up and Support for Central American Subversion.* Washington, D.C.: Department of State and Department of Defense, July 18, 1984.

———. *News Briefing on Intelligence Information on External Support of the Guerrillas in El Salvador.* Washington, D.C.: Office of Public Diplomacy for Latin America and the Caribbean, Department of State, August 8, 1984.

———. *The Sandinista Military Build-up.* Washington, D.C.: Department of State and Department of Defense, revised May 1985.

———. *The Soviet-Cuban Connection in Central America and the Caribbean.* Washington, D.C.: Department of State and Department of Defense, March 1985.

U.S. Department of Transportation, Maritime Administration. *United States Oceanborne Foreign Trade Routes.* Washington, D.C.: Department of Transportation, September 1984.

———. *U.S. Oceanborne Foreign Trade Routes.* Washington, D.C.: Department of Transportation, April 1983.

U.S. General Accounting Office. "Actions Needed to Promote a Stable

Supply of Strategic and Critical Minerals and Materials." Report No. EMD-82-69. Washington, D.C.: GAO, June 3, 1982. Mimeographed.

————. "Political and Economic Factors Influencing Economic Support Fund Programs." Report No. GAO/ID-83-43. Washington, D.C.: GAO, April 18, 1983.

————. "Providing Effective Economic Assistance to El Salvador and Honduras: A Formidable Task." Report No. GAO/NSIAD-85-82. Washington, D.C.: GAO, July 3, 1985.

————. "U.S. Security and Military Assistance: Programs and Related Activities." Report No. GAO/ID-82-40. Washington, D.C.: GAO, June 1, 1982.

————. "U.S. Security and Military Assistance: Programs and Related Activities—An Update." Report No. GAO/NSIAD-85-158. Washington, D.C.: GAO, September 30, 1985.

————. "U.S. and Soviet Bloc Training of Latin American and Caribbean Students: Considerations in Developing Future U.S. Programs." Report No. GAO/NSIAD-84-109. Washington, D.C.: GAO, August 16, 1984.

U.S. Joint Chiefs of Staff. "United States Military Posture for FY1982." Printed in U.S. Congress, House, Committee on Armed Services, *Hearings on Military Posture and H.R. 2614 and H.R. 2970.* 97th Cong., 1st Sess., 1981, pt. 1, pp. 78-203.

U.S. National Bipartisan Commission on Central America (Kissinger Commission). "Report of the National Bipartisan Commission on Central America." Washington, D.C.: Department of State (?), January 1984. Mimeographed.

U.S. National Security Council. "U.S. Objectives and Courses of Action with Respect to Latin America." NSC Report 144, March 6, 1953. Mimeographed.

Vacs, Aldo César. *Discreet Partners: Argentina and the USSR since 1917.* Pittsburgh: University of Pittsburgh Press, 1984.

Vaky, Viron P. "A Central American Tragedy." In *Central America and the Western Alliance,* edited by Joseph Cirincione, pp. 57-60. New York: Holmes and Meier, 1985.

————. "Hemispheric Relations: 'Everything is Part of Everything Else.'" *Foreign Affairs* 59 (1981): 617-647.

Vincent, R. J. *Nonintervention and International Order.* Princeton, N.J.: Princeton University Press, 1974.

Volgy, Thomas. "Toward an Exploration of Comparative Foreign Policy Distance between the United States and Latin America: A Research Note." *International Studies Quarterly* 20 (March 1971): 143-166.

Wagner, R. Harrison. *United States Policy toward Latin America: A Study in Domestic and International Politics.* Stanford: Stanford University Press, 1970.

Walker, Thomas W., ed. *Nicaragua: The First Five Years.* New York: Praeger, 1985.

Walters, Vernon A. *Silent Missions*. Garden City, N.Y.: Doubleday, 1978.

Walzer, Michael. "On Failed Totalitarianism." *Dissent* 30 (Summer 1983): 297-306.

Weisband, Edward. *The Ideology of American Foreign Policy: A Paradigm of Lockian Liberalism*. Beverly Hills, Ca.: Sage Publications, 1973.

Welles, Sumner. *The Time for Decision*. New York: Harper and Brothers, 1944.

Wesson, Robert, ed. *U.S. Influence in Latin America in the 1980s*. New York: Praeger, 1982.

White, Robert. "There Is No Military Solution in El Salvador." *The Center Magazine* 14 (July-August 1981): 5-13.

Whitehead, Laurence. "Explaining Washington's Central American Policies." *Journal of Latin American Studies* 15 (November 1983): 321-363.

Wiarda, Howard J. "Conceptual and Political Dimensions of the Crisis in U.S.-Latin American Relations: Toward a New Policy Formulation." In *The Crisis in Latin America: Strategic, Economic, and Political Dimensions*, edited by Howard J. Wiarda, pp. 22-32. Washington, D.C.: American Enterprise Institute, 1984.

————. *In Search of Policy: The United States and Latin America*. Washington, D.C.: American Enterprise Institute, 1984.

————, ed. *Rift and Revolution: The Central American Imbroglio*. Washington, D.C.: American Enterprise Institute for Public Policy Research, 1984.

Wijetilleke, Lakdasa, and Ody, Anthony J. "World Refinery Industry: Need for Restructuring." World Bank Technical Paper No. 32. Washington, D.C.: World Bank, 1984.

Wit, Joel S. "Advances in Antisubmarine Warfare." *Scientific American* 244 (February 1981): 31-41.

Wood, Bryce. *The Dismantling of the Good Neighbor Policy*. Austin: University of Texas Press, 1985.

Woods, Randall Bennett. *The Roosevelt Foreign Policy Establishment and the Good Neighbor: The United States and Argentina, 1941-1945*. Lawrence: Regents Press of Kansas, 1979.

World Bank. *The Assault on World Poverty: Problems of Rural Development, Education and Health*. Baltimore: The Johns Hopkins University Press, 1975.

————. *Guatemala: Economic and Social Position and Prospects*. Washington, D.C.: World Bank, August 1978.

————. "Land Reform." World Bank Paper—Rural Development Series. Washington, D.C.: World Bank, July 1974.

————. *World Development Report 1985*. New York: Oxford University Press, 1985.

Wyman, Thomas S. "Petroleum Imports and National Security." *Proceedings of the U.S. Naval Institute* 103 (September 1977): 33-37.

INDEX

367

LIBRARY OF CONGRESS CATALOGING-IN-PUBLICATION DATA

Schoultz, Lars.
National security and United States policy toward Latin America.
Bibliography: p.
Includes index.
1. United States—National security. 2. Latin America—Strategic
aspects. 3. United States—Military relations—Latin America.
4. Latin America—Military relations—United States. I. Title.

UA23.S36 1987   355'.033073   87-2287
ISBN 0-691-07741-X (alk. paper)   ISBN 0-691-02267-4 (pbk.)